DATA REDUCTION AND ERROR ANALYSIS FOR THE PHYSICAL SCIENCES

DATA REDUCTION AND ERROR ANALYSIS FOR THE PHYSICAL SCIENCES

THIRD EDITION

Philip R. Bevington

Late Associate Professor of Physics
Case Western Reserve University

D. Keith Robinson

Emeritus Professor of Physics
Case Western Reserve University

Boston Burr Ridge, IL Dubuque, IA Madison, WI New York
San Francisco St. Louis Bangkok Bogotá Caracas Kuala Lumpur
Lisbon London Madrid Mexico City Milan Montreal New Delhi
Santiago Seoul Singapore Sydney Taipei Toronto

McGraw-Hill Higher Education
*A Division of The **McGraw-Hill** Companies*

DATA REDUCTION AND ERROR ANALYSIS FOR THE PHYSICAL SCIENCES
THIRD EDITION

Published by McGraw-Hill, a business unit of The McGraw-Hill Companies, Inc., 1221 Avenue of the Americas, New York, NY 10020.

This book is printed on acid-free paper.

14 15 16 17 18 19 QVS/QVS 20 19 18 17 16

ISBN-13: 978-0-07-247227-1
ISBN-10: 0-07-247227-8

Publisher: *Kent A. Peterson*
Sponsoring editor: *Daryl Bruflodt*
Developmental editor: *Spencer J. Cotkin, Ph.D.*
Marketing manager: *Debra B. Hash*
Senior project manager: *Mary E. Powers*
Senior production supervisor: *Laura Fuller*
Senior media project manager: *Stacy A. Patch*
Lead media technology producer: *Judi David*
Coordinator of freelance design: *Rick D. Noel*
Cover designer: *John Rokusek/Rokusek Design*
Cover diagrams provided by: *D. Keith Robinson*
Compositor: *GAC—Indianapolis*
Typeface: *10/12 Times Roman*
Printer: *Quad/Graphics*

Library of Congress Cataloging-in-Publication Data

Bevington, Philip R., 1933-1980.
Data reduction and error analysis for the physical sciences / Philip R. Bevington, D. Keith Robinson.—3rd ed.
p. cm.
Includes bibliographical references and index.
ISBN 0-07-247227-8
1. Multivariate analysis. 2. Error analysis (Mathematics). 3. Least squares. 4. Data reduction. I. Robinson, D. Keith. II. Title.

QA278 .B48 2003
511'.43—dc21 2002070896
CIP

www.mhhe.com

CONTENTS

PREFACE TO THE THIRD EDITION

In his 1969 Preface to the first edition of this book, the late Philip Bevington aptly stated his purpose, " to provide an introduction to the techniques of data reduction and error analysis commonly employed by individuals doing research in the physical sciences and to present them in sufficient detail and breadth to make them useful for students throughout their undergraduate and graduate studies. The presentation is developed from a practical point of view, including enough derivation to justify results, but emphasizing the methods more than the theory." This third edition continues Phil's original mission, updated to reflect the ready availability of modern computers.

The first four chapters introduce the concepts of measuring uncertainties, error analysis, and probability distributions, with a new section on probabilities in low-statistics experiments. Chapter 5 provides an introduction to Monte Carlo methods for simulating experimental data, methods that are applied in later chapters to generate data for examples and to study and evaluate the statistical significance of experimental results. In chapters 6 through 9, the least-squares method is applied to problems of increasing complexity, from analytic straight-line fits to nonlinear fits that require iterative solutions. Chapter 10 provides an introduction to the direct application of the maximum-likelihood method, and chapter 11 includes a discussion of χ^2-probability, confidence intervals, and correlation coefficients. Exercises at the end of the chapters range in complexity from simple statistical calculations to minor projects such as least-squares fitting and Monte Carlo calculations. Answers to selected exercises are provided.

The appendixes from previous editions have been retained. Appendix A includes a brief section on basic differential calculus but is devoted mainly to numerical methods that are useful in analyzing data on the computer. Determinants and matrices are discussed in appendix B. Appendix C provides tables and graphs of statistical functions, augmented by computer routines on the website for calculating probabilities. Appendix D sets forth some guidelines for the preparation of effective graphs. Appendix E provides listings of computer routines that illustrate the text.

COMPUTER ROUTINES

Simple, illustrative computer routines that were a useful feature of the original book have been retained and are listed in Fortran77 in appendix E. Fortran was chosen because it has proved to be the most durable of languages over many decades. (Pascal, which was provided in the second edition, has vanished, displaced by C++.) With the help of the comments at the beginning of appendix E, students should be able to read the Fortran programs and follow their logic without special expertise in the language. To simplify the listed routines and to clarify their main objectives, we have deleted most of the calls to graphics routines.

Computer routines and programs are available for downloading in both Fortran and C++ from the www.mhhe.com/bevington website, along with supporting routines to facilitate the construction of complete programs for Monte Carlo generation, least-squares fitting, and probability calculations. A "Read Me" file on the site describes the organization of the programs and provides instructions for using them.

ACKNOWLEDGMENTS

I am most indebted to the late Philip R. Bevington for his original book, which formed the basis for these revisions. I am grateful to the Case Western Reserve Physics Department for its support, and to my undergraduate laboratory students for providing several of the examples and much inspiration. I thank Spencer Cotkin and his colleagues at McGraw-Hill for their encouragement.

I would also like to thank readers and, in particular, the following reviewers, for their many helpful comments and suggestions: Jingsong Zhang, University of California, Riverside; Gary Schmidt, University of Arizona; Herbert Strauss, University of California, Berkeley; Daniel Suson, Texas A&M, Kingsville.

Finally, I wish to thank my wife Margi for her remarkable patience and support.

D. Keith Robinson

ABOUT THE AUTHORS

The late Philip R. Bevington was a professor of physics at Case Western Reserve University. He graduated from Harvard University in 1954 and received his Ph.D. from Duke University in 1960. He taught at Duke University for five years and was an assistant professor at Stanford University from 1963 to 1968 before coming to Case Western Reserve University. He was involved in research in nuclear structure physics with Van de Graaff accelerators. While at Stanford he was active in computer applications for nuclear physics and was responsible for development of the SCANS system.

D. Keith Robinson is an emeritus professor of physics at Case Western Reserve University in Cleveland, Ohio. He received his B.Sc. in physics from Dalhousie University in Canada in 1954 and his D.Phil. from the University of Oxford in 1960. He was a member of the staff at Brookhaven National Laboratory from 1960 until 1966 when he joined CWRU. His research in experimental particle physics has included studies of boson resonances, K-meson properties, antiproton-proton interactions, and the radiative decay of hyperons. He has been strongly involved in developing computer-based laboratories for the introductory physics courses at CWRU.

CHAPTER 1

UNCERTAINTIES IN MEASUREMENTS

1.1 MEASURING ERRORS

It is a well-established fact of scientific investigation that the first time an experiment is performed the results often bear all too little resemblance to the "truth" being sought. As the experiment is repeated, with successive refinements of technique and method, the results gradually and asymptotically approach what we may accept with some confidence to be a reliable description of events. We may sometimes feel that nature is loath to give up her secrets without a considerable expenditure of effort on our part, and that first steps in experimentation are bound to fail. Whatever the reason, it is certainly true that for all physical experiments, errors and uncertainties exist that must be reduced by improved experimental techniques and repeated measurements, and those errors remaining must always be estimated to establish the validity of our results.

Error is defined by Webster as "the difference between an observed or calculated value and the true value." Usually we do not know the "true" value; otherwise there would be no reason for performing the experiment. We may know approximately what it should be, however, either from earlier experiments or from theoretical predictions. Such approximations can serve as a guide but we must always determine in a systematic way from the data and the experimental conditions themselves how much confidence we can have in our experimental results.

There is one class of error that we can deal with immediately: errors that originate from mistakes or blunders in measurement or computation. Fortunately, these errors are usually apparent either as obviously incorrect data points or as results that are not reasonably close to expected values. They are classified as *illegitimate errors* and generally can be corrected by carefully repeating the operations. Our interest is in *uncertainties* introduced by random fluctuations in our measurements, and *systematic errors* that limit the precision and accuracy of our results in more or less well-defined ways. Generally, we refer to the uncertainties as the *errors* in our results, and the procedure for estimating them as *error analysis.*

Accuracy Versus Precision

It is important to distinguish between the terms *accuracy* and *precision.* The accuracy of an experiment is a measure of how close the result of the experiment is to the true value; the precision is a measure of how well the result has been determined, without reference to its agreement with the true value. The precision is also a measure of the reproducibility of the result in a given experiment. The distinction between accuracy and precision is illustrated by the two sets of measurements in Figure 1.1 where the straight line on each graph shows the expected relation between the dependent variable y and the independent variable x. In both graphs, the scatter of the data points is a reflection of uncertainties in the measurements, consistent with the error bars on the points. The data in Figure 1.1(a) have been measured to a high degree of precision as illustrated by the small error bars, and are in excellent agreement with the expected variation of y with x, but are clearly inaccurate, deviating from the line by a constant offset. On the other hand, the data points in Figure 1.1(b) are rather imprecise as illustrated by the large error bars, but are scattered about the predicted distribution.

It is obvious that we must consider the accuracy and precision simultaneously for any experiment. It would be a waste of time and energy to determine a result with high precision if we knew that the result would be highly inaccurate. Conversely, a

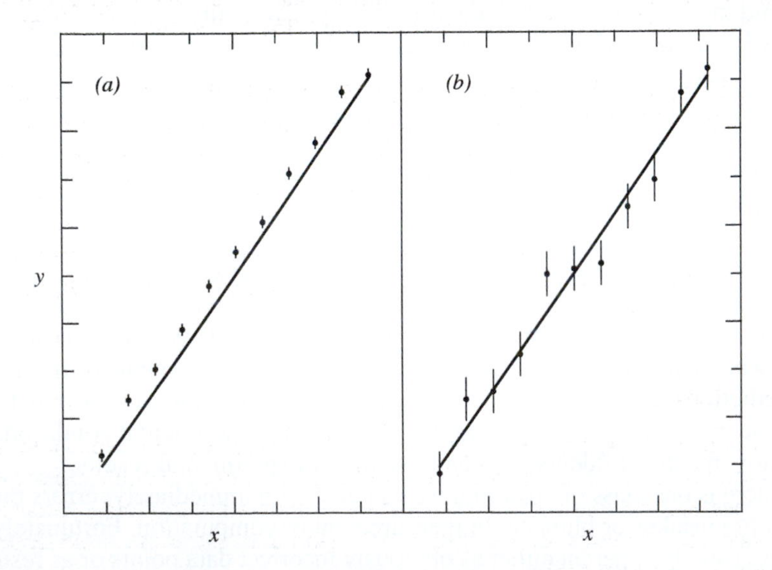

FIGURE 1.1
Illustration of the difference between precision and accuracy. (*a*) Precise but inaccurate data. (*b*) Accurate but imprecise data. True values are represented by the straight lines.

result cannot be considered to be extremely accurate if the precision is low. In general, when we quote the *uncertainty* or *error* in an experimental result, we are referring to the precision with which that result has been determined. *Absolute* precision indicates the magnitude of the uncertainty in the result in the same units as the result; *relative* precision indicates the uncertainty in terms of a fraction of the value of the result.

Systematic Errors

The accuracy of an experiment, as we have defined it, is generally dependent on how well we can control or compensate for systematic errors, errors that will make our results different from the "true" values with reproducible discrepancies. Errors of this type are not easy to detect and not easily studied by statistical analysis. They may result from faulty calibration of equipment or from bias on the part of the observer. They must be estimated from an analysis of the experimental conditions and techniques. A major part of the planning of an experiment should be devoted to understanding and reducing sources of systematic errors.

EXAMPLE 1.1 A student measures a table top with a steel meter stick and finds that the average of his measurements yields a result of (1.982 ± 0.001)m for the length of the table. He subsequently learns that the meter stick was calibrated at 25 °C and has an expansion coefficient of 0.0005 $°C^{-1}$. Because his measurements were made at a room temperature of 20°C, they are systematically too small. To correct for this effect, he multiplies his results by $1 + 0.0005 \times (20 - 25) = 0.9975$ so that his new determination of the length is 1.977 m.

When the student repeats the experiment, he discovers a second systematic error, his technique for reading the meter stick was faulty in that he did not always read the divisions from directly above. By experimentation he determines that this consistently resulted in a reading that was 2 mm short. The corrected result is 1.979 m.

In this example, the first result was given with a fairly high precision, approximately 1 part in 2000. The corrections to this result were meant to improve the accuracy by compensating for known sources of deviation of the first result from the best estimate possible. These corrections did not improve the precision at all, but did in fact worsen it, because the corrections were themselves only estimates of the exact corrections. Before quoting his final result, the student must reexamine his error analysis and take account of any additional uncertainties that may have been introduced by these corrections.

Random Errors

The precision of an experiment depends upon how well we can overcome random errors, fluctuations in observations that yield different results each time the experiment is repeated, and thus require repeated experimentation to yield precise results. A given accuracy implies an equivalent precision and, therefore, also depends to some extent on random errors.

The problem of reducing random errors is essentially one of improving the experimental method and refining the techniques, as well as simply repeating the experiment. If the random errors result from instrumental uncertainties, they may be reduced by using more reliable and more precise measuring instruments. If the random errors result from statistical fluctuations in a limited number of measurements, they may be reduced by making more measurements. There are practical limits to these improvements. In the measurement of the length of the table of Example 1.1, the student might attempt to improve the precision of his measurements by using a magnifying glass to read the scale, or he might attempt to reduce statistical fluctuations in his measurements by repeating the measurement several times. In neither case would it be useful to reduce the random errors much below the systematic errors, such as those introduced by the calibration of the meter stick or the correction for his initial faulty reading of the scale. The limits imposed by systematic errors are important considerations in planning and performing experiments.

Significant Figures and Roundoff

The precision of an experimental result is implied by the number of digits recorded in the result, although generally the uncertainty should be quoted specifically as well. The number of *significant figures* in a result is defined as follows:

1. The leftmost nonzero digit is the most significant digit.
2. If there is no decimal point, the rightmost nonzero digit is the least significant digit.
3. If there is a decimal point, the rightmost digit is the least significant digit, even if it is a 0.
4. All digits between the least and most significant digits are counted as significant digits.

For example, the following numbers each have four significant digits: 1234, 123,400, 123.4, 1001, 1000., 10.10, 0.0001010, 100.0. If there is no decimal point, there are ambiguities when the rightmost digit is 0. Thus, the number 1010 is considered to have only three significant digits even though the last digit might be physically significant. To avoid ambiguity, it is better to supply decimal points or to write such numbers in *scientific notation*, that is, as an argument in decimal notation multiplied by the appropriate power of 10. Thus, our example of 1010 would be written as 1010. or 1.010×10^3 if all four digits are significant.

When quoting an experimental result, the number of significant figures should be approximately one more than that dictated by the experimental precision. The reason for including the extra digit is to avoid errors that might be caused by rounding errors in later calculations. If the result of the measurement of Example 1.1 is $L = 1.979$ m with an uncertainty of 0.012 m, this result could be quoted as $L = (1.979 \pm 0.012)$ m. However, if the first digit of the uncertainty is large, such as 0.082 m, then we should probably quote $L = (1.98 \pm 0.08)$ m. In other words, we let the uncertainty define the precision to which we quote our result.

When insignificant digits are dropped from a number, the last digit retained should be rounded off for the best accuracy. To round off a number to fewer significant digits than were specified originally, we truncate the number as desired and treat the excess digits as a decimal fraction. Then:

1. If the fraction is greater than ½, increment the new least significant digit.
2. If the fraction is less than ½, do not increment.
3. If the fraction equals ½, increment the least significant digit only if it is odd.

The reason for rule 3 is that a fractional value of ½ may result from a previous rounding up of a fraction that was slightly less than ½ or a rounding down of a fraction that was slightly greater than ½. For example, 1.249 and 1.251 both round to three significant figures as 1.25. If we were to round again to two significant figures, both would yield the same value, either 1.2 or 1.3, depending on our convention. Choosing to round up if the resulting last digit is odd and to round down if the resulting last digit is even, reduces systematic errors that would otherwise be introduced into the average of a group of such numbers. Note that it is generally advisable to retain all available digits in intermediate calculations and round only the final results.

1.2 UNCERTAINTIES

Uncertainties in experimental results can be separated into two categories: those that result from fluctuations in measurements, and those associated with the theoretical description of our result. For example, if we measure the length of a rectangular table along one edge, we know that any uncertainties, aside from systematic errors, are associated with the fluctuations of our measurements from trial to trial. With an infinite number of measurements we might be able to estimate the length very precisely, but with a finite number of trials there will be a finite uncertainty. If we were to measure the length of the table at equally spaced positions across the table, the measurements would show additional fluctuations corresponding to irregularities in the table itself, and our result could be expressed as the mean length. If, however, we were to describe the shape of an oval table, we would be faced with uncertainties both in the measurement of position of the edge of the table at various points and in the form of the equation to be used to describe the shape, whether it be circular, elliptical, or whatever. Thus, we shall be concerned in the following chapters with a comparison of the distribution of measured data points with the distribution predicted on the basis of a theoretical model. This comparison will help to indicate whether our method of extracting the results is valid or needs modification.

The term *error* suggests a deviation of the result from some "true" value. Usually we cannot know what the true value is, and can only estimate the errors inherent in the experiment. If we repeat an experiment, the results may well differ from those of the first attempt. We express this difference as a *discrepancy* between the two results. Discrepancies arise because we can determine a result only with a given *uncertainty*. For example, when we compare different measurements of a standard physical constant, or compare our result with the accepted value, we should refer to the differences as *discrepancies*, not errors or uncertainties.

Because, in general, we shall not be able to quote the actual error in a result, we must develop a consistent method for determining and quoting the estimated error. A study of the distribution of the results of repeated measurements of the same quantity can lead to an understanding of these errors so that the quoted error is a measure of the spread of the distribution. However, for some experiments it may not be feasible to repeat the measurements and experimenters must therefore attempt to estimate the errors based on an understanding of the apparatus and their own skill in using it. For example, if the student of Example 1.1 could make only a single measurement of the length of the table, he should examine his meter stick and the table, and try to estimate how well he could determine the length. His estimate should be consistent with the result expected from a study of repeated measurements; that is, to quote an estimate for the *standard error*, he should try to estimate a range into which he would expect repeated measurements to fall about seven times out of ten. Thus, he might conclude that with a fine steel meter stick and a well-defined table edge, he could measure to about ± 1 mm or ± 0.001 m. He should resist the temptation to increase this error estimate, "just to be sure."

We must also realize that the model from which we calculate theoretical parameters to describe the results of our experiment may not be the correct model. In the following chapters we shall discuss hypothetical parameters and probable distributions of errors pertaining to the "true" states of affairs, and we shall discuss methods of making experimental estimates of these parameters and the uncertainties associated with these determinations.

Minimizing Uncertainties and Best Results

Our preoccupation with error analysis is not confined just to the determination of the precision of our results. In general, we shall be interested in obtaining the maximum amount of useful information from the data on hand without being able either to repeat the experiment with better equipment or to reduce the statistical uncertainties by making more measurements. We shall be concerned, therefore, with the problem of extracting from the data the best estimates of theoretical parameters and of the random errors, and we shall want to understand the effect of these errors on our results, so that we can determine what confidence we can place in our final results. It is reasonable to expect that the most reliable results we can calculate from a given set of data will be those for which the estimated errors are the smallest. Thus, our development of techniques of error analysis will help to determine the optimum estimates of parameters to describe the data.

It must be noted, however, that even our best efforts will yield only *estimates* of the quantities investigated.

1.3 PARENT AND SAMPLE DISTRIBUTIONS

If we make a measurement x_1 of a quantity x, we expect our observation to approximate the quantity, but we do not expect the experimental data point to be exactly equal to the quantity. If we make another measurement, we expect to observe a discrepancy between the two measurements because of random errors, and we do not expect either determination to be exactly correct, that is, equal to x. As we make more and more measurements, a pattern will emerge from the data. Some of the measurements will be too large, some will be too small. On the average, however, we expect them to be distributed around the correct value, assuming we can neglect or correct for systematic errors.

If we could make an infinite number of measurements, then we could describe exactly the distribution of the data points. This is not possible in practice, but we can hypothesize the existence of such a distribution that determines the probability of getting any particular observation in a single measurement. This distribution is called the *parent distribution.* Similarly, we can hypothesize that the measurements we have made are samples from the parent distribution and they form the *sample distribution.* In the limit of an infinite number of measurements, the sample distribution becomes the parent distribution.

EXAMPLE 1.2 In a physics laboratory experiment, students drop a ball 50 times and record the time it takes for the ball to fall 2.00 m. One set of observations, corrected for systematic errors, ranges from about 0.59 s to 0.70 s, and some of the observations are identical. Figure 1.2 shows a histogram or frequency plot of these measurements. The height of a data bar represents the number of measurements that fall between the two values indicated by the upper and lower limits of the bar on the abscissa of the plot. (See Appendix D.)

If the distribution results from random errors in measurement, then it is very likely that it can be described in terms of the *Gaussian* or *normal error distribution*, the familiar bell-shaped curve of statistical analysis, which we shall discuss in Chapter 2. A Gaussian curve, based on the mean and standard deviation of these measurements, is plotted as the solid line in Figure 1.2. This curve summarizes the data of the sample distribution in terms of the Gaussian model and provides an estimate of the parent distribution.

The measured data and the curve derived from them clearly do not agree exactly. The coarseness of the experimental histogram distinguishes it at once from the smooth theoretical Gaussian curve. We might imagine that, if the students were to make a great many measurements or combine several sets of measurements so that they could plot the histogram in finer and finer bins, under ideal circumstances the histogram would eventually approach a smooth Gaussian curve. If they were to calculate the parameters from such a large sample, they could determine the parent distribution represented by the dotted curve in Figure 1.2.

It is convenient to think in terms of a *probability density function* $p(x)$, normalized to unit area (*i.e.*, so that the integral of the entire curve is equal to 1) and defined such that in the limit of a very large number N of observations, the number ΔN of observations of the variable x between x and $x + \Delta x$ is given by $\Delta N = Np(x)\Delta x$. The solid and dashed curves in Figure 1.2 have been scaled in this way so that the ordinate values correspond directly to the numbers of observations expected in any range Δx from a 50-event sample and the area under each curve corresponds to the total area of the histogram.

Notation

A number of parameters of the parent distribution have been defined by convention. We use Greek letters to denote them, and Latin letters to denote experimental estimates of them.

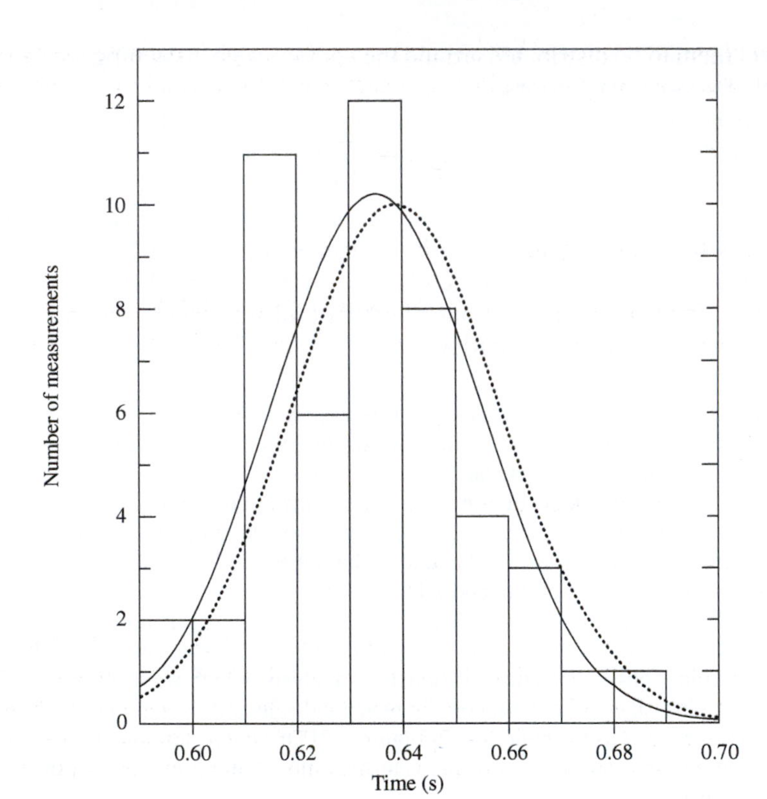

FIGURE 1.2
Histogram of measurements of the time for a ball to fall 2.00 m. The solid Gaussian curve was calculated from the mean ($\overline{T} = 0.635$ s) and standard deviation ($s = 0.020$ s) estimated from these measurements. The dashed curve was calculated from the parent distribution with mean $\mu = 0.639$ s and standard deviation $\sigma = 0.020$ s.

In order to determine the parameters of the parent distribution, we assume that the results of experiments asymptotically approach the parent quantities as the number of measurements approaches infinity; that is, the parameters of the experimental distribution equal the parameters of the parent distribution *in the limit of an infinite number of measurements.* If we specify that there are N observations in a given experiment, then we can denote this by

$$(\text{parent parameter}) = \lim_{N\to\infty} (\text{experimental parameter})$$

If we make N measurements and label them x_1, x_2, x_3, and so forth, up to a final measurement x_N, then we can identify the sum of all these measurements as

$$\sum_{i=1}^{N} x_i \equiv x_1 + x_2 + x_3 + \cdots + x_N$$

where the left-hand side is interpreted as the sum of the observations x_i over the index i from $i = 1$ to $i = N$ inclusive. Because we shall be making frequent use of the sum over N measurements of various quantities, we simplify the notation by omitting the index whenever we are considering a sum where the index i runs from 1 to N;

$$\sum x_i \equiv \sum_{i=1}^{N} x_i$$

Mean, Median, and Mode

With the preceding definitions, the *mean* $\bar{x}$ of the experimental distribution is given as the sum of N determinations x_i of the quantity x divided by the number of determinations

$$\bar{x} \equiv \frac{1}{N} \sum x_i \tag{1.1}$$

and the mean μ of the parent population is defined as the limit

$$\mu \equiv \lim_{N \to \infty} \left(\frac{1}{N} \sum x_i \right) \tag{1.2}$$

The mean is therefore equivalent to the centroid or *average* value of the quantity x.

The *median* of the parent population $\mu_{1/2}$ is defined as that value for which, in the limit of an infinite number of determinations x_i, half the observations will be less than the median and half will be greater. In terms of the parent distribution, this means that the probability is 50% that any measurement x_i will be larger or smaller than the median

$$P(x_i < \mu_{1/2}) = P(x_i > \mu_{1/2}) = 1/2 \tag{1.3}$$

so that the median line cuts the area of the probability density distribution in half. Because of inconvenience in computation, the median is not often used as a statistical parameter.

The *mode*, or *most probable value* μ_{max}, of the parent population is that value for which the parent distribution has the greatest value. In any given experimental measurement, this value is the one that is most likely to be observed. In the limit of a large number of observations, this value will probably occur most often

$$P(\mu_{max}) \geq P(x \neq \mu_{max}) \tag{1.4}$$

The relationship of the mean, median, and most probable value to one another is illustrated in Figure 1.3. For a symmetrical distribution these parameters would all be equal by the symmetry of their definitions. For an asymmetric distribution such as that of Figure 1.3, the median generally falls between the most probable value and the mean. The most probable value corresponds to the peak of the distribution, and the areas on either side of the median are equal.

Deviations

The *deviation* d_i of any measurement x_i from the mean μ of the parent distribution is defined as the difference between x_i and μ:

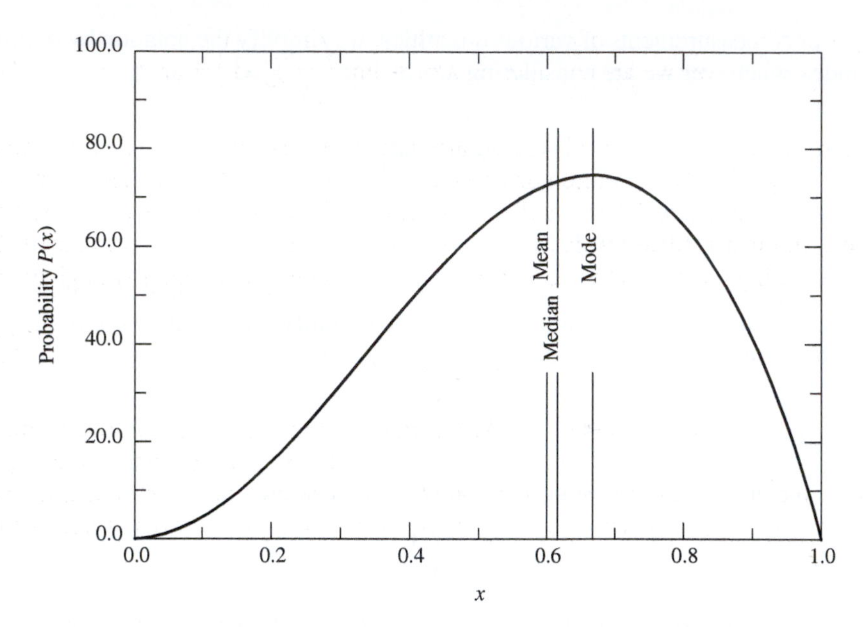

FIGURE 1.3
Asymmetric distribution illustrating the positions of the mean, median, and mode of the variable.

$$d_i \equiv x_i - \mu \tag{1.5}$$

For computational purposes, deviations are generally defined with respect to the mean, rather than the median or most probable value. If μ is the true value of the quantity, d_i is also the true error in x_i.

The average of the deviations $\bar{d}$ must vanish by virtue of the definition of the mean in Equation (1.2):

$$\lim_{N\to\infty} \bar{d} = \lim_{N\to\infty}\left[\frac{1}{N}\sum(x_i - \mu)\right] = \lim_{N\to\infty}\left(\frac{1}{N}\sum x_i\right) - \mu = 0 \tag{1.6}$$

The *average deviation* α, therefore, is defined as the average of the absolute values of the deviations:

$$\alpha \equiv \lim_{N\to\infty}\left[\frac{1}{N}\sum |x_i - \mu|\right] \tag{1.7}$$

The average deviation is a measure of the *dispersion* of the expected observations about the mean. The presence of the absolute value sign makes its use inconvenient for statistical analysis.

A parameter that is easier to use analytically and that can be justified fairly well on theoretical grounds to be a more appropriate measure of the dispersion of the observations is the *standard deviation* σ. The *variance* σ^2 is defined as the limit of the average of the squares of the deviations from the mean μ:

$$\sigma^2 \equiv \lim_{N\to\infty}\left[\frac{1}{N}\sum (x_i - \mu)^2\right] = \lim_{N\to\infty}\left(\frac{1}{N}\sum x_i^2\right) - \mu^2 \tag{1.8}$$

and the standard deviation σ is the square root of the variance. Note that the second form of Equation (1.8) is often described as "the average of the squares minus the square of the average." The standard deviation is the root mean square of the deviations, and is associated with the *second moment* of the data about the mean. The corresponding expression for the variance s^2 of the sample population is given by

$$s^2 \equiv \frac{1}{N-1}\sum (x_i - \bar{x})^2 \tag{1.9}$$

where the factor $N - 1$, rather than N, is required in the denominator to account for the fact that the parameter $\bar{x}$ has been determined from the data and not independently. We note that the symbol σ (instead of s) is often used to represent the best estimate of the standard deviation of the parent distribution determined from a sample distribution.

Significance

The mean μ and the standard deviation, as well as the median, the most probable value, and the average deviation, are all parameters that characterize the information we are seeking when we perform an experiment. Often we wish to describe our distribution in terms of just the mean and standard deviation. The mean may not be exactly equal to the datum in question if the parent distribution is not symmetrical about the mean, but it should have the same characteristics. If a more detailed description is desired, it may be useful to compute higher moments about the mean.

In general, the best we can say about the mean is that it is one of the parameters that specifies the probability distribution: It has the same units as the "true" value and, in accordance with convention, we shall consider it to be the best estimate of the "true" value under the prevailing experimental conditions.

The variance s^2 and the standard deviation s characterize the uncertainties associated with our experimental attempts to determine the "true" values. For a given number of observations, the uncertainty in determining the mean of the parent distribution is proportional to the standard deviation of that distribution. The standard deviation s is, therefore, an appropriate measure of the uncertainty due to fluctuations in the observations in our attempt to determine the "true" value.

Although, in general, the distribution resulting from purely statistical errors can be described well by the two parameters, the mean and the standard deviation, we should be aware that, at distances of a few standard deviations from the mean of an experimental distribution, nonstatistical errors may dominate. In especially severe cases, it may be preferable to describe the spread of the distribution in terms of the average deviation, rather than the standard deviation, because the latter tends to deemphasize measurements that are far from the mean. There are also distributions for which the variance does not exist. The average deviation or some other quantity must be used as a parameter to indicate the spread of the distribution in such cases.

In the following sections, however, we shall be concerned mainly with distributions that result from statistical errors and for which the variance exists.

1.4 MEAN AND STANDARD DEVIATION OF DISTRIBUTIONS

We can define the mean μ and the standard deviation σ in terms of the distribution $p(x)$ of the parent population. The probability density $p(x)$ is defined such that in the limit of a very large number of observations, the fraction dN of observations of the variable x that yield values between x and $x + dx$ is given by $dN = Np(x)\,dx$.

The mean μ is the *expectation value* $\langle x \rangle$ of x, and the variance σ^2 is the expectation value $\langle (x - \mu)^2 \rangle$ of the square of deviations of x from μ. The expectation value $\langle f(x) \rangle$ of any function of x is defined as the weighted average of $f(x)$, over all possible values of the variable x, with each value of $f(x)$ weighted by the probability density distribution $p(x)$.

Discrete Distributions

If the probability function is a discrete function $P(x)$ of the observed value x, we replace the sum over the individual observations Σx_i in Equation (1.2) by a sum over the values of the possible observations multiplied by the number of times these observations are expected to occur. If there are n possible different observable values of the quantity x, which we denote by x_j (where the index j runs from 1 to n with no two values of x_j equal), we should expect from a total of N observations to obtain each observable $NP(x_j)$ times. The mean can then be expressed as

$$\mu = \lim_{N\to\infty} \frac{1}{N} \sum_{i=1}^{N} x_i = \lim_{N\to\infty} \frac{1}{N} \sum_{j=1}^{N} [x_j NP(x_j)]$$

$$= \lim_{N\to\infty} \sum [x_j P(x_j)] \tag{1.10}$$

Similarly, the variance σ in Equation (1.8) can be expressed in terms of the probability function $P(x)$:

$$\sigma^2 = \lim_{N\to\infty} \sum_{j=1}^{n} [(x_j - \mu)^2 P(x_j)] = \lim_{N\to\infty} \sum_{j=1}^{n} [x_j^2 P(x_j)] - \mu^2 \tag{1.11}$$

In general, the expectation value of any function of $f(x)$ is given by

$$\langle f(x) \rangle = \sum_{j=1}^{n} [f(x_j)P(x_j)] \tag{1.12}$$

Continuous Distributions

If the probability density function is a continuous smoothly varying function $p(x)$ of the observed value x, we replace the sum over the individual observations by an integral over all values of x multiplied by the probability $p(x)$. The mean μ becomes the first moment of the parent distribution

$$\mu = \int_{-\infty}^{\infty} xp(x)\,dx \tag{1.13}$$

and the variance σ^2 becomes the second central product moment

$$\sigma^2 = \int_{-\infty}^{\infty} (x - \mu)^2 p(x)dx = \int_{-\infty}^{\infty} x^2 p(x)dx - \mu^2 \tag{1.14}$$

The expectation value of any function of x is

$$\langle f(x) \rangle = \int_{-\infty}^{\infty} f(x)p(x)dx \tag{1.15}$$

What is the connection between the probability distribution of the parent population and an experimental sample we obtain? We have already seen that the uncertainties of the experimental conditions preclude a determination of the "true" values themselves. As a matter of fact, there are three levels of abstraction between the data and the information we seek:

1. From our experimental data points we can determine a sample frequency distribution that describes the way in which these particular data points are distributed over the range of possible data points. We use $\bar{x}$ to denote the mean of the data and s^2 to denote the sample variance. The shape and magnitude of the sample distribution vary from sample to sample.
2. From the parameters of the sample probability distribution we can estimate the parameters of the probability distribution of the parent population of possible observations. Our best estimate for the mean μ is the mean of the sample distribution $\bar{x}$, and the best estimate for the variance σ^2 is the sample variance s^2. Even the shape of this parent distribution must be estimated or assumed.
3. From the estimated parameters of the parent distribution we estimate the results sought. In general, we shall assume that the estimated parameters of the parent distribution are equivalent to the "true" values, but the estimated parent distribution is a function of the experimental conditions as well as the "true" values, and these may not necessarily be separable.

Let us refer again to Figure 1.2, which shows a histogram of time interval measurements and two Gaussian curves, a solid curve based on the parameters $\overline{T} = 0.635$ s and $s = 0.020$ s, which were determined experimentally from the data displayed in the histogram, and a dotted curve based on the parameters $\mu = 0.639$ s and $\sigma = 0.020$ s of the parent distribution. (Although, in general we don't know the properties of the parent distribution, they could have been estimated to high precision in another experiment involving many more measurements.) Comparing the two curves, we observe a slight difference between the experimental mean $\overline{T}$ and the "true" mean μ, and between s and σ.

By considering the data to be a *sample* from the parent population with the values of the observations distributed according to the parent population, we can estimate the shape and dispersion of the parent distribution to obtain useful information on the precision and reliability of our results. Thus, we consider the sample mean $\overline{T}$ to be our best estimate from the data of the mean μ, and we consider the sample variance s^2 to be our best estimate from the data of the variance σ^2, from which we can estimate the uncertainty in our estimate of μ.

SUMMARY

Errors: Difference between measured and "true" values. Generally applied to the uncertainty in a measurement. Not *blunders* or *mistakes.*
Systematic error: Reproducible inaccuracy introduced by faulty equipment, calibration, or technique.
Random error: Indefiniteness of result introduced by finite precision of measurement or statistical variations. Measure of fluctuation after repeated experimentation.
Uncertainty: Magnitude of error that is estimated to have been made in determination of results.
Accuracy: Measure of how close the result of an experiment comes to the "true" value.
Precision: Measure of how carefully the result is determined without reference to any "true" value.
Significant figures:

1. The leftmost nonzero digit is the most significant digit.
2. If there is no decimal point, the rightmost nonzero digit is the least significant digit.
3. If there is a decimal point, the rightmost digit is the least significant digit, even if it is zero.
4. All digits between the least and most significant digits are counted as significant digits.

Roundoff: Truncate the number to the specified number of significant digits and treat the excess digits as a decimal fraction.

1. If the fraction is greater than ½, increment the new least significant digit.
2. If the fraction is less than ½, do not increment.
3. If the fraction equals ½, increment the least significant digit only if it is odd.

Parent population: Hypothetical infinite set of data points of which the experimental data points are assumed to be a random sample.
Parent distribution: Probability distribution of the parent population from which the sample data are chosen.
Expectation value $f(x)$: Weighted average of a function $f(x)$ over all values of x:

$$\langle f(x) \rangle = \lim_{N \to \infty} \left[\frac{1}{N} \sum f(x_i) \right] = \sum_{i=1}^{n} [f(x)P(x_j)] = \int_{-\infty}^{\infty} f(x)P(x)\,dx$$

Median $\mu_{1/2}$: $P(x_i < \mu_{1/2}) = P(x > \mu_{1/2}) = ½$

Most probable value μ_{max}: $P(\mu_{max}) \geq P(x \neq \mu_{max})$

Mean: $\mu \equiv \langle x \rangle$

Average deviation: $\alpha \equiv \langle |x_i - \mu| \rangle$

Variance: $\sigma^2 \equiv \langle (x_i - \mu)^2 \rangle = \langle x^2 \rangle - \mu^2$

Standard deviation: $\sigma = \sqrt{\sigma^2}$

Sample mean: $\bar{x} = (1/N)\Sigma x_i$

Sample variance: $s^2 = \dfrac{1}{(N-1)} \Sigma (x_i - \bar{x})^2$

EXERCISES

1.1. How many significant features are there in the following numbers?
(*a*) 976.45 (*b*) 84,000 (*c*) 0.0094 (*d*) 301.07
(*e*) 4.000 (*f*) 10 (*g*) 5280 (*h*) 400.
(*i*) 4.00×10^2 (*j*) 3.010×10^4

1.2. What is the most significant figure in each of the numbers in Exercise 1.1? What is the least significant?

1.3. Round off each of the numbers in Exercise 1.1 to two significant digits.

1.4. Find the mean, median, and most probable value of x for the following data (from rolling dice).

i	x_i	i	x_i	i	x_i	i	x_i	i	x_i
1	3	6	8	11	12	16	6	21	5
2	7	7	9	12	8	17	7	22	10
3	3	8	7	13	6	18	8	23	8
4	7	9	5	14	6	19	9	24	8
5	12	10	7	15	7	20	8	25	8

1.5. Find the mean, median, and most probable grade from the following set of grades. Group them to find the most probable value.

i	x_i	i	x_i	i	x_i	i	x_i
1	73	11	73	21	69	31	56
2	91	12	46	22	70	32	94
3	72	13	64	23	82	33	51
4	81	14	61	24	90	34	79
5	82	15	50	25	63	35	63
6	46	16	89	26	70	36	87
7	89	17	91	27	94	37	54
8	75	18	82	28	44	38	100
9	62	19	71	29	100	39	72
10	58	20	76	30	88	40	81

1.6. Calculate the standard deviation of the data of Exercise 1.4.

1.7. Calculate the standard deviation of the data of Exercise 1.5.

1.8. Justify the second equality in Equations (1.8) and (1.14).

1.9. Carefully measure in centimeters the length of the cover of this book along the bound edge. Estimate the uncertainty in your measurement. Quote your answer with its uncertainty in decimal form and in scientific notation.

CHAPTER 2

PROBABILITY DISTRIBUTIONS

Of the many probability distributions that are involved in the analysis of experimental data, three play a fundamental role: the *binomial distribution*, the *Poisson distribution*, and the *Gaussian distribution*. Of these, the Gaussian, or normal error, distribution is undoubtedly the most important in statistical analysis of data. Practically, it is useful because it seems to describe the distribution of random observations for many experiments, as well as describing the distributions obtained when we try to estimate the parameters of most other probability distributions.

The Poisson distribution is generally appropriate for counting experiments where the data represent the number of items or events observed per unit interval. It is important in the study of random processes such as those associated with the radioactive decay of elementary particles or nuclear states, and is also applied to data that have been sorted into ranges to form a frequency table or a histogram.

The binomial distribution is generally applied to experiments in which the result is one of a small number of possible final states, such as the number of "heads" or "tails" in a series of coin tosses, or the number of particles scattered forward or backward relative to the direction of the incident particle in a particle physics experiment. Because both the Poisson and the Gaussian distributions can be considered as limiting cases of the binomial distribution, we shall devote some attention to the derivation of the binomial distribution from basic considerations.

2.1 BINOMIAL DISTRIBUTION

Suppose we toss a coin in the air and let it land. There is a 50% probability that it will land heads up and a 50% probability that it will land tails up. By this we mean that if we continue tossing a coin repeatedly, the fraction of times that it lands with heads up will asymptotically approach ½, indicating that there was a probability of

½ of doing so. For any given toss, the probability cannot determine whether or not it will land heads up; it can only describe how we should expect a large number of tosses to be divided into two possibilities.

Suppose we toss two coins at a time. There are now four different possible permutations of the way in which they can land: both heads up, both tails up, and two mixtures of heads and tails depending on which one is heads up. Because each of these permutations is equally probable, the probability for any choice of them is ¼ or 25%. To find the probability for obtaining a particular mixture of heads and tails, without differentiating between the two kinds of mixtures, we must add the probabilities corresponding to each possible kind. Thus, the total probability of finding either head up and the other tail up is ½. Note that the sum of the probabilities for all possibilities (¼ + ¼ + ¼ + ¼) is always equal to 1 because *something* is bound to happen.

Let us extrapolate these ideas to the general case. Suppose we toss n coins into the air, where n is some integer. Alternatively, suppose that we toss one coin n times. What is the probability that exactly x of these coins will land heads up, without distinguishing which of the coins actually belongs to which group? We can consider the probability $P(x; n)$ to be a function of the number n of coins tossed and of the number x of coins that land heads up. For a given experiment in which n coins are tossed, this probability $P(x; n)$ will vary as a function of x. Of course, x must be an integer for any physical experiment, but we can consider the probability to be smoothly varying with x as a continuous variable for mathematical purposes.

Permutations and Combinations

If n coins are tossed, there are 2^n different possible ways in which they can land. This follows from the fact that the first coin has two possible orientations, for each of these the second coin also has two such orientations, for each of these the third coin also has two, and so on. Because each of these possibilities is equally probable, the probability for any one of these possibilities to occur at any toss of n coins is $1/2^n$.

How many of these possibilities will contribute to our observations of x coins with heads up? Imagine two boxes, one labeled "heads" and divided into x slots, and the other labeled "tails." We shall consider first the question of how many permutations of the coins result in the proper separation of x in one box and $n - x$ in the other; then we shall consider the question of how many combinations of these permutations should be considered to be different from each other.

In order to enumerate the number of *permutations* $Pm(n, x)$, let us pick up the coins one at a time from the collection of n coins and put x of them into the "heads" box. We have a choice of n coins for the first one we pick up. For our second selection we can choose from the remaining $n - 1$ coins. The range of choice is diminished until the last selection of the xth coin can be made from only $n - x + 1$ remaining coins. The total number of choices for coins to fill the x slots in the "heads" box is the product of the numbers of individual choices:

$$Pm(n, x) = n(n-1)(n-2)\cdots(n-x+2)(n-x+1) \tag{2.1}$$

This expansion can be expressed more easily in terms of factorials

$$Pm(n, x) = \frac{n!}{(n - x)!} \tag{2.2}$$

So far we have calculated the number of permutations $Pm(n, x)$ that will yield x coins in the "heads" box and $n - x$ coins in the "tails" box, with the provision that we have identified which coin was placed in the "heads" box first, which was placed in second, and so on. That is, we have *ordered* the x coins in the "heads" box. In our computation of 2^n different possible permutations of the n coins, we are only interested in which coins landed heads up or heads down, not which landed first. Therefore, we must consider contributions different only if there are different coins in the two boxes, not if the x coins within the "heads" box are permuted into different time orderings.

The number of different *combinations* $C(n, x)$ of the permutations in the preceding enumeration results from combining the $x!$ different ways in which x coins in the "heads" box can be permuted within the box. For every $x!$ permutations, there will be only one new combination. Thus, the number of different combinations $C(n, x)$ is the number of permutations $Pm(n, x)$ divided by the degeneracy factor $x!$ of the permutations:

$$C(n, x) = \frac{Pm(n, x)}{x!} = \frac{n!}{x!(n - x)!} = \binom{n}{x} \tag{2.3}$$

This is the number of different possible combinations of n items taken x at a time, commonly referred to as $\binom{n}{x}$ or "n over x."

Probability

The probability $P(x; n)$ that we should observe x coins with heads up and $n - x$ with tails up is the product of the number of different combinations $C(n, x)$ that contribute to that set of observations multiplied by the probability for each of the combinations to occur, which we have found to be $(\frac{1}{2})^n$.

Actually, we should separate the probability for each combination into two parts: one part is the probability $p^x = (\frac{1}{2})^x$ for x coins to be heads up; the other part is the probability $q^{n-x} = (1 - \frac{1}{2})^{n-x} = (\frac{1}{2})^{n-x}$ for the other $n - x$ coins to be tails up. For symmetrical coins, the product of these two parts $p^x q^{n-x} = (\frac{1}{2})^n$ is the probability of the combination with x coins heads up and $n - x$ coins tails up. In the general case, the probability p of success for each item is not equal in magnitude to the probability $q = 1 - p$ for failure. For example, when tossing a die, the probability that a particular number will show is $p = 1/6$, while, the probability of its not showing is $q = 1 - 1/6 = 5/6$ so that $p^x q^{n-x} = (1/6)^x \times (5/6)^n$.

With these definitions of p and q, the probability $P_B(x; n, p)$ for observing x of the n items to be in the state with probability p is given by the *binomial distribution*

$$P_B(x; n, p) = \binom{n}{x} p^x q^{n-x} = \frac{n!}{x!(n - x)!} p^x (1 - p)^{n-x} \tag{2.4}$$

where $q = 1 - p$. The name for the binomial distribution comes from the fact that the coefficients $P_B(x; n, p)$ are closely related to the binomial theorem for the expansion of a power of a sum. According to the binomial theorem,

$$(p + q)^n = \sum_{x=0}^{n} \left[\binom{n}{x} p^x q^{n-x} \right] \tag{2.5}$$

The $(j + 1)$th term, corresponding to $x = j$, of this expansion, therefore, is equal to the probability $P_B(j; n, p)$. We can use this result to show that the binomial distribution coefficients $P_B(x; n, p)$ are normalized to a sum of 1. The right-hand side of Equation (2.5) is the sum of probabilities over all possible values of x from 0 to n and the left-hand side is just $1^n = 1$.

Mean and Standard Deviation

The mean of the binomial distribution is evaluated by combining the definition of μ in Equation (1.10) with the formula for the probability function of Equation (2.4):

$$\mu = \sum_{x=0}^{n} \left[x \frac{n!}{x!(n-x)!} p^x (1-p)^{n-x} \right] = np \tag{2.6}$$

We interpret this to mean that if we perform an experiment with n items and observe the number x of successes, after a large number of repeated experiments the average $\bar{x}$ of the number of successes will approach a mean value μ given by the probability for success of each item p times the number of items n. In the case of coin tossing where $p = \frac{1}{2}$, we should expect on the average to observe half the coins land heads up, which seems eminently reasonable.

The variance σ^2 of a binomial distribution is similarly evaluated by combining Equations (1.11) and (2.4):

$$\sigma^2 = \sum_{x=0}^{n} \left[(x - \mu)^2 \frac{n!}{x!(n-x)!} p^x (1-p)^{n-x} \right] = np(1-p) \tag{2.7}$$

The evaluation of these sums is left as an exercise. We are mainly interested in the results, which are remarkably simple.

If the probability for a single success p is equal to the probability for failure $p = q = \frac{1}{2}$, then the distribution is symmetric about the mean μ, and the median $\mu_{1/2}$ and the most probable value are both equal to the mean. In this case, the variance σ^2 is equal to half the mean: $\sigma^2 = \mu/2$. If p and q are not equal, the distribution is asymmetric with a smaller variance.

Example 2.1. Suppose we toss 10 coins into the air a total of 100 times. With each coin toss we observe the number of coins that land heads up and denote that number by x_i, where i is the number of the toss; i ranges from 1 to 100 and x_i can be any integer from 0 to 10. The probability function governing the distribution of the observed values of x is given by the binomial distribution $P_B(x; n, p)$ with $n = 10$ and $p = \frac{1}{2}$. This is the parent distribution and is not affected by the number N of repeated procedures in the experiment.

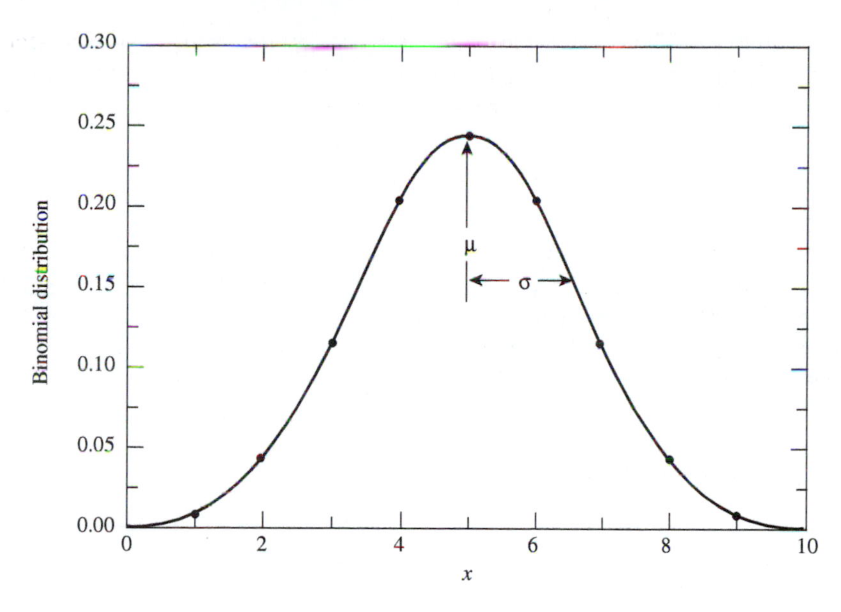

FIGURE 2.1
Binomial distribution for $\mu = 5.0$ and $p = \frac{1}{2}$ shown as a continuous curve although the function is only defined at the discrete points indicated by the round dots.

The parent distribution $P_B(x; 10, \frac{1}{2})$ is shown in Figure 2.1 as a smooth curve drawn through discrete points. The mean μ is given by Equation (2.6):

$$\mu = np = 10(\tfrac{1}{2}) = 5$$

the standard deviation σ is given by Equation (2.7):

$$\sigma = \sqrt{np(1-p)} = \sqrt{10(\tfrac{1}{2})(\tfrac{1}{2})} = \sqrt{2.5} \simeq 1.58$$

The curve is symmetric about its peak at the mean so that approximately 25% of the throws yield five heads and five tails, about 20% yield four heads and six tails, and the same fraction yields six heads and four tails. The magnitudes of the points are such that the sum of the probabilities over all ten points is equal to 1.

Example 2.2. Suppose we roll ten dice. What is the probability that x of these dice will land with the 1 up? If we throw one die, the probability of its landing with 1 up is $p = \frac{1}{6}$. If we throw ten dice, the probability for x of them to land with 1 up is given by the binomial distribution $P_B(x; n, p)$ with $n = 10$ and $p = \frac{1}{6}$:

$$P_B\left(x; 10, \frac{1}{6}\right) = \frac{10!}{x!(10-x)!}\left(\frac{1}{6}\right)^x\left(\frac{5}{6}\right)^{10-x}$$

This distribution is illustrated in Figure 2.2 as a smooth curve drawn through discrete points. The mean and standard deviation are

$$\mu = 10/6 \simeq 1.67$$

and

$$\sigma = \sqrt{10(1/6)(5/6)} \simeq 1.18$$

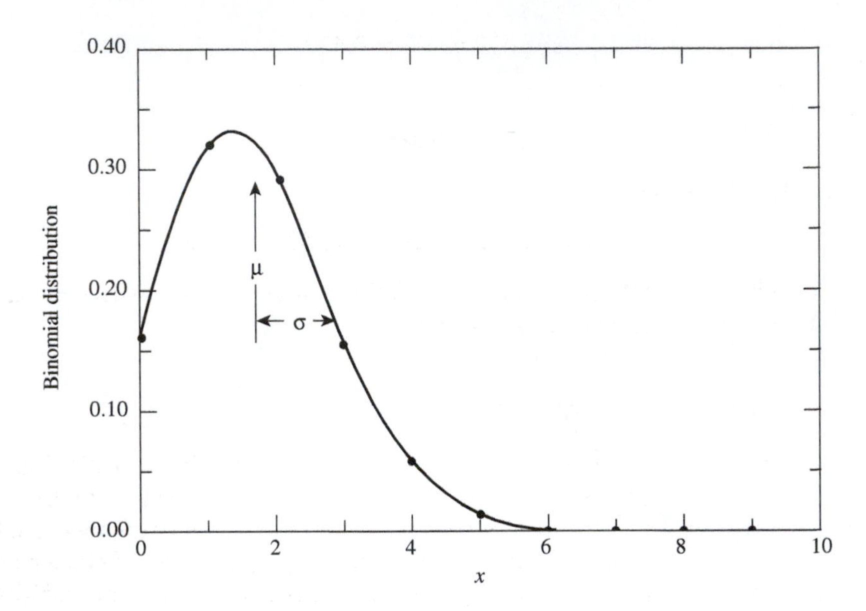

FIGURE 2.2
Binomial distribution for $\mu = 10/6$ and $p = 1/6$ shown as a continuous curve.

The distribution is not symmetric about the mean or about any other point. The most probable value is $x = 1$, but the peak of the smooth curve occurs for a slightly larger value of x.

Example 2.3 A particle physicist makes some preliminary measurements of the angular distribution of K mesons scattered from a liquid hydrogen target. She knows that there should be equal numbers of particles scattered forward and backward in the center-of-mass system of the particles. She measures 1000 interactions and finds that 472 scatter forward and 528 backward. What uncertainty should she quote in these numbers?

The uncertainty is given by the standard deviation from Equation (2.7),

$$\sigma = \sqrt{np(1-p)} = \sqrt{1000(\tfrac{1}{2})(\tfrac{1}{2})} = \sqrt{250} \simeq 15.8$$

Thus, she could quote

$$f_F = (472 \pm 15.8)/1000 = 0.472 \pm 0.15$$

for the fraction of particles scattered in the forward direction and

$$f_B = (528 \pm 15.8)/1000 = 0.528 \pm 0.15$$

for the fraction scattered backward.

Note that the uncertainties in the numbers scattering forward and backward must be the same because losses from one group must be made up in the other.

If the experimenter did not know the a priori probabilities of scattering forward and backward, she would have to estimate p and q from her measurements; that is,

$$p \simeq 472/1000 = 0.472$$

and

$$q \simeq 528/1000 = 0.528$$

She would then calculate

$$s \simeq \sqrt{1000\,(0.472)(0.528)} = \sqrt{249.2} = 15.8$$

For probability p near 50%, the standard deviation is relatively insensitive to uncertainties in the experimental determination of p.

2.2 POISSON DISTRIBUTION

The Poisson distribution represents an approximation to the binomial distribution for the special case where the average number of successes is much smaller than the possible number; that is, when $\mu \ll n$ because $p \ll 1$. For such experiments the binomial distribution correctly describes the probability $P_B(x; n, p)$ of observing x events per time interval out of n possible events, each of which has a probability p of occurring, but the large number n of possible events makes exact evaluation from the binomial distribution impractical. Furthermore, neither the number n of possible events nor the probability p for each is usually known. What may be known instead is the average number of events μ expected in each time interval or its estimate $\bar{x}$. The Poisson distribution provides an analytical form appropriate to such investigations that describes the probability distribution in terms of just the variable x and the parameter μ.

Let us consider the binomial distribution in the limiting case of $p \ll 1$. We are interested in its behavior as n becomes infinitely large while the mean $\mu = np$ remains constant. Equation (2.4) for the probability function of the binomial distribution can be written as

$$P_B(x; n, p) = \frac{1}{x!}\frac{n!}{(n-x)!}p^x(1-p)^{-x}(1-p)^n \tag{2.8}$$

If we expand the second factor

$$\frac{n!}{(n-x)!} = n(n-1)(n-2)\cdots(n-x-2)(n-x-1) \tag{2.9}$$

we can consider it to be the product of x individual factors, each of which is very nearly equal to n because $x \ll n$ in the region of interest. The second factor in Equation (2.8) thus asymptotically approaches n^x. The product of the second and third factors then becomes $(np)^x = \mu^x$. The fourth factor is approximately equal to $1 + px$, which tends to 1 as p tends to 0.

The last factor can be rearranged by substituting μ/p for n and expanding the expression to show that it asymptotically approaches $e^{-\mu}$:

$$\lim_{p\to 0}(1-p)^x = \lim_{p\to 0}[(1-p)^{1/p}]^\mu = \left(\frac{1}{e}\right)^\mu = e^{-\mu} \tag{2.10}$$

Combining these approximations, we find that the binomial distribution probability function $P_B(x;\ n,\ p)$ asymptotically approaches the *Poisson distribution* $P_P(x;\ \mu)$ as p approaches 0:

$$\lim_{p\to 0} P_B(x;\ n, p) = P_P(x;\ \mu) \equiv \frac{\mu^x}{x!}e^{-\mu} \tag{2.11}$$

Because this distribution is an approximation to the binomial distribution for $p \ll 1$, the distribution is asymmetric about its mean μ and will resemble that of Figure 2.2. Note that $P_p(x;\ \mu)$ does not become 0 for $x = 0$ and is not defined for negative values of x. This restriction is not troublesome for counting experiments because the number of counts per unit time interval can never be negative.

Derivation

The Poisson distribution can also be derived for the case where the number of events observed is small compared to the total possible number of events.[1] Assume that the average rate at which events of interest occur is constant over a given interval of time and that event occurrences are randomly distributed over that interval. Then, the probability dP of observing no events in a time interval dt is given by

$$dP(0;\ t, \tau) = -P(0;\ t, \tau)\frac{dt}{\tau} \tag{2.12}$$

where $P(x;\ t, \tau)$ is the probability of observing x events in the time interval dt, τ is a constant proportionality factor that is associated with the mean time between events, and the minus sign accounts for the fact that increasing the differential time interval dt decreases the probability proportionally. Integrating this equation yields the probability of observing no events within a time t to be

$$P(0;\ t, \tau) = P_0 e^{-t/\tau} \tag{2.13}$$

where P_0, the constant of integration, is equal to 1 because $P(0;\ t, \tau) = 1$ at $t = 0$.

The probability $P(x;\ t, \tau)$ for observing x events in the time interval τ can be evaluated by integrating the differential probability

$$d^x P(x;\ t, \tau) = \frac{e^{-t/\tau}}{x!}\prod_{i=1}^{x}\frac{dt_i}{\tau} \tag{2.14}$$

which is the product of the probabilities of observing each event in a different interval dt_i and the probability $e^{-t/\tau}$ of not observing any other events in the remaining time. The factor of $x!$ in the denominator compensates for the ordering implicit in the probabilities $dP_i(1, t, \tau)$ as discussed in the preceding section on permutations and combinations.

Thus, the probability of observing x events in the time interval t is obtained by integration

$$P_P(x;\ \mu) = P(x;\ t, \tau) = \frac{e^{-t/\tau}}{x!}\left(\frac{t}{\tau}\right)^x \tag{2.15}$$

---

[1]This derivation follows that of Orear (1958), pages 21–22.

or

$$P_P(x; \mu) = \frac{\mu^x}{x!} e^{-\mu} \tag{2.16}$$

which is the expression for the Poisson distribution, where $\mu = t/\tau$ is the average number of events observed in the time interval t. Equation (2.16) represents a normalized probability function; that is, the sum of the function evaluated at each of the allowed values of the variable x is unity:

$$\sum_{x=0}^{\infty} P_P(x, \mu) = \sum_{x=0}^{\infty} \frac{\mu^x}{x!} e^{-\mu} = e^{-\mu} \sum_{x=0}^{\infty} \frac{\mu^x}{x!} = e^{-\mu} e^{\mu} = 1 \tag{2.17}$$

Mean and Standard Deviation

The Poisson distribution, like the binomial distribution, is a *discrete* distribution. That is, it is defined only at integral values of the variable x, although the parameter μ is a positive, real number. The mean of the Poisson distribution is actually the parameter μ that appears in the probability function $P_P(x; \mu)$ of Equation (2.16). To verify this, we can evaluate the expectation value $\langle x \rangle$ of x:

$$\langle x \rangle = \sum_{x=0}^{\infty} \left(x \frac{\mu^x}{x!} e^{-\mu} \right) = \mu e^{-\mu} \sum_{x=1}^{\infty} \frac{\mu^{x-1}}{(x-1)!} = \mu e^{-\mu} \sum_{y=0}^{\infty} \frac{\mu^y}{y!} = \mu \tag{2.18}$$

To find the standard deviation σ, the expectation value of the square of the deviations can be evaluated:

$$\sigma^2 = \langle (x - \mu)^2 \rangle = \sum_{x=0}^{\infty} \left[(x - \mu)^2 \frac{\mu^x}{x!} e^{-\mu} \right] = \mu \tag{2.19}$$

Thus, the standard deviation σ is equal to the square root of the mean μ and the Poisson distribution has only a single parameter, μ.

Computation of the Poisson distribution by Equation (2.16) can be limited by the factorial function in the denominator. The problem can be avoided by using logarithms or by using the recursion relations

$$P(0; \mu) = e^{-\mu} \qquad P(x; \mu) = \frac{\mu}{x} P(x - 1; \mu) \tag{2.20}$$

This form has the disadvantage that, in order to calculate the function for particular values of x and μ, the function must be calculated at all lower values of x as well. However, if the function is to be summed from $x = 0$ to some upper limit to obtain the summed probability or to generate the distribution for a Monte Carlo calculation (Chapter 5), the function must be calculated at all lower values of x anyway.

Example 2.4 As part of an experiment to determine the mean life of radioactive isotopes of silver, students detected background counts from cosmic rays. (See Example 8.1.) They recorded the number of counts in their detector for a series of 100 2-s intervals, and found that the mean number of counts was 1.69 per interval. From the mean they estimated the standard deviation to be $\sigma = \sqrt{1.69} = 1.30$, compared to $s = 1.29$ from a direct calculation with Equation (1.9).

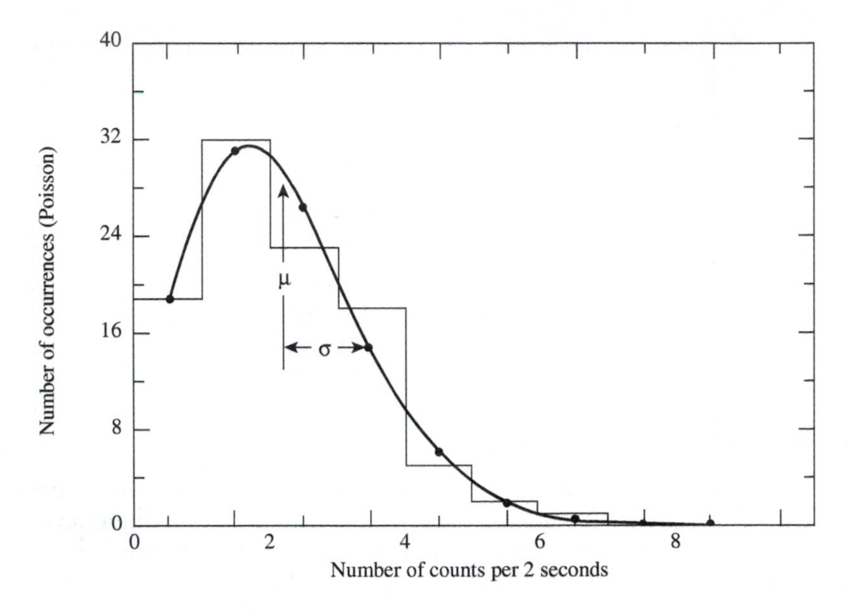

FIGURE 2.3
Histogram of counts in a cosmic ray detector. The Poisson distribution is an estimate of the parent distribution based on the measured mean $\bar{x} = 1.69$. It is shown as a continuous curve although the function is only defined at the discrete points indicated by the round dots.

The students then repeated the exercise, this time recording the number of counts in 15-s intervals for 60 intervals, obtaining a mean of 11.48 counts per interval, with standard deviations $\sigma = \sqrt{11.48} = 3.17$ and $s = 3.39$.

Histograms of the two sets of data are shown in Figures 2.3, and. 2.4. The calculated mean in each case was used as an estimate of the mean of the parent distribution to calculate a Poisson distribution for each data set. The distributions are shown as continuous curves, although only the points at integral values of the abscissa are physically significant.

The asymmetry of the distribution in Figure 2.3 is obvious, as is the fact that the mean μ does not coincide with the most probable value of x at the peak of the curve. The curve of Figure 2.4, on the other hand, is almost symmetric about its mean and the data are consistent with the curve. As μ increases, the symmetry of the Poisson distribution increases and the distribution becomes indistinguishable from the Gaussian distribution.

Summed Probability

We may want to know the probability of obtaining a sample value of x between limits x_1 and x_2 from a Poisson distribution with mean μ. This probability is obtained by summing the values of the function calculated at the integral values of x between the two integral limits x_1 and x_2,

$$S_P(x_1, x_2; \mu) = \sum_{x_1}^{x_2} P_P(x; \mu) \tag{2.21}$$

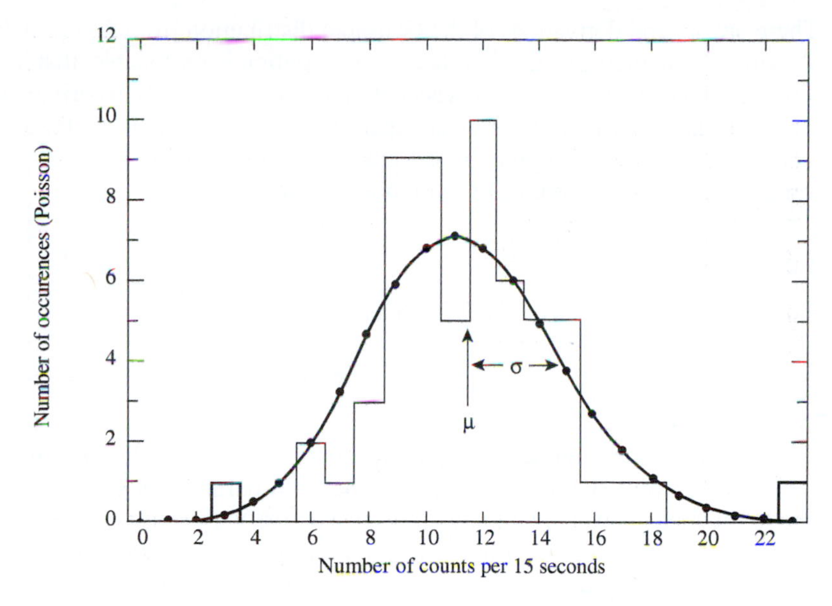

FIGURE 2.4
Histogram of counts in a cosmic ray detector. The Poisson distribution, shown as a continuous curve, is an estimate of the parent distribution based on the measured mean $\bar{x} = 11.48$. Only the calculated points indicated by the round dots are defined.

More likely, we may want to find the probability of recording n or more events in a given interval when the mean number of events is μ. This is just the sum

$$S_P(n, \infty; \mu) = \sum_{x=n}^{\infty} P_P(x; \mu) = 1 - \sum_{x=0}^{n-1} P_P(x; \mu) = 1 - e^{-\mu} \sum_{x=0}^{n-1} \frac{\mu^x}{x!} \tag{2.22}$$

In Example 2.4, the mean number of counts recorded in a 15-s time interval was $\bar{x} = 11.48$. In one of the intervals, 23 counts were recorded. From Equation (2.22), the probability of collecting 23 or more events in a single 15-s time interval is ~ 0.0018, and the probability of this occurring in any one of 60 15-s time intervals is just the complement of the joint probability that 23 or more counts *not* be observed in any of the 60 time intervals, or $p \simeq 1 - (1 - 0.0018)^{60} \simeq 0.10$, or about 10%.

For large values of μ, the probability sum of Equation (2.22) may be approximated by an integral of the Gaussian function.

2.3 GAUSSIAN OR NORMAL ERROR DISTRIBUTION

The Gaussian distribution is an approximation to the binomial distribution for the special limiting case where the number of possible different observations n becomes infinitely large and the probability of success for each is finitely large so $np \gg 1$. It is also, as we observed, the limiting case for the Poisson distribution as μ becomes large.

There are several derivations of the Gaussian distribution from first principles, none of them as convincing as the fact that the distribution is reasonable, that it has a fairly simple analytic form, and that it is accepted by convention and experimentation to be the most likely distribution for most experiments. In addition, it has the satisfying characteristic that the most probable estimate of the mean μ from a random sample of observations x is the average of those observations $\bar{x}$.

Characteristics

The Gaussian probability density is defined as

$$p_G = \frac{1}{\sigma\sqrt{2\pi}} \exp\left[-\frac{1}{2}\left(\frac{x-\mu}{\sigma}\right)^2\right] \tag{2.23}$$

This is a continuous function describing the probability of obtaining the value x in a random observation from a parent distribution with parameters μ and σ, corresponding to the mean and standard deviation, respectively. Because the distribution is continuous, we must define an interval in which the value of the observation x will fall. The probability density function is properly defined such that the probability $dP_G(x; \mu, \sigma)$ that the value of a random observation will fall within an interval dx around x is given by

$$dP_G(x; \mu, \sigma) = p_G(x; \mu, \sigma)dx \tag{2.24}$$

considering dx to be an infinitesimal differential, and the probability density function to be normalized, so that

$$\int_{x=-\infty}^{x=\infty} dP_G(x; \mu, \sigma) = \int_{x=-\infty}^{x=\infty} p_G(x; \mu, \sigma)dx \tag{2.25}$$

The width of the curve is determined by the value of σ, such that for $x = \mu + \sigma$, the height of the curve is reduced to $e^{-1/2}$ of its value at the peak:

$$p_G(x; \mu \pm \sigma, \sigma) = e^{-1/2} p_G(\mu; \mu, \sigma) \tag{2.26}$$

The shape of the Gaussian distribution is shown in Figure 2.5. The curve displays the characteristic bell shape and symmetry about the mean μ.

We can characterize a distribution by its *full-width at half maximum* Γ, often referred to as the *half-width*, defined as the range of x between values at which the probability $p_G(x; \mu, \sigma)$ is half its maximum value:

$$p_G(\mu \pm \tfrac{1}{2}\Gamma, \mu, \sigma) = \tfrac{1}{2} p_G(\mu; \mu, \sigma) \tag{2.27}$$

With this definition, we can determine from Equation (2.23) that

$$\Gamma = 2.354\,\sigma \tag{2.28}$$

As illustrated in Figure 2.5, tangents drawn along a portion of steepest descent of the curve intersect the curve at the $e^{-1/2}$ points $x = \mu \pm \sigma$ and intersect the x axis at the points $x = \mu \pm 2\sigma$.

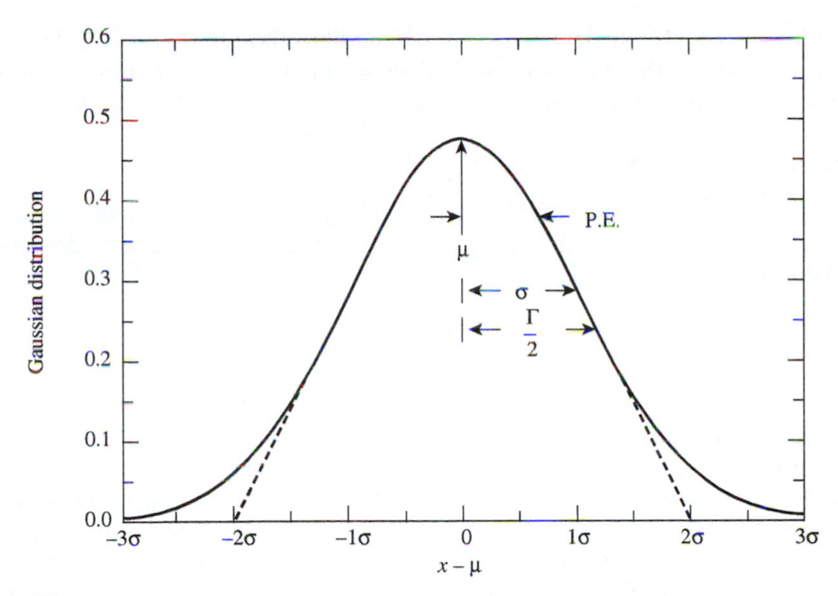

FIGURE 2.5
Gaussian probability distribution illustrating the relation of μ, σ, Γ, and P.E. to the curve. The curve has unit area.

Standard Gaussian Distribution

It is generally convenient to use a standard form of the Gaussian equation obtained by defining the dimensionless variable $z = (x - \mu)/\sigma$, because with this change of variable, we can write

$$p_G(z)\,dz = \frac{1}{\sqrt{2\pi}}\exp\left(-\frac{z^2}{2}\right)dz \qquad (2.29)$$

Thus, from a single computer routine or a table of values of $p_G(z)$, we can find the Gaussian probability function $p_G(x; \mu, \sigma)$ for all values of the parameters μ and σ by changing the variable and scaling the function by $1/\sigma$ to preserve the normalization.

Mean and Standard Deviation

The parameters μ and σ in Equation (2.23) for the Gaussian probability density distribution correspond to the mean and standard deviation of the function. This equivalence can be verified by calculating μ and σ with Equations (1.13) and (1.14) as the expectation values for the Gaussian function of x and $(x - \mu)^2$, respectively.

For a finite data sample, which is expected to follow the Gaussian probability density distribution, the mean and standard deviation can be calculated directly from Equations (1.1) and (1.9). The resulting values of $\bar{x}$ and s will be estimates of the mean μ and standard deviation σ. Values of $\bar{x}$ and s, obtained in this way from the original 50 time measurements in Example 1.2, were used as estimates of μ and

σ in Equation (2.23) to calculate the solid Gaussian curve in Figure 1.2. The curve was scaled to have the same area as the histogram. The curve represents our estimate of the parent distribution based on our measurements of the sample.

Integral Probability

We are often interested in knowing the probability that a measurement will deviate from the mean by a specified amount Δx or greater. The answer can be determined by evaluating numerically the integral

$$P_G(\Delta x, \mu, \sigma) = \frac{1}{\sigma\sqrt{2\pi}} \int_{\mu-\Delta x}^{\mu+\Delta x} \exp\left[-\frac{1}{2}\left(\frac{x-\mu}{\sigma}\right)^2\right] dx \tag{2.30}$$

which gives the probability that any random value of x will deviate from the mean by less than $\pm\Delta x$. Because the probability function $P_G(x; \mu, \sigma)$ is normalized to unity, the probability that a measurement will deviate from the mean by *more* than Δx is just $1 - P_G(\Delta x; \mu, \sigma)$. Of particular interest are the probabilities associated with deviations of σ, 2σ, and so forth from the mean, corresponding to 1, 2, and so on standard deviations. We may also be interested in the probable error (σ_{pe}), defined to be the absolute value of the deviation $|x - \mu|$ such that the probability for the deviation of any random observation $|x_i - \mu|$ is less than ½. That is, half the observations of an experiment would be expected to fall within the boundaries denoted by $\mu \pm \sigma_{pe}$.

If we use the standard form of the Gaussian distribution of Equation (2.29), we can calculate the integrated probability $P_G(z)$ in terms of the dimensionless variable $z = (x - \mu)/\sigma$,

$$P_G(z) = \frac{1}{\sqrt{2\pi}} \int_{-\Delta z}^{\Delta z} e^{-z^2/2}\, dz \tag{2.31}$$

where $\Delta z = \Delta x/\sigma$ measures the deviation from the mean in units of the standard deviation σ.

The integral of Equation (2.31) cannot be evaluated analytically, so in order to obtain the probability $P_G(\Delta x; \mu, \sigma)$ it is necessary either to expand the Gaussian function in a Taylor's series and integrate the series term by term, or to integrate numerically. With modern computers, numerical integration is fast and accurate, and reliable results can be obtained from a simple quadratic integration (Appendix A.3).

Tables and Graphs

The Gaussian probability density function $p_G(z)$ and the integral probability $P_G(z)$ are tabulated and plotted in Tables C.1 and C.2, respectively. From the integral probability Table C.2, we note that the probabilities are about 68% and 95% that a given measurement will fall within 1 and 2 standard deviations of the mean, respectively. Similarly, by considering the 50% probability limit we can see that the probable error is given by $\sigma_{pe} = 0.6745\sigma$.

Comparison of Gaussian and Poisson Distributions

A comparison of the Poisson and Gaussian curves reveals the nature of the Poisson distribution. It is the appropriate distribution for describing experiments in which the possible values of the data are strictly bounded on one side but not on the other. The Poisson curve of Figure 2.3 exhibits the typical Poisson shape. The Poisson curve of Figure 2.4 differs little from the corresponding Gaussian curve of Figure 2.5, indicating that for large values of the mean μ, the Gaussian distribution becomes an acceptable description of the Poisson distribution. Because, in general, the Gaussian distribution is more convenient to calculate than the Poisson distribution, it is often the preferred choice. However, one should remember that the Poisson distribution is only defined at 0 and positive integral values of the variable x, whereas the Gaussian function is defined at all values of x.

2.4 LORENTZIAN DISTRIBUTION

There are many other distributions that appear in scientific research. Some are phenomenological distributions, created to parameterize certain data distributions. Others are well grounded in theory. One such distribution in the latter category is the Lorentzian distribution, similar but unrelated to the binomial distribution. The Lorentzian distribution is an appropriate distribution for describing data corresponding to resonant behavior, such as the variation with energy of the cross section of a nuclear or particle reaction or absorption of radiation in the Mössbauer effect.

The *Lorentzian probability density* function $P_L(x; \mu, \Gamma)$, also called the *Cauchy distribution*, is defined as

$$p_L(x; \mu, \Gamma) = \frac{1}{\pi} \frac{\Gamma/2}{(x - \mu)^2 + (\Gamma/2)^2} \tag{2.32}$$

This distribution is symmetric about its mean μ with a width characterized by its half-width Γ. The most striking difference between it and the Gaussian distribution is that it does not diminish to 0 as rapidly; the behavior for large deviations is proportional to the inverse square of the deviation, rather than exponentially related to the square of the deviation.

As with the Gaussian distribution, the Lorentzian distribution function is a continuous function, and the probability of observing a value x must be related to the interval within which the observation may fall. The probability $dP_L(x; \mu, \Gamma)$ for an observation to fall within an infinitesimal differential interval dx around x is given by the product of the probability density function $p_L(x; \mu, \Gamma)$ and the size of the interval dx:

$$dP_L(x; \mu, \Gamma) = p_L(x; \mu, \Gamma)\, dx \tag{2.33}$$

The normalization of the probability density function $p_L(x; \mu, \Gamma)$ is such that the integral of the probability over all possible values of x is unity:

$$\int_{-\infty}^{\infty} p_L(x; \mu, \Gamma)\, dx = \frac{1}{\pi} \int_{-\infty}^{\infty} \frac{1}{1 + z^2}\, dz = 1 \tag{2.34}$$

where $z = (x - \mu)/(\Gamma/2)$.

Mean and Half-Width

The mean μ of the Lorentzian distribution is given as one of the parameters in Equation (2.32). It is obvious from the symmetry of the distribution that μ must be equal to the mean as well as to the median and to the most probable value.

The standard deviation is not defined for the Lorentzian distribution as a consequence of its slowly decreasing behavior for large deviations. If we attempt to evaluate the expectation value for the square of the deviations

$$\sigma^2 = \langle (x - \mu)^2 \rangle = \frac{1}{\pi} \frac{\Gamma^2}{4} \int_{-\infty}^{\infty} \frac{z^2}{1 + z^2}\, dz \tag{2.35}$$

we find that the integral is unbounded: the integral does not converge for large deviations. Although it is possible to calculate a *sample standard deviation* by evaluating the average value of the square of the deviations from the sample mean, this calculation has no meaning and will not converge to a fixed value as the number of samples increases.

The width of the Lorentzian distribution is instead characterized by the *full-width at half maximum* Γ, generally called the *half-width*. This parameter is defined such that when $x = \mu \pm \Gamma/2$, the probability density function is equal to one-half its maximum value, or $p(\mu \pm \Gamma/2; \mu, \Gamma) = \frac{1}{2} p(\mu; \mu, \Gamma)$. Thus, the half-width Γ is the full width of the curve measured between the levels of half maximum probability. We can verify that this identification of Γ with the full-width at half maximum is correct by substituting $x = \mu \pm \Gamma/2$ into Equation (2.32).

The Lorentzian and Gaussian distributions are shown for comparison in Figure 2.6, for $\mu = 10$ and $\Gamma = 2.354$ (corresponding to $\sigma = 1$ for the Gaussian function). Both distributions are normalized to unit area according to their definitions in Equations (2.23) and (2.32). For both curves, the value of the maximum probability is inversely proportional to the half-width Γ. This results in a peak value of $2/\pi\Gamma \simeq 0.270$ for the Lorentzian distribution and a peak value of $1/\sigma\sqrt{2\pi} \simeq 0.399$ for the Gaussian distribution.

Except for the normalization, the Lorentzian distribution is equivalent to the dispersion relation that is used, for example, in describing the cross section of a nuclear reaction for a Breit-Wigner resonance:

$$\sigma = \pi\lambda^2 \frac{\Gamma_1 \Gamma_2}{(E - E_0)^2 + (\Gamma/2)^2} \tag{2.36}$$

SUMMARY

Binomial distribution: Describes the probability of observing x successes out of n tries when the probability for success in each try is p:

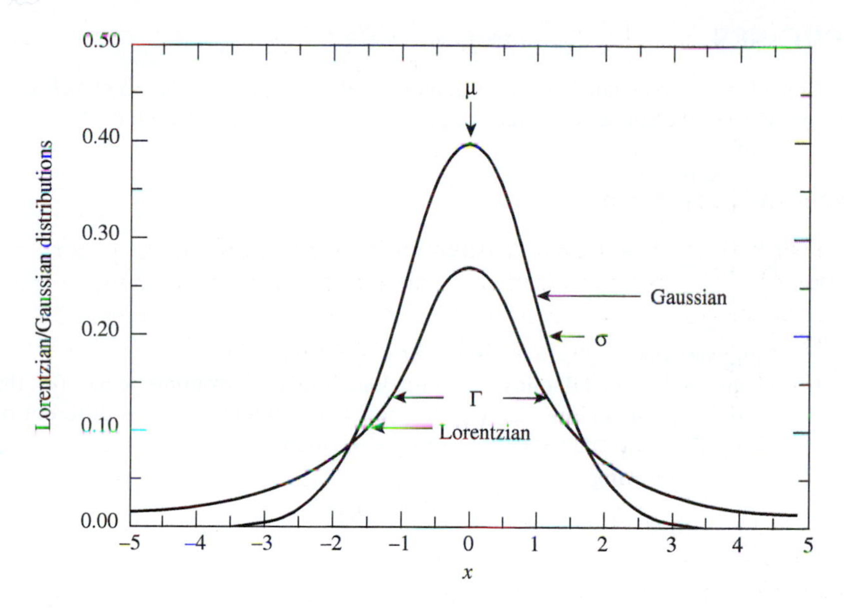

FIGURE 2.6
Comparison of normalized Lorentzian and Gaussian distributions, with $\Gamma = 2.354\sigma$.

$$p_B(x; n, p) = \binom{n}{x} p^x q^{n-x} = \frac{n!}{x!(n-x)!} p^x (1-p)^{n-x}$$

$$\mu = np \qquad \sigma^2 = np(1-p)$$

Poisson distribution: Limiting case of the binomial distribution for large n and constant μ; appropriate for describing small samples from large populations.

$$P_P(x; \mu) = \frac{\mu^x}{x!} e^{-\mu}, \qquad \sigma^2 = \mu$$

Gaussian distribution: Limiting case of the binomial distribution for large n and finite p; appropriate for smooth symmetric distributions.

$$p_G(x; \mu, \sigma) = \frac{1}{\sigma\sqrt{2\pi}} \exp\left[-\frac{1}{2}\left(\frac{x-\mu}{\sigma}\right)^2\right]$$

Half-width $\Gamma = 2.354\sigma$; probable error P.E. $= 0.6745\sigma$.
Standard form:

$$p_G(z)\, dz = \frac{1}{\sqrt{2\pi}} \exp\left(-\frac{z}{2}\right)^2 dz$$

Lorentzian distribution: Describes resonant behavior

$$p_L(x; \mu, \Gamma) = \frac{1}{\pi} \frac{\Gamma/2}{(x-\mu)^2 + (\Gamma/2)^2}$$

EXERCISES

2.1. Consider five coins labeled *a*, *b*, *c*, *d*, and *e*. Let x = *number of heads showing*.

(*a*) Manually count and tabulate *all* possible permutations for each of the following configurations:

i. $x = 0$
ii. $x = 1$
iii. $x = 2$
iv. $x = 3$
v. $x = 4$
vi. $x = 5$

Compare your results to those given by Equation (2.2).

(*b*) Manually delete all duplicate permutations from each example of part (*a*), that is, cross out permutations that repeat a previous combination in a different order. Compare your results to those given by Equation (2.3).

2.2. Evaluate the following:

(*a*) $\binom{6}{3}$ (*b*) $\binom{4}{2}$ (*c*) $\binom{10}{3}$ (*d*) $\binom{52}{4}$

2.3. Evaluate the binomial distribution $P_B(x; n, p)$ for $n = 6$, $p = \frac{1}{2}$, and $x = 0$ to 6. Sketch the distribution and identify the mean and standard deviation. Repeat for $p = \frac{1}{6}$.

2.4. The probability distribution of the sum of the points showing on a pair of dice is given by

$$\begin{aligned} P(x) &= \frac{x-1}{36} \qquad 2 \le x \le 7 \\ &= \frac{13-x}{36} \qquad 7 \le x \le 12 \end{aligned}$$

Find the mean, median, and standard deviation of the distribution.

2.5. Show that the sum in Equation (2.6) reduces to $\mu = np$. *Hint*: Define $y = x - 1$ and $m = n - 1$ and use the fact that

$$\sum_{y=0}^{m}\left[\frac{m!}{y!(m-y)!}p^y(1-p)^{m-y}\right] = \sum_{y=0}^{m} P_B(y; m, p) = 1$$

2.6. On a certain kind of slot machine there are 10 different symbols that can appear in each of three windows. The machine pays off different amounts when either one, two, or three lemons appear. What should be the payoff ratio for each of the three possibilities if the machine is honest and there is no cut for the house?

2.7. Show that the sum in Equation (2.7) reduces to $\sigma^2 = np(1 - p)$. *Hint*: Define $y = x - 1$ and $m = n - 1$ and use the results of Exercise 2.5.

2.8. At rush hour on a typical day, 25.0% of the cars approaching a fork in the street turn left and 75.0% turn right. On a particular day, 283 cars turned left and 752 turned right. Find the predicted uncertainty in these numbers and the probability that these measurements were not made on a "typical day"; that is, find the probability of obtaining a result that is as far or farther from the mean than the result measured on the particular day.

2.9. In a certain physics course, 7.3% of the students failed and 92.7% passed, averaged over many semesters.

(*a*) What is the expected number of failures in a particular class of 32 students, drawn from the same population?

(*b*) What is the probability that five or more students will fail?

2.10. Evaluate and plot the two Poisson distributions of Example 2.4. Plot on each graph the corresponding Gaussian distribution with the same mean and standard deviation.

2.11. Verify that, for the Poisson distribution, if μ is an integer, the probability for $x = \mu$ is equal to the probability for $x = \mu - 1$, $P_P(\mu, \mu) = P_P(\mu - 1; \mu)$.

2.12. Show that the sum in Equation (2.19) reduces to $\sigma^2 = \mu$. *Hint*: Use Equation (2.18) to simplify the expression. Define $y = x - 1$ and show that the sum reduces to $\mu\langle y + 1\rangle = \mu^2$.

2.13. Members of a large collaboration that operated a giant proton-decay detector in a salt mine near Cleveland, Ohio, detected a burst of 8 neutrinos in their apparatus coincident with the optical observation of the explosion of the Supernova 1987A.

(*a*) If the average number of neutrinos detected in the apparatus is 2 per day, what is the probability of detecting a fluctuation of 8 or more in one day?

(*b*) In fact, the 8 neutrinos were all detected within a 10-min period. What is the probability of detecting a fluctuation of 8 or more neutrinos in a 10-min period if the average rate is 2 per 24 hours?

2.14. In a scattering experiment to measure the polarization of an elementary particle, a total of $N = 1000$ particles was scattered from a target. Of these, 670 were observed to be scattered to the right and 330 to the left. Assume that there is no uncertainty in $N = N_R + N_L$.

(*a*) Based on the experimental estimate of the probability, what is the uncertainty in N_R? In N_L?

(*b*) The asymmetry parameter is defined as $A = (N_R - N_L)/(N_R + N_L)$. Calculate the experimental asymmetry and its uncertainty.

(*c*) Assume that the asymmetry has been predicted to be $A = 0.400$ and recalculate the uncertainties in (*a*) and (*b*) using the predicted probability.

2.15. A problem arises when recording data with electronic counters in that the system may saturate when rates are very high, leading to a "dead time." For example, after a particle has passed through a detector, the equipment will be "dead" while the detector recovers and the electronics stores away the results. If a second particle passes through the detector in this time period, it will not be counted.

(*a*) Assume that a counter has a dead time of 200 ns (200×10^{-9} s) and is exposed to a beam of 1×10^6 particles per second so that the mean number of particles hitting the counter in the 200-ns time slot is $\mu = 0.2$. From the Poisson probability for this process, find the efficiency of the counter, that is, the ratio of the average number of particles counted to the average number that pass through the counter in the 200-ns time period.

(*b*) Repeat the calculation for beam rates of 2, 4, 6, 8, and 10×10^6 particles per second, and plot a graph of counter efficiency as a function of beam rate.

2.16. Show by numerical calculation that, for the Gaussian probability distribution, the full-width at half maximum Γ is related to the standard deviation by $\Gamma = 2.354\sigma$ [Equation (2.28)].

2.17. The probability that an electron is at a distance r from the center of the nucleus of a hydrogen atom is given by

$$dP(r) = Cr^2 e^{-r/R} dr$$

Find the mean radius $\bar{r}$ and the standard deviation. Find the value of the constant C.

2.18. Show that a tangent to the Gaussian function is steepest at $x = \mu \pm \sigma$, and therefore intersects the curve at the $e^{-1/2}$ points. Show also that these tangents intersect the x axis at $x = \mu \pm 2\sigma$.

CHAPTER 3

ERROR ANALYSIS

In Chapter 1 we discussed methods for extracting from a set of data points estimates of the mean and standard deviation that describe, respectively, the desired result and the uncertainties in the results. In this chapter we shall further consider how to estimate uncertainties in our measurements, the sources of the uncertainties, and how to combine uncertainties in separate measurements to find the error in a result calculated from those measurements.

3.1 INSTRUMENTAL AND STATISTICAL UNCERTAINTIES

Instrumental Uncertainties

If the quantity x has been measured with a physical instrument, the uncertainty in the measurement generally comes from fluctuations in readings of the instrumental scale, either because the settings are not exactly reproducible due to imperfections in the equipment, or because of human imprecision in observing settings, or a combination of both. Such uncertainties are called *instrumental* because they arise from a lack of perfect precision in the measuring instruments (including the observer). We can include in this category experiments that deal with measurements of such characteristics as length, mass, voltage, current, and so forth. These uncertainties are often independent of the actual value of the quantity being measured.

Instrumental uncertainties are generally determined by examining the instruments and considering the measuring procedure to estimate the reliability of the measurements. In general, one should attempt to make readings to a fraction of the smallest scale division on the instrument. For example, with a good mercury thermometer, it is often easy to estimate the level of the mercury to a least count of one-half of the smallest scale division and possibly even to one-fifth of a division. The measurement is generally quoted to plus or minus one-half of the least count, and this number represents an estimate of the standard deviation of a single measurement. Recalling that, for a Gaussian distribution, there is a 68% probability that a random measurement will lie within 1 standard deviation of the mean, we observe that our object in estimating errors is not to place outer limits on the range of the measurement, which is impossible, but to set a particular *confidence level* that a repeated measurement of the quantity will fall this close to the mean or closer. Often we choose the standard deviation, the 68% confidence level, but other levels are used as well. We shall discuss the concept of confidence levels in Chapter 11.

Digital instruments require special consideration. Generally, manufacturers specify a *tolerance;* for example, the tolerance of a digital multimeter may be given as $\pm 1\%$. At any rate, the precision cannot be better than half the last digit on the display. The manufacturer's quoted tolerances may require interpretation as to whether the uncertainty must be treated as a systematic effect or a statistical effect. For example, if a student uses a resistor with a stated 1% tolerance in an experiment, he can expect the stated uncertainty in the resistance to make a systematic contribution to all experiments with that resistor. On the other hand, when he combines his results with those of the other students in the class, each of whom used a different resistor, the uncertainties in the individual resistances contribute in a statistical manner to the variation of the combined sample.

If it is possible to make repeated measurements, then an estimate of the standard deviation can be calculated from the spread of these measurements as discussed in Chapter 1. The resulting estimate of the standard deviation corresponds to the expected uncertainty in a single measurement. In principle, this *internal* method of determining the uncertainty should agree with that obtained by the *external* method of considering the equipment and the experiment itself, and in fact, any significant discrepancy between the two suggests a problem, such as a misunderstanding of some aspect of the experimental procedure. However, when reasonable agreement is achieved, then the standard deviation calculated internally from the data generally provides the better estimate of the uncertainties.

Statistical Uncertainties

If the measured quantity x represents the number of counts in a detector per unit time interval for a random process, then the uncertainties are called *statistical* because they arise not from a lack of precision in the measuring instruments but from overall statistical fluctuations in the collections of finite numbers of counts over finite intervals of time. For statistical fluctuations, we can estimate analytically the standard deviation for each observation, without having to determine it experimentally. If we were to make the same measurement repeatedly, we should find that the observed values were distributed about their mean in a Poisson distribution (as discussed in Section 2.2) instead of a Gaussian distribution. We can justify the use of this distribution intuitively by considering that we should expect a distribution that is related to the binomial distribution, but that is consistent with our boundary conditions that we can collect any positive number of counts, but no fewer than zero counts, in any time interval.

The Poisson distribution and statistical uncertainties do not apply solely to experiment where counts are recorded in unit time intervals. In any experiment in which data are grouped in bins according to some criterion to form a histogram or frequency plot, the number of events in each individual bin will obey Poisson statistics and fluctuate with statistical uncertainties.

One immediate advantage of the Poisson distribution is that the standard deviation is automatically determined:

$$\sigma = \sqrt{\mu} \tag{3.1}$$

The relative uncertainty, the ratio of the standard deviation to the average rate, $\sigma/\mu = 1/\sqrt{\mu}$, decreases as the number of counts received per interval increases. Thus relative uncertainties are smaller when counting rates are higher.

The value for μ to be used in Equation (3.1) for determining the standard deviation σ is, of course, the value of the mean counting rate from the parent population, of which each measurement x is only an approximate sample. In the limit of an infinite number of determinations, the average of all the measurements would very closely approximate the parent value, but often we cannot make more than one measurement of each value of x, much less an infinite number. Thus, we are forced to use $\sqrt{x}$ as an estimate of the standard deviation of a single measurement.

Example 3.1. Consider an experiment in which we count gamma rays emitted by a strong radioactive source. We cannot determine the counting rate instantaneously because no counts will be detected in an infinitesimal time interval. But we can determine the number of counts x detected over a time interval Δt, and this should be representative of the average counting rate over that interval. Assume that we have recorded 5212 counts in a 1-s time interval. The distribution of counts is random in time and follows the Poisson probability function, so our estimate of the standard deviation of the distribution is $\sigma = \sqrt{5212}$. Thus, we should record our result for the number of counts x in the time interval Δt as 5212 ± 72 and the relative error is

$$\frac{\sigma_x}{x} = \frac{\sqrt{x}}{x} = \frac{1}{\sqrt{x}} \simeq \frac{1}{72} = 0.014 = 1.4\%$$

There may also be instrumental uncertainties contributing to the overall uncertainties. For example, we can determine the time intervals with only finite precision. However, we may have some control over these uncertainties and can often organize our experiment so that the statistical errors are dominant. Suppose that the major instrumental error in our example is the uncertainty $\sigma_t = 0.01$ s in the time interval $\Delta t = 1.00$ s. The relative uncertainty in the time interval is thus

$$\frac{\sigma_t}{\Delta t} = \frac{0.01}{1.00} = 0.01 = 1.\%$$

This relative instrumental error in the time interval will produce a 1.% relative error in the number of counts x. Because the instrumental uncertainty is comparable to the statistical uncertainty, it might be wise to attempt a more precise measurement of the interval or to increase its length. If we increase the counting time interval from 1 s to 4 s, the number of counts x will increase by about a factor of 4 and the relative statistical error will therefore decrease by a factor of 2 to about 0.7%, whereas the relative instrumental uncertainty will decrease by a factor of 4 to 0.25%, as long as the instrumental uncertainty σ_t remains constant at 0.01 s.

3.2 PROPAGATION OF ERRORS

We often want to determine a dependent variable x that is a function of one or more different measured variables. We must know how to propagate or carry over the uncertainties in the measured variables to determine the uncertainty in the dependent variable.

Example 3.2. Suppose we wish to find the volume V of a box of length L, width W, and height H. We can measure each of the three dimensions to be L_0, width W_0, and height H_0 and combine these measurements to yield a value for the volume:

$$V_0 = L_0 W_0 H_0 \tag{3.2}$$

How do the uncertainties in the estimates L_0, W_0, and H_0, affect the resulting uncertainties in the final result V_0?

If we knew the actual errors, $\Delta L = L - L_0$ and so forth, in each dimension, we could obtain an estimate of the error in the final result V_0 by expanding V about the point (L_0, W_0, H_0) in a Taylor series. The first term in the Taylor expansion gives

$$V \simeq V_0 + \Delta L\left(\frac{\partial V}{\partial L}\right)_{W_0 H_0} + \Delta W\left(\frac{\partial V}{\partial W}\right)_{L_0 H_0} + \Delta H\left(\frac{\partial V}{\partial H}\right)_{L_0 W_0} \tag{3.3}$$

from which we can find $\Delta V = V - V_0$. The terms in parentheses are the partial derivatives of V, with respect to each of the dimensions, L, W, and H, evaluated at the point L_0, W_0, H_0. They are the proportionality constants between changes in V and infinitesimally small changes in the corresponding dimensions. The partial derivative of V with respect to L, for example, is evaluated with the other variables W and H held fixed at the values W_0 and H_0 as indicated by the subscript. This approximation neglects higher-order terms in the Taylor expansion, which is equivalent to neglecting the fact that the partial derivatives are not constant over the ranges of L, W, and H given by their errors. If the errors are large, we must include in this definition at least second partial derivatives ($\partial V^2/\partial L^2$, etc.) and partial cross derivatives ($\partial^2 V/\partial L\, \partial W$, etc.), but we shall omit these from the discussion that follows.

For our example of $V = LWH$, Equation (3.3) gives

$$\Delta V \simeq W_0 H_0 \Delta L + L_0 H_0 \Delta W + L_0 W_0 \Delta H \tag{3.4}$$

which we could evaluate if we knew the uncertainties ΔL, ΔW, and ΔH.

Uncertainties

In general, however, we do not know the actual errors in the determination of the dependent variables (or if we do, we should make the necessary corrections). Instead, we may be able to estimate the error in each measured quantity, or to estimate some characteristic, such as the standard deviation σ, of the probability distribution of the measured qualities, How can we combine the standard deviation of the individual measurements to estimate the uncertainty in the result?

Suppose we want to determine a quantity x that is a function of at least two measured variables, u and v. We shall determine the characteristics of x from those of u and v and from the fundamental dependence

$$x = f(u, v, \ldots) \tag{3.5}$$

Although it may not always be exact, we shall assume that the most probable value for x is given by

$$\bar{x} = f(\bar{u}, \bar{v}, \ldots) \tag{3.6}$$

The uncertainty in the resulting value for x can be found by considering the spread of the values of x resulting from combining the individual measurements u_i, v_i, ... into individual results x_i:

$$x_i = f(u_i, v_i, \ldots) \tag{3.7}$$

In the limit of an infinite number of measurements, the mean of the distribution will coincide with the average $\bar{x}$ given in Equation (3.6) and we can use the definition of Equation (1.8) to find the variance σ_x^2 (which is the square of the standard deviation σ_x):

$$\sigma_x^2 = \lim_{N\to\infty}\left[\frac{1}{N}\sum (x_i - \bar{x})^2\right] \tag{3.8}$$

Just as we expressed the deviation of V in Equation (3.4) as a function of the deviations in the dimensions L, W, and H, so we can express the deviations $x_i - \bar{x}$ in terms of the deviations $u_i - \bar{u}$, $v_i - \bar{v}$, ... of the observed parameters

$$x_i - \bar{x} \simeq (u_i - \bar{u})\left(\frac{\partial x}{\partial u}\right) + (v - \bar{v})\left(\frac{\partial x}{\partial v}\right) + \cdots \tag{3.9}$$

where we have omitted specific notation of the fact that each of the partial derivatives is evaluated with all the other variables fixed at their mean values.

Variance and Covariance

Combining Equations (3.8) and (3.9) we can express the variance σ_x^2 for x in terms of the variances σ_u^2, σ_v^2, ... for the variables u, v, ..., which were actually measured:

$$\begin{aligned}\sigma_x^2 &\simeq \lim_{N\to\infty}\frac{1}{N}\sum\left[(u_i - \bar{u})\left(\frac{\partial x}{\partial u}\right) + (v_i - \bar{v})\left(\frac{\partial x}{\partial v}\right) + \cdots\right]^2 \\ &\simeq \lim_{N\to\infty}\frac{1}{N}\sum\left[(u_i - \bar{u})^2\left(\frac{\partial x}{\partial u}\right)^2 + (v_i - \bar{v})^2\left(\frac{\partial x}{\partial v}\right)^2\right. \\ &\qquad \left. + 2(u_i - \bar{u})(v_i - \bar{v})\left(\frac{\partial x}{\partial u}\right)\left(\frac{\partial x}{\partial v}\right) + \cdots\right]\end{aligned} \tag{3.10}$$

The first two terms of Equation (3.10) can be expressed in terms of the variances σ_u^2 and σ_v^2 given by Equation (1.8):

$$\sigma_u^2 = \lim_{N\to\infty}\left[\frac{1}{N}\sum (u_i - \bar{u}_i)^2\right] \qquad \sigma_v^2 = \lim_{N\to\infty}\left[\frac{1}{N}\sum (v_i - \bar{v}_i)^2\right] \tag{3.11}$$

In order to express the third term of Equation (3.10) in a similar form, we introduce the *covariances* σ_{uv}^2 between the variables u and v defined analogous to the variances of Equation (3.11):

$$\sigma_{uv}^2 \equiv \lim_{N\to\infty}\left[\frac{1}{N}\sum [(u_i - \bar{u})(v_i - \bar{v})]\right] \tag{3.12}$$

With these definitions, the approximation for the variance σ_x^2 for x given in Equation (3.10) becomes

$$\sigma_x^2 \simeq \sigma_u^2\left(\frac{\partial x}{\partial u}\right)^2 + \sigma_v^2\left(\frac{\partial x}{\partial v}\right)^2 + \cdots + 2\sigma_{uv}^2\left(\frac{\partial x}{\partial u}\right)\left(\frac{\partial x}{\partial v}\right) + \cdots \tag{3.13}$$

Equation (3.13) is known as the *error propagation equation.*

The first two terms in the equation are averages of squares of deviations weighted by the squares of the partial derivatives, and may be considered to be the averages of the squares of the deviations in x produced by the uncertainties in u and in v, respectively. In general, these terms dominate the uncertainties. If there are additional variables besides u and v in the determination of x, their contributions to the variance of x will have similar terms.

The third term is the average of the cross terms involving products of deviations in u and v weighted by the product of the partial derivatives. If the fluctuations in the measured quantities u and $v, \ldots$ are uncorrelated, then, on the average, we should expect to find equal distributions of positive and negative values for this term, and we should expect the term to vanish in the limit of a large random selection of observations. This is often a reasonable approximation and Equation (3.13) then reduces to

$$\sigma_x^2 \simeq \sigma_u^2\left(\frac{\partial x}{\partial u}\right)^2 + \sigma_v^2\left(\frac{\partial x}{\partial v}\right)^2 + \cdots \tag{3.14}$$

with similar terms for additional variables. In general, we use Equation (3.14) for determining the effects of measuring uncertainties on the final result and neglect the covariant terms. However, as we shall see in Chapter 7, the covariant terms often make important contributions to the uncertainties in parameters determined by fitting curves to data by the least-squares method.

3.3 SPECIFIC ERROR FORMULAS

The expressions of Equations (3.13) and (3.14) were derived for the general relationship of Equation (3.5) giving x as an arbitrary function of u and $v, \ldots$. In the following specific cases of functions $f(u, v, \ldots)$, the parameters a and b are defined as constants and u and v are variables.

Simple Sums and Differences

If the dependent variable x is related to a measured quantity u by the relation

$$x = u + a \tag{3.15}$$

then the partial derivative $\partial x/\partial u = 1$ and the uncertainty in x is just

$$\sigma_x = \sigma_u \tag{3.16}$$

and the relative uncertainty is given by

$$\frac{\sigma_x}{x} = \frac{\sigma_u}{x} = \frac{\sigma_u}{u + a} \tag{3.17}$$

Note that if we are dealing with a small difference between u and a, the uncertainty in x might be greater than the magnitude of x, even for a small relative uncertainty in u.

Example 3.3. In an experiment to count particles emitted by a decaying radioactive source, we measure $N_1 = 723$ counts in a 15-s time interval at the beginning of the experiment and $N_2 = 19$ counts in a 15-s time interval later in the experiment. The events are random and obey Poisson statistics so that we know that the uncertainties in N_1 and N_2 are just their square roots. Assume that we have made a very careful measurement of the background counting rate in the absence of the radioactive source and obtained a value $B = 14.2$ counts with negligible error for the same time interval Δt. Because we have averaged over a long time period, the mean number of background counts in the 15-s interval is not an integral number.

For the first time interval, the corrected number of counts is

$$x_1 = N_1 - B = 723 - 14.2 = 708.8 \text{ counts}$$

The uncertainty in x_1 is given by

$$\sigma_{x_1} = \sigma_{N_1} = \sqrt{723} \simeq 26.9 \text{ counts}$$

and the relative uncertainty is

$$\frac{\sigma_x}{x} = \frac{26.9}{708} = 0.038 \simeq 3.8\%$$

For the second time interval, the corrected number of events is

$$x_2 = N_2 - B = 19 - 14.2 \simeq 4.8 \text{ counts}$$

The uncertainty in x is given by

$$\sigma_{x_2} = \sigma_{N_2} = \sqrt{19} \simeq 4.4 \text{ counts}$$

and the relative uncertainty in x is

$$\frac{\sigma_x}{x} \simeq \frac{4.4}{4.8} = 0.91$$

Weighted Sums and Differences

If x is the weighted sum of u and v,

$$x = au + bv \tag{3.18}$$

the partial derivatives are simply the constants

$$\left(\frac{\partial x}{\partial u}\right) = a \qquad \left(\frac{\partial x}{\partial v}\right) = b \tag{3.19}$$

and we obtain

$$\sigma_x^2 = a^2\sigma_u^2 + b^2\sigma_v^2 + 2ab\sigma_{uv}^2 \tag{3.20}$$

Note the possibility that the variance σ_x^2 might vanish if the covariance σ_{uv}^2 has the proper magnitude and sign. This could happen in the unlikely event that the fluctuations were completely correlated so that each erroneous observation of u was exactly compensated for by a corresponding erroneous observation of v.

Example 3.4. Suppose that, in the previous example, the background counts B were not averaged over a long time period but were simply measured for 15 s to give $B = 14$ with standard deviation $\sigma_B = \sqrt{14} \simeq 3.7$ counts. Then the uncertainty in x would be given by

$$\sigma_x^2 = \sigma_N^2 + (-\sigma_B)^2 = N + B$$

because the uncertainties in N and B are equal to their square roots.

For the first time interval, we would calculate

$$x_1 = (723 - 14) \pm \sqrt{723 + 14} = 709 \pm 27.1 \text{ counts}$$

and the relative uncertainty would be

$$\frac{\sigma_x}{x} = \frac{27.1}{709} \simeq 0.038$$

For the second time interval, we would calculate

$$x_2 = (19 - 14) \pm \sqrt{19 + 14} = 5 \pm 5.7 \text{ counts}$$

and the relative uncertainty would be

$$\frac{\sigma_x}{x} = \frac{5.7}{5} \simeq 1.1$$

Multiplication and Division

If x is the weighted product of u and v,

$$x = auv \tag{3.21}$$

the partial derivatives of each variable are functions of the other variable,

$$\left(\frac{\partial x}{\partial u}\right) = av \qquad \left(\frac{\partial x}{\partial v}\right) = au \tag{3.22}$$

and the variance of x becomes

$$\sigma_x^2 = (av\sigma_u)^2 + (au\sigma_v)^2 + 2a^2uv\sigma_{uv}^2 \tag{3.23}$$

which can be expressed more symmetrically as

$$\frac{\sigma_x^2}{x^2} = \frac{\sigma_u^2}{u^2} + \frac{\sigma_v^2}{v^2} + 2\frac{\sigma_{uv}^2}{uv} \tag{3.24}$$

Similarly, if x is obtained through division,

$$x = \frac{au}{v} \tag{3.25}$$

the relative variance for x is given by

$$\frac{\sigma_x^2}{x^2} = \frac{\sigma_u^2}{u^2} + \frac{\sigma_v^2}{v^2} - 2\frac{\sigma_{uv}^2}{uv} \tag{3.26}$$

Example 3.5. The area of a triangle is equal to half the product of the base times the height $A = bh/2$. If the base and height have values $b = 5.0 \pm 0.1$ cm and $h = 10.0 \pm 0.3$ cm, the area is $A = 25.0\ \mathrm{cm}^2$ and the relative uncertainty in the area is given by

$$\frac{\sigma_A^2}{A^2} = \frac{\sigma_b^2}{b^2} + \frac{\sigma_h^2}{h^2} \tag{3.27}$$

or

$$\begin{aligned}\sigma_A^2 &= A^2\left(\frac{\sigma_b^2}{b^2} + \frac{\sigma_h^2}{h^2}\right) \\ &= 25^2(\mathrm{cm})^4\left(\frac{0.1^2}{5^2} + \frac{0.3^2}{10^2}\right)(\mathrm{cm}^2/\mathrm{cm}^2) \\ &\simeq 0.81\ \mathrm{cm}^4\end{aligned}$$

Although the absolute uncertainty in the height is 3 times the absolute uncertainty in the base, the relative uncertainty is only 1½ times as large and its contribution to the variance of the area is only $(1½)^2$ as large.

Powers

If x is obtained by raising the variable u to a power

$$x = au^b \tag{3.28}$$

the derivative of x with respect to u is

$$\left(\frac{\partial x}{\partial u}\right) = abu^{b-1} = \frac{bx}{u} \tag{3.29}$$

and relative error in x becomes

$$\frac{\sigma_x}{x} = b\frac{\sigma_u}{u} \tag{3.30}$$

For the special cases of $b = +1$, we have

$$x = au \qquad \sigma_x = a\sigma_\mu$$

so

$$\frac{\sigma_x}{x} = \frac{\sigma_u}{u} \tag{3.31}$$

For $b = -1$, we have

$$x = \frac{a}{u} \qquad \sigma_x = -\frac{a\sigma_u}{u^2}$$

so

$$\frac{\sigma_x}{x} = -\frac{\sigma_u}{u} \tag{3.32}$$

The negative sign indicates that, in division, a positive error in u will produce a corresponding negative error in x.

Example 3.6. The area of a circle is proportional to the square of the radius $A = \pi r^2$. If the radius is determined to be $r = 10.0 \pm 0.3$ cm, the area is $A = 100.\pi$ cm^2 with an uncertainty given by

$$\frac{\sigma_A}{A} = 2\frac{\sigma_r}{r}$$

or

$$\sigma_A = 2A\frac{\sigma_r}{r} = 2\pi(10.0\text{ cm})^2(0.3\text{ cm})/(10.0\text{ cm}) = 6\pi\text{ cm}^2$$

Exponentials

If x is obtained by raising the natural base to a power proportional to u,

$$x = ae^{bu} \tag{3.33}$$

the derivative of x with respect to u is

$$\frac{\partial x}{\partial u} = abe^{bu} = bx \tag{3.34}$$

and the relative uncertainty becomes

$$\frac{\sigma_x}{x} = b\sigma_u \tag{3.35}$$

If the constant that is raised to the power is not equal to e, the expression can be rewritten as

$$\begin{aligned} x &= a^{bu} \\ &= \left(e^{\ln a}\right)^{bu} = e^{(b \ln a)u} \\ &= e^{cu} \text{ with } c = b \ln a \end{aligned} \tag{3.36}$$

where ln indicates the natural logarithm. Solving in the same manner as before we obtain

$$\frac{\sigma_x}{x} = c\sigma_u = (b \ln a)\sigma_u \tag{3.37}$$

Logarithms

If x is obtained by taking the logarithm of u,

$$x = a \ln(bu) \tag{3.38}$$

the derivative with respect to u is

$$\frac{\partial x}{\partial u} = \frac{ab}{u} \tag{3.39}$$

$$\sigma_x = ab \frac{\sigma_u}{u} \tag{3.40}$$

Angle Functions

If x is determined as a function of u, such as

$$x = a \cos(bu) \tag{3.41}$$

The derivative of x with respect to u is

$$\frac{dx}{du} = -ab \sin(bu) \tag{3.42}$$

so

$$\sigma_x = -\sigma_u ab \sin(bu) \tag{3.43}$$

Similarly, if

$$x = a \sin(bu) \tag{3.44}$$

then

$$\frac{dx}{du} = ab \cos(bu) \tag{3.45}$$

so

$$\sigma_x = \sigma_u ab \cos(bu) \tag{3.46}$$

Note that σ_u is the uncertainty in an angle and therefore must be expressed in radians.

These relations can be useful for making quick estimates of the uncertainty in a calculated quantity caused by the uncertainty in a measured variable. For a simple product or quotient of the measured variable u with a constant, a 1% error in u causes a 1% error in x. If u is raised to a power b, the resulting error in x becomes b% for a 1% uncertainty in u. Even if the complete expression for x involves other measured variables, $x = f(u, v, \ldots)$ and is considerably more complicated than these simple examples, it is often possible to use these relations to make approximate estimates of uncertainties.

3.4 APPLICATION OF ERROR EQUATIONS

Even for relatively simple calculations, such as those encountered in undergraduate laboratory experiments, blind application of the general error propagation expression

[Equation (3.14)] can lead to very lengthy and discouraging equations, especially if the final results depend on several different measured quantities. Often the error equations can be simplified by neglecting terms that make negligible contributions to the final uncertainty, but this requires a certain amount of practice.

Approximations

Students should practice making quick, approximate estimates of the various contributions to the uncertainty in the final result by considering separately the terms in Equation (3.14). A convenient rule of thumb is to neglect terms that make final contributions that are less than 10% of the largest contribution. (Like all rules of this sort, one should be wary of special cases. Several smaller contributions to the final uncertainty can sum to be as important as one larger uncertainty.)

Example 3.7. Suppose that the area of a rectangle $A = LW$ is to be determined from the following measurements of the lengths of two sides:

$$L = 22.1 \pm 0.1 \text{ cm} \qquad W = 7.3 \pm 0.1 \text{ cm}$$

The relative contribution of σ_L to the error in L will be

$$\frac{\sigma_{A_L}}{A} = \frac{\sigma_L}{L} = \frac{0.1}{22.1} = 0.005$$

and the corresponding contribution of σ_W will be

$$\frac{\sigma_{A_W}}{A} = \frac{\sigma_W}{W} = \frac{0.1}{7.3} \simeq 0.014$$

The contribution from σ_L is thus about one-third of that from σ_W. However, when the contributions are combined, we obtain

$$\sigma_A = A\sqrt{0.014^2 + 0.005^2}$$

which can be expanded to give

$$\sigma_A \simeq 0.014A\left(1 + \frac{1}{2}\left(\frac{0.005}{0.014}\right)^2\right) \simeq 0.014A(1 + 0.06) = 0.015A$$

Thus, the effective contribution from σ_L is only about 6% of the effective contribution from σ_W and could safely be neglected in this calculation.

Computer Calculation of Uncertainties

Finding analytic forms for the partial derivatives is sometimes quite difficult. One should always break Equation (3.14) into separate components and not attempt to find one complete equation that incorporates all error terms. In fact, if the analysis is being done by computer, it may not even be necessary to find the derivatives explicitly. The computer can find numerically the variations in the dependent variable caused by variations in each independent, or measured, variable.

Suppose that we have a particularly complicated equation, or set of equations, relating our final result x to the individually measured variables u, v, and so forth. Let us assume that the actual equations are programmed as a computer function CALCULATE, which returns the single variable x when called with arguments corresponding to the measured parameters

```
X = CALCULATE(U, V, W . . . )
```

We shall further assume that correlations are small so that the covariances may be ignored. Then, to find the variations of x with the measured quantities u, v, and so forth, we can make successive calls to the function of the form

```
DXU = CALCULATE(U + DU, V, W, . . . ) -X,
DXV = CALCULATE(U, V + DV, W, . . . ) - X,
DXW = CALCULATE(U, V, W + DW, . . . ) - X,
ETC.
```

where DU, DV, DW, and so forth are the standard deviations σ_u, σ_v, σ_w, and so on. The resulting contributions to the uncertainty in x are combined in quadrature as

```
DX = SQRT(SQR(DXU) + SQR(DXV) + SQR(DXW) + . . . )
```

Note that it would not be correct to incorporate all the variations into one equation such as

```
DX = CALCULATE(U + DU, V + DV, W + DW, . . . ) - X
```

because this would imply that the errors DU, DV, and so on were actually known quantities, rather than independent, estimated variations of the measured quantities, corresponding to estimates of the widths of the distributions of the measured variables.

SUMMARY

Covariance: $\sigma_{uv}^2 = \langle (u - \bar{u})(v - \bar{v}) \rangle$.

Propagation of errors: Assume $x = f(u, v)$:

$$\sigma_x^2 = \sigma_u^2 \left(\frac{\partial x}{\partial u} \right)^2 + \sigma_v^2 \left(\frac{\partial x}{\partial v} \right)^2 + 2\sigma_{uv}^2 \left(\frac{\partial x}{\partial u} \right) \left(\frac{\partial x}{\partial v} \right)$$

For u and v uncorrelated, $\sigma_{uv}^2 = 0$.

Specific formulas:

$$x = au + bv \qquad \sigma_x^2 = a^2\sigma_u^2 + b^2\sigma_v^2 + 2ab\sigma_{uv}^2$$

$$x = auv \qquad \frac{\sigma_x^2}{x^2} = \frac{\sigma_u^2}{u^2} + \frac{\sigma_v^2}{v^2} + 2\frac{\sigma_{uv}^2}{uv}$$

$$x = \frac{au}{v} \qquad \frac{\sigma_x^2}{x^2} = \frac{\sigma_u^2}{u^2} + \frac{\sigma_v^2}{v^2} - 2\frac{\sigma_{uv}^2}{uv}$$

$$x = au^b \qquad \frac{\sigma_x}{x} = b\frac{\sigma_u}{u}$$

$$x = ae^{bu} \qquad \frac{\sigma_x}{x} = b\sigma_u$$

$$x = a^{bu} \qquad \frac{\sigma_x}{x} = (b \ln a)\sigma_u$$

$$x = a \ln(bu) \qquad \sigma_x = ab\frac{\sigma_u}{u}$$

$$x = a \cos(bu) \qquad \sigma_x = -\sigma_u ab \sin(bu)$$

$$x = a \sin(bu) \qquad \sigma_x = \sigma_u ab \cos(bu)$$

EXERCISES

3.1. Find the uncertainty σ_x in x as a function of the uncertainties σ_u and σ_v in u and v for the following functions:
(*a*) $x = 1/2(u + v)$ (*b*) $x = 1/2(u - v)$ (*c*) $x = 1/u^2$
(*d*) $x = uv^2$ (*e*) $x = u^2 + v^2$

3.2. If the diameter of a round table is determined to within 1%, how well is its area known? Would it be better to determine its radius to within 1%?

3.3. The resistance R of a cylindrical conductor is proportional to its length L and inversely proportional to its cross-sectional area $A = \pi r^2$. Which should be determined with higher precision, r or L, to optimize the determination of R? How much higher?

3.4. The initial activity N_0 and the mean life τ of a radioactive source are known with uncertainties of 1% each. The activity follows the exponential distribution $N = N_0 e^{-t/\tau}$. The uncertainty in the initial activity N_0 dominates at small t; the uncertainty in the mean life τ dominates at large t ($t \gg \tau$). For what value of t/τ do the uncertainties in N_0 and τ contribute equally to the uncertainty in N? What is the resulting uncertainty in N?

3.5. Snell's law relates the angle of refraction θ_2 of a light ray traveling in a medium of index of refraction n_2 to the angle of incidence θ_1 of a ray traveling in a medium of index n_1 through the equation $n_2 \sin \theta_2 = n_1 \sin \theta_1$. Find n_2 and its uncertainty from the following measurements:

$$\theta_1 = (22.03 \pm 0.2)° \qquad \theta_2 = (14.45 \pm 0.2)° \qquad n_1 = 1.0000$$

3.6. The change in frequency produced by the Doppler shift when a sound source of frequency f is moving with velocity v toward a fixed observer is given by $\Delta f = fv/(u - v)$, where u is the velocity of sound. From the following values of u, f, and v and their uncertainties, calculate Δf and its uncertainty. Which, if any, of the uncertainties make a negligible contribution to the uncertainty in Δf?

$$u_1 = (332 \pm 8) \text{ m/s}; \; f = (1000 \pm 1)\text{Hz}; \; \text{and } v = (0.123 \pm 0.003) \text{ m/s}.$$

3.7. The radius of a circle can be calculated from measurements of the length L of a chord and the distance h from the chord to the circumference of the circle from the equation $R = L^2/2h + h/2$. Calculate the radius and its uncertainty from the following values of L and h.
(*a*) $L = (125.0 \pm 5.0)$ cm, $h = (0.51 \pm 0.22)$ cm
(*b*) $L = (125.0 \pm 5.0)$ cm, $h = (57.4 \pm 1.2)$ cm
Was it necessary to use the second term to calculate R in both (*a*) and (*b*)? Explain.

3.8. Students measure the speed of sound in the laboratory by creating a sound pulse that travels down a 1-m tube and reflects back so that both the initial and reflected pulses are detected by the same microphone. The signals are recorded by computer and the pulse amplitudes versus time are displayed on the monitor. The students measure the time intervals for ten such pairs of pulses on the monitor and record the following transit times in milliseconds:

Trial	1	2	3	4	5	6	7	8	9	10
Transit times	5.77	5.78	5.74	5.80	5.78	5.83	5.76	5.78	5.76	5.78

(*a*) Examine the data and try to estimate the spread of the data, that is, their standard deviation.
(*b*) Calculate the mean transit time, the standard deviation of the sample, and the standard error (error in the mean).
(*c*) One of the transit time measurements differs from the mean by more than 2 standard deviations. In a ten-event sample, how many measurements are predicted by Gaussian statistics to differ from the mean by 2 or more standard deviations? Refer to Table C.2.
(*d*) Calculate the speed of sound and its uncertainty from the data.

3.9. Students in the undergraduate laboratory recorded the following counts in 1-min intervals from a radioactive source. The nominal mean decay rate from the source is 3.7 decays per minute.

Decays per minute	0	1	2	3	4	5	6	7	8	9	10
Frequencey of occurrence	1	9	20	24	19	11	11	0	3	1	1

(*a*) Find the mean decay rate and its standard deviation. Compare the standard deviation to the value expected from the Poisson distribution for the mean value that you obtained.
(*b*) Plot a histogram of the data and show Poisson curves of both the parent and observed distributions.

3.10. Find by numerical integration the probability of observing a value from the Gaussian distribution that is:
(*a*) More than 1 standard deviation (σ) from the mean.
(*b*) More than 2 standard deviations from the mean.
(*c*) More than 3 standard deviations from the mean.

3.11. Find by numerical integration the probability of observing a value from the Lorentzian distribution that is:
(*a*) More than 1 half-width ($\Gamma/2$) from the mean.
(*b*) More than 2 half-widths from the mean.
(*c*) More than 3 half-widths from the mean.

CHAPTER 4

ESTIMATES OF MEAN AND ERRORS

4.1 METHOD OF LEAST SQUARES

In Chapter 2 we defined the mean μ of the parent distribution and noted that the most probable estimate of the mean μ of a random set of observations is the average $\bar{x}$ of the observations. The justification for that statement is based on the assumption that the measurements are distributed according to the Gaussian distribution. In general, we expect the distribution of measurements to be either Gaussian or Poisson, but because these distributions are indistinguishable for most physical situations we can assume the Gaussian distribution is obeyed.

Method of Maximum Likelihood

Assume that, in an experiment, we have observed a set of N data points that are randomly selected from the infinite set of the parent population, distributed according to the parent distribution. If the parent distribution is Gaussian with mean μ and standard deviation σ, the probability dP_i for making any single observation x_i within an interval dx is given by

$$dP_i = p_i dx \tag{4.1}$$

with probability function $p_i = p_G(x_i; \mu, \sigma)$ [see Equation(2.23)]. For simplicity, we shall denote the probability P_i for making an observation x_i by

$$P_i = \frac{1}{\sigma\sqrt{2\pi}} \exp\left[-\frac{1}{2}\left(\frac{x_i - \mu}{\sigma}\right)^2\right] \tag{4.2}$$

Because, in general, we do not know the mean μ of the distribution for a physical experiment, we must estimate it from some experimentally derived parameter. Let us call the estimate μ'. What formula for deriving μ' from the data will yield the maximum likelihood that the parent distribution had a mean equal to μ?

If we hypothesize a trial distribution with a mean μ' and standard deviation $\sigma' = \sigma$, the probability of observing the value x_i is given by the probability function

$$P_i(\mu') = \frac{1}{\sigma\sqrt{2\Pi}} \exp\left[-\frac{1}{2}\left(\frac{x_i - \mu'}{\sigma}\right)^2\right] \tag{4.3}$$

Considering the entire set of N observations, the probability for observing that particular set is given by the product of the individual probability functions, $P_i(\mu')$,

$$P(\mu') = \prod_{i=1}^{N} P_i(\mu') \tag{4.4}$$

where the symbol Π denotes the product of the N probabilities $P_i(\mu')$.

The product of the constants multiplying the exponential in Equation (4.3) is the same as the product to the Nth power, and the product of the exponentials is the same as the exponential of the sum of the arguments. Therefore, Equation (4.4) reduces to

$$P(\mu') = \left(\frac{1}{\sigma\sqrt{2\pi}}\right)^N \exp\left[-\frac{1}{2}\sum\left(\frac{x_i - \mu'}{\sigma}\right)^2\right] \tag{4.5}$$

According to the *method of maximum likelihood,* if we compare the probabilities $P(\mu')$ of obtaining our set of observations from various parent populations with different means μ' but with the same standard deviation $\sigma' = \sigma$, the probability is greatest that the data were derived from a population with $\mu' = \mu$; that is, the most likely population from which such a set of data might have come is assumed to be the correct one.

Calculation of the Mean

The method of maximum likelihood states that the most probable value for μ' is the one that gives the maximum value for the probability $P(\mu')$ of Equation (4.5). Because this probability is the product of a constant times an exponential to a negative argument, maximizing the probability $P(\mu')$ is equivalent to minimizing the argument X of the exponential,

$$X = -\frac{1}{2}\sum\left(\frac{x_i - \mu'}{\sigma}\right)^2 \tag{4.6}$$

To find the minimum value of a function X we set the derivative of the function to 0,

$$\frac{dX}{d\mu'} = -\frac{d}{d\mu'}\frac{1}{2}\sum\left(\frac{x_i - \mu'}{\sigma}\right)^2 = 0 \tag{4.7}$$

and obtain

$$\frac{dX}{d\mu'} = -\frac{1}{2}\sum \frac{d}{d\mu'}\left(\frac{x_i - \mu'}{\sigma}\right)^2 = \sum \left(\frac{x_i - \mu'}{\sigma^2}\right) = 0 \tag{4.8}$$

which, because σ is a constant, gives

$$\mu' = \bar{x} \equiv \frac{1}{N}\sum x_i \tag{4.9}$$

Thus, the maximum likelihood method for estimating the mean by maximizing the probability $P(\mu')$ of Equation (4.5) shows that the most probable value of the mean is just the average $\bar{x}$ as defined in Equation (1.1).

Estimated Error in the Mean

What uncertainty σ is associated with our determination of the mean μ' in Equation (4.9)? We have assumed that all data points x_i were drawn from the same parent distribution and were thus obtained with an uncertainty characterized by the same standard deviation σ. Each of these data points contributes to the determination of the mean μ' and therefore each data point contributes some uncertainty to the determination of the final results. A histogram of our data points would follow the Gaussian shape, peaking at the value μ' and exhibiting a width corresponding to the standard deviation σ. Clearly we are able to determine the mean to much better than $\pm\sigma$, and our determination will improve as we increase the number of measured points N and are thus able to improve the agreement between our experimental histogram and the smooth Gaussian curve.

In Chapter 3 we developed the error propagation equation [see Equation (3.13)] for finding the contribution of the uncertainties in several terms contributing to a single result. Applying this relation to Equation (4.9) to find the variance σ_μ^2 of the mean μ', we obtain

$$\sigma_\mu^2 = \sum\left[\sigma_i^2\left(\frac{\partial \mu'}{\partial x_i}\right)^2\right] \tag{4.10}$$

where the variance σ_i^2 in each measured data point x_i is weighted by the square of the effect $\partial\mu'/\partial x_i$, that that data point has on the result. This approximation neglects correlations between the measurements x_i as well as second- and higher-order terms in the expansion of the variance σ_μ^2, but it should be a reasonable approximation as long as none of the data points contributes a major portion of the final result.

If the uncertainties of the data points are all equal $\sigma_i = \sigma$, the partial derivatives in Equation (4.10) are simply

$$\frac{\partial \mu'}{\partial x_i} = \frac{\partial}{\partial x_i}\left(\frac{1}{N}\sum x_i\right) = \frac{1}{N} \tag{4.11}$$

and combining Equations (4.10) and (4.11), we obtain

$$\sigma_\mu^2 = \sum\left[\sigma_i^2\left(\frac{1}{N}\right)^2\right] = \frac{\sigma^2}{N} \tag{4.12}$$

for the estimated error in the mean σ_μ. Thus, the standard deviation of our determination of the mean μ' and, therefore, the precision of our estimate of the quantity μ, improves as the square root of the number of measurements.

The standard deviation σ of the parent population can be estimated from a consideration of the measuring equipment and conditions, or internally from the data, according to Equation (1.8):

$$\sigma \simeq s = \sqrt{\frac{1}{N-1}\sum(x_i - \bar{x})^2} \tag{4.13}$$

which gives for the uncertainty σ_μ in the determination of the mean

$$\sigma_\mu = \frac{\sigma}{\sqrt{N}} \simeq \frac{s}{\sqrt{N}} \tag{4.14}$$

where σ_μ is referred to as the standard deviation of the mean, or the *standard error.* In principle, the value of σ obtained from Equation (4.13) should be consistent with the estimate made from the experimental equipment.

Example 4.1 We return to the student's measurement of the dropped ball (Example 1.2). Let us assume that the time for the ball to fall 2.00 m had been established previously by careful measurements to be $T_{est} = 0.639$ s. The student drops the ball 50 times and concludes, from a consideration of the electronic timer and the experimental arrangement that the uncertainty in each of his individual measurements is ± 0.020 s, consistent with the standard deviation determined from the data. This finite precision of the apparatus results in a spread of observations grouped around the established time as illustrated by the histogram of the data in Figure 1.2.

Because the uncertainties in all the data points are equal ($s_i = s$), the student calculates from his measurements and Equation (4.9) that his estimate of the mean time is $\mu \simeq \bar{T} = 0.635$s, with a standard deviation from Equation (4.13) of $\sigma \simeq s = 0.020$ s. From Equation (4.14), he estimates the uncertainty in his determination of the mean to be $\sigma_\mu \simeq s/\sqrt{N} = 0.020/\sqrt{50}$ or $\sigma_\mu \simeq 0.0028$. He quotes his experimental result as $T_{exp} = (0.635 \pm 0.003)$ s.

To compare his experimental value T_{exp} to the established value T_{est}, the student calculates the number of standard deviations by which the two differ, $n = |T_{exp} - T_{est}|/\sigma_\mu = 1.4$. From the integral of the Gaussian probability equation in Table C.2, we observe that we might expect a measurement to be within 1.4 standard deviations in about 83.8% of repeated experiments, or to exceed 1.4 standard deviations in about 16.2% of the cases.

It is important to realize that the standard deviation of the *data* does not decrease with repeated measurement; it just becomes better determined. On the other hand, the standard deviation of the *mean* decreases as the square root of the number of measurements, indicating the improvement in our ability to estimate the mean of the distribution. Graphically we could illustrate this improvement by plotting a histogram of the data and noting that our ability to determine the peak of the distribution improves as the number of measurements increases and the distribution becomes smoother.

A Warning About Statistics

Equation (4.12) might suggest that the error in the mean of a set of measurements x_i can be reduced indefinitely by repeated measurements of x_i. We should be aware of the limitations of this equation before assuming that an experimental result can be improved to any desired degree of accuracy if we are willing to do enough work. There are three main limitations to consider: those of available time and resources, those imposed by systematic errors, and those imposed by nonstatistical fluctuations.

The first of these limitations is a very practical one. It may not be possible to take enough repeated measurements to make a significant improvement in the standard deviation of the result. The student of Example 1.2 may be able to make 50 measurements of the time, but might not have the patience to make four times as many measurements to cut the uncertainty by a factor of 2. Similarly, an experiment at a particle accelerator may be assigned 1000 hours of beam time. It may not be possible to increase the allocation to 16,000 hours to improve the precision of the result by a factor of 4.

All experiments are subject to systematic errors at some level. Even after every possible effort has been made to understand the experimental equipment and correct for all known defects and errors of calibration, there comes a point at which further knowledge is unobtainable. For instance, any error in the placement of the detectors that measure times at the beginning and ending of the ball's fall in Example 1.2 will lead to a systematic uncertainty in the time (or in the distance through which the ball fell) and thus in the final result of the experiment.

The phrase "nonstatistical fluctuations" can hide a multitude of sins, or at least problems, in our experiments. It is a rare experiment that follows the Gaussian distribution beyond 3 or 4 standard deviations. More likely, some unexplained data points, or *outliers,* may appear in our data sample, far from the mean. Such points may imply the existence of other contaminating points within the central probability region, masked by the large body of good points. A thorough study of background effects and sources of possible contaminating is obviously required, but at some level, these effects are bound to limit the accuracy of the experiment.

What are we to make of those unexpected points that appear in our data plots well beyond their level of probability? Some may arise from a chance careless measurement. Did our attention wander at the instant when we should have recorded the data point? Did we accidentally interchange two digits in writing down our measurement? Perhaps we can understand and make corrections for some of these effects. Other anomalies in the data may be caused by equipment malfunction. Did our electronic detector respond to a particularly striking clash of metal from the local all-powerful rock radio station? Did our trusty computer decide to check e-mail rather than respond to an urgent data interrupt? And was the distribution that we chose to represent our data the correct one for this experiment?

We may be able to make corrections for these problems, once we are aware of their existence, but there are always others. At some level, things will happen that we cannot understand, and for which we cannot make corrections, and these "things" will cause data to appear where statistically no data should exist, and data points to vanish that should have been there. The moral is, be aware and do not trust statistics in the tails of the distributions.

Elimination of Data Points

There will be occasions when we feel justified in eliminating or correcting outlying data points. For example, suppose that among the time measurements in Example 1.2, the student had recorded one as 0.86s. The student would likely conclude that he had meant to write 0.68s and either ignore or correct the point. What if one measurement had been recorded as 0.72s? Should any action be taken? The point is about 4 standard deviations away from the mean of all the data points, and referring to Table C.2 we see that there is about a 0.06% probability of obtaining in a single measurement a value that is that far from the mean. Thus, in a sample of 50 such measurements we should expect to collect about $50 \times 0.00006 = 0.003$ such events.

The established condition for discarding data in such circumstances is known as Chauvenet's criterion, which states that we should discard a data point if we expect less than half an event to be farther from the mean than the suspect point. If our sample point satisfies this requirement and, as long as we are convinced that our data do indeed follow the Gaussian distribution, we may discard the point with reasonable confidence and recalculate the mean and standard deviation. Thus, for the two examples cited in the preceding paragraph, it would be permissible under Chauvenet's criterion to discard both the 0.86s and the 0.72s data points.

Removing an outlying point has a greater effect on the standard deviation than on the mean of a data sample, because the standard deviation depends on the squares of the deviations from the mean. Deleting one such point will lead to a smaller standard deviation and perhaps another point or two will now become candidates for rejection. We should be very cautious about changing data unless we are confident that we understand the source of the problem we are seeking to correct, and repeated point deletion is generally not recommended. The importance of keeping good records of any changes to the data sample must also be emphasized.

Weighting the Data–Nonuniform Uncertainties

In developing the probability $P(\mu')$ of Equation (4.5) from the individual probabilities $P_i(\mu')$ of Equation (4.3), we assumed that the data points were all extracted from the same parent population. In some circumstances, however, there will be data points that have been measured with better or worse precision than others. We can express this quantitatively by assuming parent distributions with the same mean μ but with different standard deviations σ_i.

If we assign to each data point x_i its own standard deviation σ_i representing the precision with which that particular data point was measured, Equation (4.5) for the probability $P(\mu')$ that the observed set of N data points come from parent distributions with means $\mu_i = \mu'$ and standard deviations σ_i becomes

$$P(\mu') = \prod_{i=1}^{n}\left(\frac{1}{\sigma_i\sqrt{2\pi}}\right)\exp\left[-\frac{1}{2}\sum\left(\frac{x_i-\mu'}{\sigma_i}\right)^2\right] \tag{4.15}$$

Using the method of maximum likelihood, we must maximize this probability, which is equivalent to minimizing the argument in the exponential. Setting the first derivative of the argument to 0, we obtain

$$-\frac{1}{2}\frac{d}{d\mu'}\sum\left(\frac{x_i-\mu'}{\sigma_i}\right)^2 = \sum\left(\frac{x_i-\mu'}{\sigma_i^2}\right) = 0 \tag{4.16}$$

The most probable value is therefore the weighted average of the data points

$$\mu' = \frac{\Sigma(x_i/\sigma_i^2)}{\Sigma(1/\sigma_i^2)} \tag{4.17}$$

where each data point x_i in the sum is weighted inversely by its own variance σ_i^2.

Error in the Weighted Mean

If the uncertainties of the data points are not equal, we evaluate $\partial\mu'/\partial x_i$ from the expression of Equation (4.17) for the mean μ':

$$\frac{\partial\mu'}{\partial x_i} = \frac{\partial}{\partial x_i}\frac{\Sigma(x_i/\sigma_i^2)}{\Sigma(1/\sigma_i^2)} = \frac{1/\sigma_i^2}{\Sigma(1/\sigma_i^2)} \tag{4.18}$$

Substituting this result into Equation (4.10) yields a general formula for the uncertainty of the mean σ:

$$\sigma_\mu^2 = \sum\frac{1/\sigma_i^2}{[\Sigma(1/\sigma_i^2)]^2} = \frac{1}{\Sigma(1/\sigma_i^2)} \tag{4.19}$$

Relative Uncertainties

It may be that the relative values of σ_i are known, but the absolute magnitudes are not. For example, if one set of data is acquired with one scale range and another set with a different scale range, the σ_i may be equal within each set but differ by a known factor between the two sets, as would be the case if σ_i were proportional to the scale range. In such a case, the *relative* values of the σ_i should be included as weighting factors in the determination of the mean μ and its uncertainty, and the *absolute* magnitudes of the σ_i can be estimated from the dispersion of the data points around the mean.

Let us define weighting factors w_i such that

$$kw_i = 1/\sigma_i^2 \tag{4.20}$$

where k is an unknown scaling constant and the σ_i are the standard deviations associated with each measurement. We assume that the weights w_i are known but that the absolute values of the standard deviations σ_i are not. Then, Equation (4.17) can be written

$$\mu' = \frac{\Sigma(x_i/\sigma_i^2)}{\Sigma(1/\sigma_i^2)} = \frac{\Sigma kw_ix_i}{\Sigma kw_i} = \frac{\Sigma w_ix_i}{\Sigma w_i} \tag{4.21}$$

and the result depends only on the relative weights and not on the absolute magnitudes of the σ_i.

To find the error in the estimate μ' of the mean we must calculate a weighted *average variance* of the data:

$$\sigma^2 = \frac{\Sigma w_i(x_i - \mu')^2}{\Sigma w_i} \times \frac{N}{(N-1)} = \left(\frac{\Sigma w_i x_i^2}{\Sigma w_i} - \mu'^2\right) \times \frac{N}{(N-1)} \tag{4.22}$$

where the last factor corrects for the fact that the mean μ' was itself determined from the data. We may recognize the expression in brackets as the difference between the weighted average of the squares of our measurements x_i and the square of the weighted average. The variance of the mean can then be determined by substituting the expression for σ^2 from Equation (4.22) into Equation (4.14):

$$\sigma_\mu^2 = \frac{\sigma^2}{N} \tag{4.23}$$

If they are required, the value of the scaling constant k and of the values of the separate variances σ_i can be estimated by equating the two expressions for σ_μ of Equations (4.14) and (4.19) and replacing $1/\sigma_i^2$ by kw_i to give

$$\frac{\sigma^2}{N} = \frac{1}{\Sigma(1/\sigma_i^2)} = \frac{1}{k\Sigma w_i} \tag{4.24}$$

so

$$k = \frac{N}{\sigma^2}\frac{1}{\Sigma w_i} \tag{4.25}$$

and therefore

$$\sigma_i^2 = \frac{1}{kw_i} = \frac{\sigma^2 \Sigma w_i}{N w_i} \tag{4.26}$$

Example 4.2. A student performs an experiment to determine the voltage of a standard cell. The student makes 40 measurements with the apparatus and finds a result $\bar{x}_1 = 1.022$ V with a spread $s_1 = 0.01$ V in the observations. After looking over her data she realizes that she could improve the equipment to decrease the uncertainty by a factor of 2.5 ($s_2 = 0.004$ V) so she makes 10 more measurements that yield a result $\bar{x}_2 = 1.018$ V.

The mean of all these observations is given by Equation (4.17):

$$\begin{aligned}\mu \simeq \bar{x} &= \frac{\dfrac{40(1.022)}{0.01^2} + \dfrac{10(1.018)}{0.004^2}}{\dfrac{40}{0.01^2} + \dfrac{10}{0.004^2}}\ \text{V} \\ &= \frac{4.00(1.022) + 6.25(1.018)}{4.00 + 6.25}\ \text{V} \\ &= 1.0196\ \text{V}\end{aligned}$$

The uncertainty σ_μ in the mean is given by Equation (4.19):

$$\sigma_\mu \simeq s = \left(\frac{40}{0.01^2} + \frac{10}{0.004^2}\right)^{-1/2} = 0.00099 \text{ V}$$

The result should be quoted as $\mu = (1.0196 \pm 0.0010)$V although $\mu = (1.020 \pm 0.001)$V would also be acceptable. Carrying the fourth place (which is completely undefined) after the decimal point just eliminates any possible rounding errors if these data should later be merged with data from other experiments.

The precision of the final result in Example 4.2 is better than that for either part of the experiment. The uncertainties in the estimates of the means μ_1 and μ_2 determined from the two sets of data independently are given by Equation (4.14):

$$s_2 = \frac{0.01}{\sqrt{40}} \text{ V} = 0.0016 \text{ V} \qquad s_2 = \frac{0.004}{\sqrt{10}} \text{ V} = 0.0013 \text{ V}$$

A comparison of these values illustrates the fact that taking more measurements decreases the resulting uncertainty only as the square root of the number of observations, which for this case is not so important as decreasing σ_i.

What if the student did not know the absolute uncertainties in her measurements, but only that the uncertainties had been improved by a factor of 2.5? She could obtain the estimate of the mean directly from Equation (4.21) by replacing $1/\sigma_1^2$ by the weight $w_i = 1$, and $1/\sigma_2^2$ by the weight $w_i = 2.5^2$, to give

$$\mu \simeq \frac{40(1)(1.022) \text{ V} + 10(2.5^2)(1.018) \text{ V}}{40(1) + 10(2.5)^2} = 1.0196 \text{ V}$$

To find the error in the mean the student could calculate σ from her data by Equation (4.22) and use Equation (4.23) to estimate σ_μ.

Discarding Data

Even though the student in Example 4.2 made four times as many observations at the lower precision (higher uncertainty), the high-precision contribution is over 1.5 times as effective as the low-precision data in determining the mean. The student should probably consider ignoring the low-precision data entirely and using only the high-precision data. Why should we ever throw away data that are not known to be bad? Additionally, because in this case the earlier data are weighted so as to be rather unimportant to the result, what is the point in neglecting them and thereby wasting all the effort that went into collecting those first 40 data points?

These are questions that arise again and again in experimental science as one works to find the elusive parameters of the parent distribution. The answer lies in the fact that experiments tend to be improved over time and often the earliest data-taking period is best considered a training period for the experimenters and a "shakedown" period for the equipment. Why risk contaminating the sample with data of uncertain results when they contribute so little to the final result? The relative standard deviations of the two data sets can serve as a guide. If the spread of the later distribution shows marked improvement over that of the earlier data, then we should seriously consider throwing away the earlier data unless we are certain of their reliability. There is no hard and fast rule that defines when a group of data should be ignored—common sense must be applied. However, we should make an effort to overcome the natural bias toward using all data simply to recover our investment of time and effort. Greater reliability may be gained by using the cleaner sample alone.

4.2 STATISTICAL FLUCTUATIONS

For some experiments the standard deviations σ_i can be determined more accurately from a knowledge of the estimated parent distribution than from the data or from other experiments. If the observations are known to follow the Gaussian distribution, the standard deviation σ is a free parameter and must be determined experimentally. If, however, the observations are known to be distributed according to the Poisson distribution, the standard deviation is equal to the square root of the mean.

As discussed in Chapter 2, Poisson probability is appropriate for describing the distribution of the data points in counting experiments where the observations are the numbers of events detected per unit time interval. In such experiments, there are fluctuations in the counting rate from observation to observation that result solely from the intrinsically random nature of the process and are independent of any imprecision in measuring the time interval or of any inexactness in counting the number of events occurring in the interval. Because the fluctuations in the observations result from the statistical nature of the process, they are classified as *statistical fluctuations,* and the resulting errors in the final determinations are classified as *statistical errors.*

In any given time interval there is a finite chance of observing *any* positive (or zero) integral number of events. The probability for observing any specific number of counts is given by the Poisson probability function, with mean μ_t, where the subscript t indicates that these are average values for the time interval of length Δt. Thus, if we make N measurements of the number of counts in time intervals of fixed length Δt, we expect that a histogram of the number of counts x_i recorded in each time interval would follow the Poisson distribution for mean μ_t.

Mean and Standard Deviation

For values of the mean μ_t greater than about ten, the Gaussian distribution closely approximates the shape of the Poisson distribution. Therefore, we can use the formula of Equation (4.9) for estimating the mean with the assumption that all data points were extracted from the same parent population and thus have the same uncertainties:

$$\mu_t \simeq \bar{x}_t = \frac{1}{N}\sum x_i \tag{4.27}$$

Here the x_i are the numbers of events detected in the N time intervals Δt, and the assumption that the data were all drawn from the same parent population is equivalent to assuming that the lengths of the time intervals were the same for all measurements.

According to Equation (2.19), the variance σ^2 for a Poisson distribution is equal to the mean μ:

$$\sigma_i^2 = \mu_t \simeq \bar{x}_i \tag{4.28}$$

The uncertainty in the mean $\sigma_{t\mu}$ is obtained by combining Equations (4.12) and (4.28):

$$\sigma_{t_\mu} = \frac{\sigma_t}{\sqrt{N}} = \sqrt{\frac{\mu_t}{N}} \simeq \sqrt{\frac{\bar{x}_t}{N}} \tag{4.29}$$

We usually wish to find the mean number of counts per unit time, which is just

$$\mu = \frac{\mu_t}{\Delta t} \quad \text{with} \quad \sigma_\mu = \frac{\sigma_{t_\mu}}{\Delta t} = \sqrt{\frac{\mu}{N\Delta t}} \tag{4.30}$$

As we might expect, the uncertainty in the mean number of counts per unit time σ_μ is inversely proportional to the square roots of both the time interval Δt and the number of measurements N.

In some experiments, as in Example 4.2, data may be obtained with varying uncertainties. For purely statistical fluctuations, this implies that counts were recorded in varying time intervals Δt_i. If we wish to find the mean number of counts μ per unit time from such data, there are two possible ways to proceed. If we have the raw data counts (the x_i) and we know they are all independent, then we can simply add all the x_i and divide the sum by the sum of the time intervals:

$$\mu = \frac{\Sigma x_i}{\Sigma \Delta t_i} \quad \text{and} \quad \sigma^2 = \mu$$

The more likely situation is that we know only the means μ_j and corresponding standard deviations σ_j of the means, obtained from the experiments. For example, when dealing with published experimental data, we should assume that the errors incorporate instrumental as well as statistical uncertainties. With such data, the safest procedure is to apply Equations (4.17) and (4.19) to evaluate the weighted mean μ of the individual means μ_i and the standard deviation σ_μ of the mean:

$$\mu \simeq \frac{\Sigma(\mu_j/\sigma_j^2)}{\Sigma(1/\sigma_j^2)} \quad \text{and} \quad \sigma_\mu^2 = \frac{1}{\Sigma(1/\sigma_j^2)} \tag{4.31}$$

Example 4.3. The activity of a radioactive source is measured $N = 10$ times with a time interval $\Delta t = 1$ min. The data are given in Table 4.1. The average of these data points is $\bar{x} = 15.1$ counts per minute. The spread of the data points is characterized by $\sigma = 3.9$ counts per minute calculated from the mean according to Equation (4.27). The uncertainty in the mean is calculated according to Equation (4.29) to be $\sigma_{\bar{x}} \simeq 1.2$ counts per minute.

TABLE 4.1
Experimental data for the activity of a radioactive source from the experiment of Example 4.3

Interval Δt_i (min)	Counts x_i	
1	19	
1	11	
1	24	$\bar{x} = \frac{1}{N}\Sigma x_i = 151$ counts per 10 minutes
1	16	$= 15.1$ counts per minute
1	11	
1	15	
1	22	$\sigma \simeq \sqrt{\bar{x}} = 3.9$ counts per minute
1	9	
1	9	$\sigma_{\bar{x}} \simeq \frac{\sigma}{\sqrt{N}} = 1.2$ counts per minute
1	15	
	Sum = 151	
10	147	$\sigma_{10} = \sqrt{147}$ counts per 10 minutes
		$= 1.2$ counts per minute
Total 20	298	$\bar{x}_{20} = (151 + 147)/(10 + 10)$
		$= 298/20 = 14.9$ counts per minute
		$\sigma_{20} = \sqrt{298}$ counts per 20 minutes
		$= 0.9$ counts per minute

Note: The data tabulated are the number of counts x_i detected in each time interval Δt_i.

If we were to combine the data into one observation $x' = \Sigma x_i$ from one 10-min interval, we would obtain the same result. The activity is $x' = 151$ counts per 10 minutes $= 15.1$ counts per minute as before. The uncertainty in the result is given by the standard deviation of the single data point $\sigma_{x'} = \sqrt{151} = 12.3$ counts per 10 minutes $= 1.2$ counts per minute.

Suppose that we made an additional measurement for a 10-min period and obtained $x'' \cong 147$ counts. We could combine x' and x'' exactly as before to obtain a total

$$\bar{x}_T = x' + x'' = (151 + 147)/(10 + 10) = 14.9 \text{ counts per minute}$$

with an uncertainty

$$\sigma_{\bar{x}_T} = \sqrt{298}/20 = 0.87 \text{ counts per minute}$$

which is smaller than $\sigma_{\bar{x}}$ by a factor of $\sqrt{2}$. Alternatively, we could combine the original data points according to Equation (4.17) and calculate the uncertainty in the final result σ_T by combining the uncertainties of the individual data points according to Equation (4.19).

Note that, although we could have simplified matters by recording all the data as one experimental point, $x = 298$ counts per 20 minutes, by so doing, we would lose all independent information about the shape of the distribution that could be used as a partial check on the validity of the experiment.

4.3 PROBABILITY TESTS

The object of our analysis is to obtain the best estimates, $\bar{x}$ and s_μ, of the mean μ and its uncertainty σ_μ, and to interpret the probability associated with the uncertainty as a measure of our success in determining the parent parameters. Regardless of the method used to make the measurements and analyze the data, we must always estimate the uncertainty in our results to indicate numerically our confidence in them.

Generally, we relate the uncertainty to a Gaussian probability. We have noted that approximately 68% of the measurements in a Gaussian distribution fall within ± 1 standard deviation of the mean μ. Thus, when we find the average of a large number of individual measurements, we expect the distribution of means to be Gaussian, centered on $\bar{x} \simeq \mu$ with width $s \simeq \sigma$, so that approximately 68% of our measurements of x would fall within the range $(\bar{x} - s) < x < (\bar{x} + s)$. Similarly, if we were to repeat the entire experiment many times, we should expect our individual determinations of $\bar{x}$ to form a Gaussian distribution about the mean μ, with width $s_\mu = s/\sqrt{N} \simeq \sigma/\sqrt{N}$. Again, we should expect that approximately 68% of our determinations of $\bar{x}$ should fall within the range $(\mu - s_\mu) < \bar{x} < (\mu + s_\mu)$. If we are convinced that we have made careful and unbiased measurements, we make a slight logical leap to state that there is approximately 68% probability that the true value of the mean μ lies in the range $(\bar{x} - s_\mu) < \mu < (\bar{x} + s_\mu)$ or that the specified range is the 68% *confidence interval.*

Rather than state confidence intervals in terms of 1 standard deviation, we may prefer to state a range that refers to a specific probability level. For example, we may wish to state that our result lies between two values, x_1 and x_2 with a 90% level of confidence, which would correspond to $x_1 = \bar{x} - 1.64\, s_\mu$ and $x_2 = \bar{x} + 1.64\, s_\mu$. Thus, in Example 4.1, the student may report 90% probability that the mean time is within the interval $0.635 \pm (1.64 \times 0.0028)$ s, or $\bar{T} = (0.635 \pm 0.005)$ s at a 90% confidence level. In science, it is customary to report 1 standard deviation uncertainties unless we state otherwise. In other fields, for example political polling, it is customary to report a 95% confidence level, corresponding to approximately 2 standard deviations. American polls are generally accompanied by a statement like "Poll of 1000 adults; margin of error plus or minus 3 percentage points." Canadian media would report "Poll results are likely to be accurate within 3 percentage points 19 times out of 20." If you assume a binomial distribution, you should realize that both statements have almost the same content.

Student's *t* Distribution

We should be aware that Gaussian probability may not apply to our particular data set, and even an experimental distribution that nominally follows Gaussian statistics is apt to deviate in the tails. When the data set is small, there is another consideration. Not only the mean, but also our estimate s_μ of the standard error σ_μ may be poorly determined. The probabilities that we calculate from the Gaussian distribution take no account of the latter problem.

In such cases, a better estimate of the probability can be obtained from *Student's t* distribution,[1] which describes the distribution of the parameter $t = |x - \bar{x}|/s_\mu$, where t is the number of standard deviations of the sample distribution s_μ by which x differs from $\bar{x}$.

$$p_t(t, \nu) = \frac{1}{\sqrt{(\nu\pi)}} \frac{\Gamma[(\nu + 1)/2]}{\Gamma(\nu/2)} \left(1 + \frac{t^2}{\nu}\right)^{-(\nu+1)/2}$$

where the gamma function $\Gamma(n)$ is equivalent to the factorial function n! extended to nonintegral arguments. (See Equation 11.7).

Unlike the Gaussian distribution, Student's t distribution depends upon the number of degrees of freedom ν. If $\bar{x}$ represents the mean of N numbers and x is not derived from the data, then $\nu = N - 1$. If both x and $\bar{x}$ are means, s_μ must be the joint standard deviation of x and $\bar{x}$, and ν must be the total number of degrees of freedom. In the limit of large ν, Student's t and Gaussian probability distributions agree. As with the Gaussian distribution, we are usually interested in integrated values that relate to the probability of obtaining a result within a specific range $\pm t$ standard deviations. For example, we might wish to report our estimate of the probability that the true value of μ lies within the range $(\bar{x} - ts_\mu) < \mu < (\bar{x} + ts_\mu)$ with $t = |\bar{x} - \mu|/s_\mu$.

Table C.8 lists probabilities obtained by integrating the Student's t distribution from $x = \bar{x} - ts_\mu$ to $x = \bar{x} + ts_\mu$ for specified values of t and the number of degrees of freedom ν. The corresponding values for Gaussian probability (which are independent of ν) are listed in the last column.

Consider again Example 4.1 in which the student made 50 time measurements and found that the mean of his measurements deviated by $1.4s_\mu$ from the established value. From Gaussian probability we observed that approximately 84% of experiments should yield a result that is within 1.4 standard deviations of the expected result. From Student's t distribution (Table C.8.), we observe that the probability is lower by about 0.6%. However, suppose the student made only six measurements using a more precise measuring system and again obtained a result that differed from the mean by $t = 1.4s_\mu$ (see Exercise 4.12). Small numbers of measurements are common in undergraduate laboratory experiments, where time may be short and the measurements may be tedious. What probability is implied for 5 degrees of freedom by a difference of $t = 1.4s_\mu$? The Gaussian probability is unchanged at ~84%; Student's t predicts ~78%. Thus, for experiments with only a few degrees of freedom, Gaussian probability overestimates the confidence level associated with a given range t. Another way of looking at this is to note that, for the same confidence level, Student's t probability requires a larger uncertainty estimate than does Gaussian probability.

Generally, a result is considered to be significant only at confidence levels of 95% or better. In Gaussian probability, this corresponds to a range of approximately $\pm 2\sigma$. We can observe from Table C.8 that for a sample of only three data points

---

[1] "Review of Particle Physics," *The European Physical Journal C,* vol. 15, p. 193 (2000)

($\nu = 2$), the Student's t probability for 95% confidence corresponds to a range of more than $\pm 4\sigma$.

4.4 CHI-SQUARE TESTS OF A DISTRIBUTION

Once we have calculated the mean and standard deviation from our data, we may be in a position to say even more about the parent population. If we can be fairly confident of the type of parent distribution that describes the spread of the data points (e.g., Gaussian or Poisson distribution), then we can describe the parent distribution in detail and predict the outcome of future experiments from a statistical point of view.

Because we are concerned with the behavior of the probability density function $p(x_i)$ as a function of the observed values of x_i, a complete discussion will be postponed until Chapter 11 following the development of procedures for comparing data with complex functions. Let us for now use the results of Chapter 11 without derivation. The test that we shall describe here is the χ^2 (chi-square) test for goodness of fit.

Probability Distribution

If N measurements x_i are made of the quantity x, we can truncate the data to a common least count and group the observations into frequencies of identical observations to make a histogram. Let us assume that j runs from 1 to n so there are n possible different values of x_j, and let us call the frequency of observations, or number of counts in each histogram bin, $h(x_j)$ for each different measured value of x_j. If the probability for observing the value x_j in any random measurement is denoted by $P(x_j)$, then the expected number of such observations is $y(x_j) = NP(x_j)$, where N is the total number of measurements. Figures 4.1 and 4.2 show the same six-bin histogram, drawn from a Gaussian parent distribution with mean $\mu = 5.0$ and standard deviation $\sigma = 1$, corresponding to 100 total measurements. The parent distribution, $y(x_j) = NP(x_j)$, is illustrated by the solid Gaussian curve on each histogram.

For each measured value x_j, there is a standard deviation $\sigma_j(\mathrm{h})$ associated with the uncertainty in the observed frequency $h(x_j)$. This is not the same as the uncertainty σ_i associated with the spread of the individual measurements x_i about their mean μ, but rather describes the spread of the measurements of each of the frequencies $h(x_j)$ about its mean μ_j. If we were to repeat the experiment many times to determine the distribution of frequency measurements at each value of x_j, we should find each parent distribution to be Poisson with mean $\mu_j = y(x_j)$ and variance $\sigma_j^2(y) = y(x_j)$. Thus, for each value of x_j, there is a distribution curve, $P_j(y_k)$, that describes the probability of obtaining the value of the frequency $h_k(x_j)$ in the kth trial experiment when the expected value is $y(x_j)$. It is the spread of these measurements for each value of j that is characterized by $\sigma_j(h)$. These distributions are illustrated in Figures 4.1 and 4.2 as dotted Poisson curves at each value of x_j. In Figure 4.1 the Poisson curves are centered at the observed frequencies $h(x_j)$ with standard deviations $\sigma_j(h) = \sqrt{h(x_j)}$. In principle, we should center the Poisson curves at the

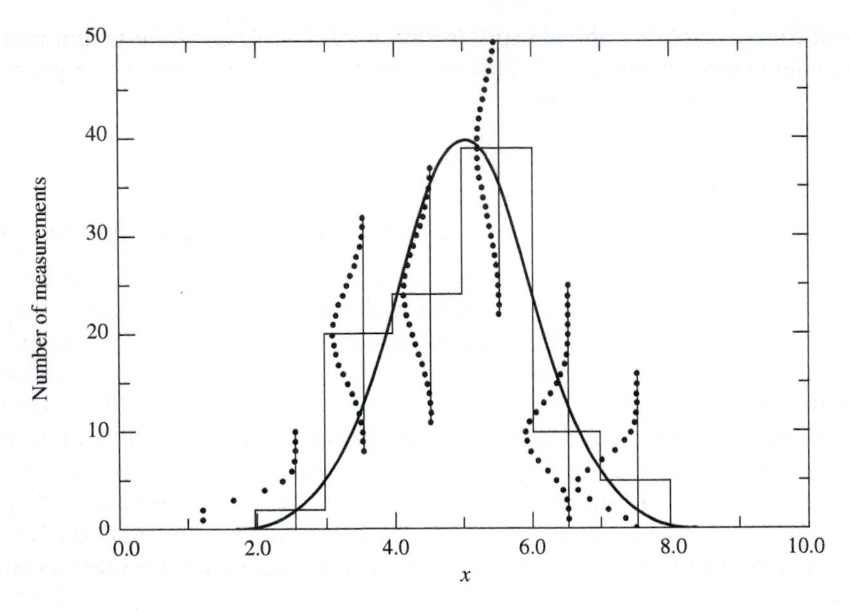

FIGURE 4.1
Histogram, drawn from a Gaussian distribution mean $\mu = 5.0$ and standard deviation $\sigma = 1$, corresponding to 100 total measurements. The parent distribution $y(x_j) = NP(x_j)$ is illustrated by the large Gaussian curve. The smaller dotted curves represent the Poisson distribution of events in each bin, based on the sample data.

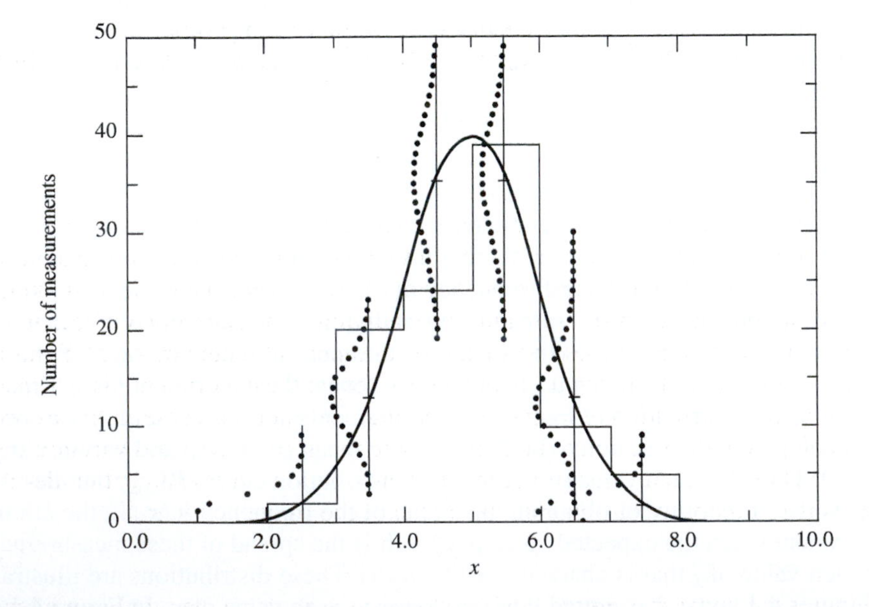

FIGURE 4.2
The same histogram as shown in Figure 4.1 with dotted curves representing the Poisson distribution of events in each bin, based on the parent distribution.

frequencies $\mu_j = y(x_j)$ with standard deviation $\sigma_j(h) = \sqrt{\mu_j}$ of the parent population as illustrated in Figure 4.2. However, in an actual experiment, we generally would not know these parameters.

Definition of χ^2

With the preceding definitions for n, N, x_j, $h(x_j)$, $P(x_j)$, and $\sigma_j(h)$, the definition of χ^2 from Chapter 11 is

$$\chi^2 \equiv \sum_{j=1}^{n} \frac{[h(x_j) - NP(x_j)]^2}{\sigma_j(h)^2} \tag{4.32}$$

In most experiments, however, we do not know the values of $\sigma_j(h)$ because we make only one set of measurements $f(x_j)$. Fortunately, these uncertainties can be estimated from the data directly without measuring them explicitly.

If we consider the data of Figure 4.2, we observe that for each value of x_j, we have extracted a proportionate random sample of the parent population for that value. The fluctuations in the observed frequencies $h(x_j)$ come from the statistical probabilities of making random selections of finite numbers of items and are distributed according to the Poisson distribution with $y(x_j)$ as mean. Although the distribution of frequencies $y(x_j)$ in Figure 4.2 is Gaussian, the probability functions for the spreads of the measurements of each frequency are Poisson distributions.

For the Poisson distribution, the variance $\sigma_j(h)^2$ is equal to the mean $y(x_j)$ of the distribution, and thus we can estimate $\sigma_j(h)$ from the data to be $\sigma_j(h) = \sqrt{NP(x_j)} \simeq \sqrt{h(x_j)}$. Equation (4.32) simplifies to

$$\chi^2 \equiv \sum_{j=1}^{n} \frac{[h(x_j) - NP(x_j)]^2}{NP(x_j)} \simeq \sum_{j=1}^{n} \frac{[h(x_j) - NP(x_j)]^2}{h(x_j)} \tag{4.33}$$

Test of χ^2

As defined in Equations (4.32) and (4.33), χ^2 is a statistic that characterizes the dispersion of the observed frequencies from the expected frequencies. If the observed frequencies were to agree exactly with the predicted frequencies $h(x_j) = NP(x_j)$, then we should find $\chi^2 = 0$. From our understanding of probability, we realize that this is not a very likely outcome of an experiment. The numerator of Equation (4.32) is a measure of the spread of the observations; the denominator is a measure of the expected spread. We might imagine that for good agreement, the average spread of the data would correspond to the expected spread, and thus we should get a contribution of about one from each frequency, or $\chi^2 \simeq n$ for the entire distribution. This is almost correct. In fact, the true expectation value for χ^2 is

$$\langle\chi^2\rangle = \nu = n - n_c \tag{4.34}$$

where ν is the number of degrees of freedom and is equal to the number n of sample frequencies minus the number n_c of constraints or parameters that have been calculated from the data to describe the probability function $NP(x_j)$. For our example, even if $NP(x_j)$ is chosen completely independently of the distribution $h(x_j)$, there is still the normalizing factor N corresponding to the total number of events in the distribution, so that the expectation value of χ^2 must at best be $\langle\chi^2\rangle = n - 1$.

In order to estimate the probability that our calculated values of χ^2 are consistent with our expected distribution of the data, we must know how χ^2 is distributed. If our value of χ^2 corresponds to a reasonable high probability, then we can have confidence in our assumed distribution.

It is convenient to define the *reduced chi-square* as $\chi^2_\nu \equiv \chi^2/\nu$, with expectation value $\langle\chi^2_\nu\rangle = 1$. Values of χ^2_ν much larger than 1 result from large deviations from the assumed distribution and may indicate poor measurements, incorrect assignment of uncertainties, or an incorrect choice of probability function. Very small values of χ^2_ν are equally unacceptable and may imply some misunderstanding of the experiment. Rather than consider the probability of obtaining any particular value of χ^2 or χ^2_ν (which is infinitesimally small), we shall use an integral test to determine the probability of observing a value of χ^2_ν equal to or greater than the one we calculated. This is similar to our consideration of the probability that a measurement of a variable deviates by more than a certain amount from the mean.

Table C.4 gives the probability that a random sample of data points drawn from the assumed probability distribution would yield a value of χ^2 as large as or larger than the observed value in a given experiment with ν degrees of freedom.

If the probability is reasonably close to 1, then the assumed distribution describes the spread of the data points well. If the probability is small, either the assumed distribution is not a good estimate of the parent distribution or the data sample is not representative of the parent distribution. There is no yes-or-no answer to the test; in fact, we should expect to find a probability of about 0.5 with $\chi^2_\nu \simeq 1$, because statistically the observed values of χ^2 should exceed the norm half the time. But in most cases, the probability is either reasonably large or unreasonably small, and the test is fairly conclusive. A further discussion of the statistical significance of the χ^2 probability function will be given in Chapter 11.

Let us consider again the data of Example 1.2 (and 4.1), which are summarized as a histogram in Figure 1.2 with the frequencies listed in Table 4.2. To test the agreement between the data and the predicted distribution, we have calculated the function $y(x_j) = NP(x_j)$ at each value of x_j from the mean and standard deviation of the parent distribution (column 3 of Table 4.2), and from the mean and standard deviation of the data, that is, from the sample distribution (column 6). The uncertainties σ_j calculated as the square roots of the values predicted by the parent distribution and by the sample distribution are listed in columns 4 and 7 respectively. The individual contributions (before squaring) to the values of χ^2, $[h(x_j) - NP(x_j)]/\sigma_j$, are listed in columns 5 and 8. The calculated values of χ^2 from the comparison between the data and each distribution are the sums of the squares of these last quantities.

For the comparison of the 11 data points with the parent distribution we have one constraint, the normalization constant N determined from the data, and therefore the expectation value of χ^2 is $\nu = 11 - 1 = 10$. We obtained $\chi^2 = 13.03$ and thus, $\chi^2_\nu = 1.30$. Interpolating in Table C.4, we observe that the corresponding probability of obtaining a value $\chi^2_\nu \geq 1.30$ with 10 degrees of freedom is ~23%. For a similar comparison with an *estimate* of the parent distribution based on the mean and standard deviation of the data, we have two additional constraints, the mean and standard deviation. Thus, for this comparison, the expectation value of χ^2 is $\nu = 11 - 3 = 8$. We obtained $\chi^2 = 7.85$ and, thus, $\chi^2_\nu = 0.98$. The corresponding probability for obtaining a value $\chi^2_\nu \geq 0.98$ with 8 degrees of freedom is ~45%.

Generalizations of the χ^2 Test

In the preceding example we knew the parent distributions and were therefore able to determine the uncertainties $\sigma_j(h)$ from the predicted probability. In most cases, where the actual parameters of the probability function are being determined in the calculation, we must use an estimate of the parent population based on these parameters and must estimate the uncertainties in the $y(x_j)$ from the data themselves. To do this we must replace the uncertainties in columns 4 and 7 of Table 4.2 with the square roots of the observed frequencies in column 2.

Furthermore, although our example was clearly based on a simple probability function, the χ^2 test is often generalized to compare data obtained in any type of experiment to the prediction of a model. The uncertainties in the measurements may be instrumental or statistical or a combination of both, and the uncertainty $\sigma_j(h)^2$ in the denominator of Equation (4.32) may represent a Gaussian error distribution rather than the Poisson distribution. In fact, several of the histogram bins in our example contained small numbers of counts, and thus, the statistical application of the test was not strictly correct, because we assume Gaussian statistics in the χ^2 calculation. However, the test still provides us with a reproducible method of evaluating

TABLE 4.2
χ^2 analysis of the data of Example 4.1

	Observed frequency h_j	From parent distribution			From sample distribution		
Time	h_j	y_j	σ_j	$\frac{y_j - h_j}{\sigma_j}$	y_j	σ_j	$\frac{y_j - h_j}{\sigma_j}$
0.595	2	0.89	0.94	−1.18	1.35	1.16	0.56
0.605	2	2.35	1.53	0.23	3.24	1.80	−0.69
0.615	11	4.85	2.20	−2.79	6.05	2.46	2.01
0.625	6	7.81	2.79	0.65	8.80	2.97	−0.94
0.635	12	9.78	3.13	−0.71	9.97	3.16	0.64
0.645	8	9.53	3.09	0.50	8.80	2.97	−0.27
0.655	4	7.24	2.69	1.20	6.05	2.46	−0.83
0.665	3	4.28	2.07	0.62	3.24	1.80	−0.13
0.675	1	1.97	1.40	0.69	1.35	1.16	−0.30
0.685	1	0.71	0.84	−0.35	0.44	0.66	0.85
0.695	0	0.20	0.44	0.44	0.11	0.33	−0.33
		$\chi^2_\nu = 13.03/10 = 1.30$			$\chi^2_\nu = 7.85/8 = 0.98$		

Note: Parameters of the parent Gaussian distribution are $\mu = 0.639$ and $\sigma = 0.020$ s; parameters estimated from the sample distribution are $\mu = 0.635$ s and $\sigma = 0.020$ s.

the quality of our data, and if we are concerned with statistical accuracy, we can merge the low-count bins to satisfy the Gaussian statistics requirement.

Another application of the chi-squared test is in comparing two sets of data to attempt to decide whether or not they were drawn from the same parent population. Suppose that we have measured two distributions, $g(x_j)$ and $h(x_j)$, and wish to determine the probability that the two sets were not drawn from the same parent probability distribution $P(x_j)$. Clearly, we could apply the χ^2 test separately to the two sets of data and determine separately χ^2 probabilities that each set was not associated with the supposed parent population $P(x_j)$. However, we can also make a direct test, independent of the parent population, by writing

$$\chi^2 = \sum_{j=1}^{n} \frac{[g(x_j) - h(x_j)]^2}{\sigma^2(g) + \sigma^2(h)} \tag{4.35}$$

The denominator $\sigma^2(g) + \sigma^2(h)$ is just the variance of the difference $g(x_j) - h(x_j)$. As in the previous examples, the expectation value of χ^2 depends on the relation between the two parts of the numerator, $g(x_j)$ and $h(x_j)$. If the two parts, corresponding to the distributions of the two data sets, were obtained completely independently of one another, then the number of degrees of freedom equals n and $\langle\chi^2\rangle = n$. If one of the distributions $g(x_j)$ or $h(x_j)$ has been normalized to the other, then the number of degrees of freedom is reduced by 1 and $\langle\chi^2\rangle = n - 1$. Again, we interpret the χ^2 probability in a negative sense. If the value of χ^2/ν is large, and therefore the probability given in Table C.4 is low, we may conclude that the two sets of data were drawn from different distributions. However, for a low value of χ^2 and therefore high probability, we cannot draw the opposite conclusion that the two data sets $g(x_j)$ and $h(x_j)$ were drawn from the same distribution. There is always the possibility that there are indeed two different but closely similar distributions and that our data are not sufficiently sensitive to detect the difference between the two.

Constraints and Degrees of Freedom

Equation (4.34) defines the number of degrees of freedom, ν, and $<\chi^2>$, the expectation value of χ^2. To clarify the relation between constraints and degrees of freedom in a χ^2 test, consider a data set that is expected to show a linear relation between the measured values x_j and h_j, that is,

$$y_j = A + Bx_j$$

Clearly, two measurements of y at two different values of x are required just to define the two parameters, A and B, of the straight line so there are two constraints ($n_c = 2$) on the system and at least three measurements ($n = 3$) must be made before a test can be applied. Under these circumstances, if we assume that points $j = 1$ and $j = 2$ are used to calculate A and B, Equation (4.32) becomes

$$\chi^2 = (h_3 - y_3)^2/\sigma_3^2(h)$$

and we should expect to find

$$\langle\chi^2\rangle = n - n_c = 3 - 2 = 1$$

Similarly, if we measure $n = 4$ points, there will be two points available for the χ^2 test or 2 degrees of freedom. Of course, in general, we would not use just two points to calculate the two parameters. Rather, we should perform a *least-squares fit* in which all measurements are treated equally (or weighted according to their uncertainties). However, the same principle holds: we impose two constraints on our calculation to define the two parameters of a straight line, leaving 2 degrees of freedom.

SUMMARY

Weighted mean:

$$\bar{x} = \frac{\Sigma(x_i/\sigma_i^2)}{\Sigma(1/\sigma_i^2)} \underset{\sigma_i=\sigma}{\longrightarrow} \frac{1}{N}\sum x_i$$

Variance of mean:

$$\sigma_\mu^2 = \frac{1}{\Sigma(1/\sigma_j^2)} \underset{\sigma_i=\sigma}{\longrightarrow} \frac{\sigma^2}{N}$$

Instrumental uncertainties: Fluctuations in measurements due to finite precision of measuring instruments:

$$\sigma^2 \simeq s^2 = \frac{1}{N-1}\sum (x_i - \bar{x})^2$$

Statistical fluctuations: Fluctuations in observations resulting from statistical probability of taking random samples of finite numbers of items:

$$\sigma^2 = \mu \simeq \bar{x}$$

χ^2 test: Comparison of observed frequency distribution $h(x_j)$ of possible observations x_j versus predicted distribution $NP(x_j)$, where N is the number of data points and $P(x_j)$ is the theoretical probability distribution:

$$\chi^2 \equiv \sum_{j=1}^{n} \frac{[h(x_j) - NP(x_j)]^2}{\sigma_j(h)^2}$$

Degrees of freedom ν: Number of data points minus the number of parameters to be determined from the data points.
Reduced χ^2: $\chi_\nu^2 = \chi^2/\nu$. For χ^2 tests, χ_ν^2 should be approximately equal to 1.
Graphs and tables of χ^2: Table C.4 gives the probability that a random sample of data when compared to its *parent distribution* would yield values of χ_ν^2 as large as or larger than the observed value.

EXERCISES

4.1. Calculate the standard deviation and the error in the mean value of x from the data of Exercise 1.4. Are the values reasonable? (See Exercise 2.4.)

4.2. Repeat Exercise 4.1 for the data of Exercise 1.5.

4.3. Read the data of Example 2.4 from Figures 2.3 and 2.4. Recalculate the curves and calculate χ^2 and χ^2_ν for the agreement between the curves and the histograms. Use only bins with five or more counts.

4.4. Work out the intermediate steps in Equation (4.19).

4.5. A student measures the period of a pendulum and obtains the following values.

Trial	1	2	3	4	5	6	7	8
Period	1.35	1.34	1.32	1.36	1.33	1.34	1.37	1.35

(*a*) Find the mean and standard deviation of the measurements and the standard deviation of the mean.

(*b*) Estimate the probability that another single measurement will fall within 0.02 s of the mean.

4.6. (*a*) Find the mean and the standard deviation of the mean of the following numbers under the assumption that they were all drawn from the same parent population.

(*b*) In fact, data points 1 through 20 were measured with uniform uncertainty σ, whereas data points 21 through 30 were measured more carefully so that the uniform uncertainty was only $\sigma/2$. Find the mean and standard deviation of the mean under these conditions.

Trial	$x(\sigma)$	**Trial**	$x(\sigma)$	**Trial**	$x(\sigma/2)$
1	2.40	11	1.94	21	2.59
2	2.45	12	1.55	22	2.65
3	2.47	13	2.12	23	2.55
4	3.13	14	2.17	24	2.07
5	2.92	15	3.06	25	2.61
6	2.85	16	1.97	26	2.61
7	2.05	17	2.23	27	2.54
8	2.52	18	3.20	28	2.76
9	2.94	19	2.24	29	2.37
10	1.89	20	2.60	30	2.57

4.7. A counter is set to count gamma rays from a radioactive source. The total number of counts, including background, recorded in each 1-min interval is listed in the accompanying table. An independent measurement of the background in a 5-min interval gave 58 counts. From these data find:

(*a*) The mean background in a 1-min interval and its uncertainty.

(*b*) The corrected counting rate from the source alone and its uncertainty.

Trial	1	2	3	4	5	6	7	8	9	10
Total counts	125	130	105	126	128	119	137	131	115	116

4.8. The *Particle Data Tables* list the following eight experimental measurements of the mean lifetime of the K_s meson with their uncertainties, in units of 10^{-10} s. Find the weighted mean of the data and the uncertainty in the mean.

0.8971 ± 0.0021 0.8941 ± 0.0014 0.8929 ± 0.0016 0.8920 ± 0.0044 0.881 ± 0.009
0.8924 ± 0.0032 0.8937 ± 0.0048 0.8958 ± 0.0045

4.9. Eleven students in an undergraduate laboratory combined their measurements of the mean lifetime of an excited state. Their individual measurements are tabulated.

Student	1	2	3	4	5	6	7	8	9	10	11
$\tau(s)$	34.3	32.2	35.4	33.5	34.7	33.5	27.9	32.0	32.4	31.0	19.8
σ_τ	1.6	1.2	1.5	1.4	1.6	1.5	1.9	1.2	1.4	1.8	1.3

Find the maximum likelihood estimate of the mean and its uncertainty.

4.10. Assume that you have a box of resistors that have a Gaussian distribution of resistances with mean value $\mu = 100\ \Omega$ and standard deviation $\sigma = 20\ \Omega$ (i.e., 20% resistors). Suppose that you wish to form a subgroup of resistors with $\mu = 100\ \Omega$ and standard deviation of $5\ \Omega$ (i.e., 5% resistors) by selecting all resistors with resistance between the two limits $r_1 = \mu - a$ and $r_2 = \mu + a$.

(*a*) Find the value of a.

(*b*) What fraction of the resistors should satisfy the condition?

(*c*) Find the standard deviation of the remaining sample.

4.11. Suppose that 1000 adults responded to a poll about a current bill in Congress, and that 622 approved, while 378 disapproved.

(*a*) Assume that there was 50% a priori probability of obtaining either answer and calculate the standard deviation of the result. Find the "margin of error," that is, the uncertainty that corresponds to a 95% confidence interval. (Use Gaussian probability. Justify this.)

(*b*) Assume the probabilities implied by the observed numbers of votes in each category and repeat the calculation. Note the insensitivity of the standard deviation of the binomial distribution to variations in probability near 50%.

(*c*) Refer to the two statements about polling reports in Section 4.3 and show that they are approximately equivalent.

4.12. Six measurements of the length of a wooden block yielded the following values: 20.3, 20.4, 19.8, 20.4, 19.9, 20.7.

(*a*) From these numbers, calculate the mean, standard deviation, and standard error. Assume that the actual mean length has been established by previous measurements to be 20.00 cm and calculate t, the number of standard errors by which the calculated mean differs from the established value.

Refer to the tables in Appendix C to find the limits on the 95% confidence level for both Gaussian and Student's t probabilities.

(*b*) The experiment was repeated to obtain a total of 25 data sets of six measurements each from which the following 25 values of the mean were calculated.

20.25 20.10 20.02 20.12 20.00 19.73 19.73 20.13 20.22 20.22 20.27 19.83 20.00
19.77 20.10 20.28 19.97 19.88 20.32 19.98 20.05 20.23 19.92 19.97 19.77

Find the mean of these "means" and calculate their standard deviation. Compare this standard deviation to the standard error calculated in (*a*).

4.13. The following data represent the frequency distribution of 200 variables drawn from a parent Gaussian population with mean $\mu = 26.00$ and standard deviation $\sigma = 5.00$. The bins are two units wide and the lower edge of the first bin is at $x = 14$.

4 8 11 20 26 31 29 22 26 13 5 2 3

(*a*) Plot a histogram of these data.

(*b*) From the mean μ and standard deviation σ, calculate the Gaussian function that represents the parent distribution, normalized to the area of the histogram. Your first point should be calculated at $x = 15$, the midpoint of the first bin.

(*c*) Calculate χ^2 to test the agreement between the data and the theoretical curve.

(*d*) What is the expectation value of χ^2?

(*e*) Refer to Table C.4 to find the χ^2 probability of the fit, that is, the probability of drawing a random sample from the parent population that will yield a value of χ^2 as large as or larger than your calculated value.

4.14. Plot a histogram in ten-point bins of the course grades listed in Exercise 1.5. Plot a Gaussian curve based on the mean and standard deviation of the data, normalized to the area of the histogram. Apply the χ^2 test and check the associated probability from Table C.4.

CHAPTER 5

MONTE CARLO TECHNIQUES

5.1 INTRODUCTION

We saw in Chapter 4 the importance of probability distributions in the analysis of data samples, and observed that we are usually interested in the integrals or sums of such distributions over specified ranges. Although we have considered only experiments that are described by a single distribution, most experiments involve a combination of many different probability distributions. Consider, for example, a simple scattering experiment to measure the angular distribution of particles scattered from protons in a fixed target. The magnitude and direction of the momentum vector of the incident particles, the probability that a particle will collide with a proton in the target, and the resulting momentum vectors of the scattered particles can all be described in terms of probability distributions. The final experimental result can be treated in terms of a multiple integration over all these distributions.

Analytical evaluation of such an integral is rarely possible, so numerical methods must be used. However, even the simplest first-order numerical integration can become very tedious for a multidimensional integral. A one-dimensional integral of a function can be determined efficiently by evaluating the function N times on a regular grid, where the number of samples N depends on the structure of the function and the required accuracy. (See Appendix A.3.) A two-dimensional integral requires sampling in two dimensions and, for accuracy comparable to that of the corresponding one-dimensional problem, requires something like N^2 samples. A three-dimensional integral requires something like N^3 samples. For integrals with many dimensions, the number of grid points at which the function must be calculated becomes excessively large.

Before we continue with methods of extracting parameters from data, let us look at the Monte Carlo method, a way of evaluating these multiple integrals that depends on random sampling from probability density distributions, rather than regular grid-based sampling techniques. The Monte Carlo method provides the experimental scientist with one of the most powerful tools available for planning experiments and analyzing data. Basically, Monte Carlo is a method of calculating multiple integrals by random sampling. Practically, it provides a method of simulating experiments and creating models of experimental data. With a Monte Carlo calculation, we can test the statistical significance of data with relatively simple calculations that require neither a deep theoretical understanding of statistical analysis nor sophisticated programming techniques.

The name *Monte Carlo* comes from the city on the Mediterranean with its famous casino, and a Monte Carlo calculation implies a statistical method of studying problems based on the use of random numbers, similar to those generated in the casino games of chance. One might reasonably ask whether the study of science can be aided by such associations, but in fact, with Monte Carlo techniques, very complicated scientific and mathematical problems can be solved with considerable ease and precision.

Example 5.1. Suppose that we wish to find the area of a circle of radius r_c but have forgotten the equation. We might inscribe the circle within a square of known area A_s and cover the surface of the square uniformly with small markers, say grains of rice. We find the ratio of the number of grains that lie within the circle to those that cover the square, and determine the area of the circle A_c from the relation

$$A_c = A_s N_c / N_s \tag{5.1}$$

where N_c and N_s are the numbers of grains of rice within the boundaries of the circle and of the square, respectively.

What would be the accuracy of this determination; that is, how close should we expect our answer to agree with the true value for the area of a circle? Clearly it would depend on the number and size of the rice grains relative to the size of the square, and on the uniformity of both the grains and their distribution over the square. What if we decided that instead of attempting to cover the square uniformly, we would be content with a random sampling obtained by tossing the rice grains from a distance so that they landed randomly on the square, with every location equally probable? Then we would obtain an interesting result: Our problem would reduce to a simple binomial calculation as long as we did not overpopulate the square but kept the density of rice grains low so that position of any grain on the square was not influenced by the presence of other grains. We should find that, for a fixed number of grains N_s thrown onto the square, the uncertainty σ in the measurement of the circular area would be given by the standard deviation for the binomial distribution with probability $p = A_c/A_s$,

$$\sigma = \sqrt{N_s p(1-p)} = \sqrt{N_c(1-p)} \tag{5.2}$$

Thus, if we were to increase the number of rice grains N_c by a factor of 4, the relative error in our determination of the area of the circle would decrease by a factor of 2.

Replacing the tossed rice grains by a set of computer generated random numbers is an obvious improvement. Let us inscribe our circle of unit radius in a square of side length 2, and generate $N = 100$ pairs of random numbers between -1 and $+1$ to determine the area. Then the probability of a "hit" is just the ratio of the area of the circle to the area of a square, or $p = \pi/4$, so in 100 tries, the mean number of hits will be $\mu = 100p = 78.5$, and the standard deviation, from Equation (5.2), will be $\sigma = \sqrt{Np(1-p} = \sqrt{100(\pi/4)(1-\pi/4)} = 4.1$. For our measurements of the area of the circle with 100 tries we should expect to obtain from Equation (5.1) $A_c = A_s \times N_c/N_s = (78.5 \pm 4.1) \times 2^2/100 = 3.14 \pm 0.16$.

Figure 5.1 shows a typical distribution of hits from one "toss" of 100 pairs of random numbers. In this example there were 73 hits, so we should estimate the area and its uncertainty from Equations (5.1) and (5.2) to be $A = 2.92 \pm 0.18$. To determine the uncertainty, we assumed that we did not know the a priori probability $p = \pi/4$ and, therefore, we used our experimental estimate $p \simeq 73/100$.

Figure 5.2 shows a histogram of the circle area estimates obtained in 100 independent Monte Carlo runs, each with 100 pairs of random numbers (or a total of 10,000 "tosses"). The Gaussian curve was calculated from the mean, $A = 3.127$, and standard deviation, $\sigma = 0.156$, of the 100 estimated areas.

Obviously, the area determination problem of Example 5.1 is much too simple to require a Monte Carlo calculation. However, for problems involving integrations of many variables and for those with complicated integration limits, the Monte

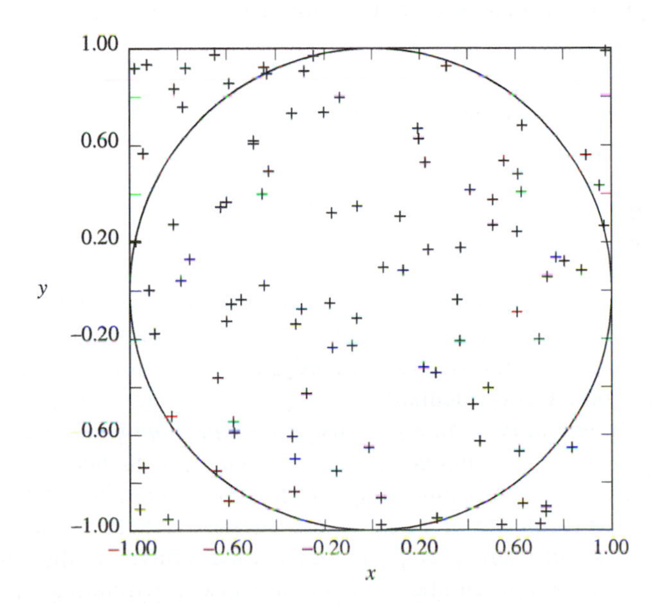

FIGURE 5.1
Estimation of the area of a circle by the Monte Carlo method. The plot illustrates a typical distribution of hits from one "toss" of 100 pairs of random numbers uniformly distributed between −1.00 and +1.00.

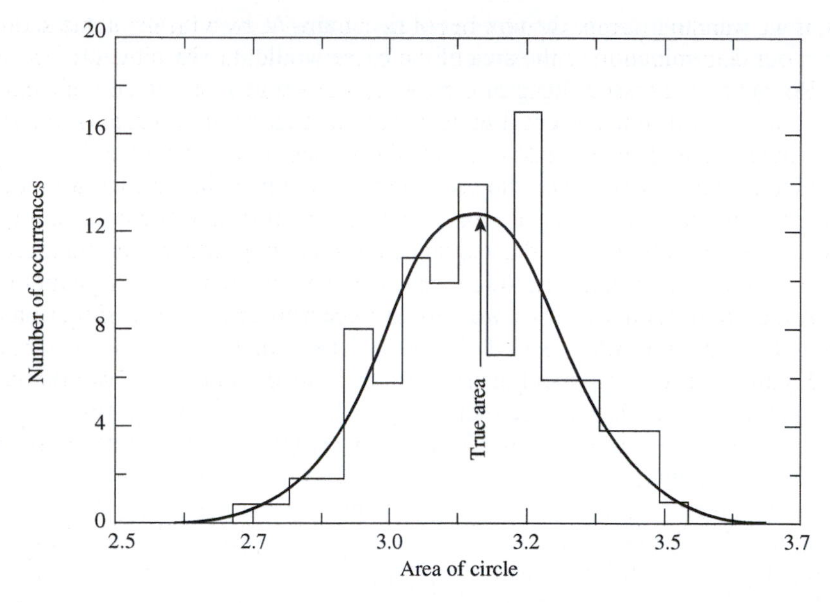

FIGURE 5.2
Histogram of the circle area estimates obtained in 100 independent Monte Carlo runs, each with 100 pairs of random numbers. The Gaussian curve was calculated from the mean $A = 3.127$ and standard deviation $\sigma = 0.156$ of the 100 estimated areas.

Carlo technique is invaluable, with its straightforward sampling and its relatively simple determination of the uncertainties.

5.2 RANDOM NUMBERS

A successful Monte Carlo calculation requires a reliable set of *random numbers*, but truly random numbers for use in calculations are hard to obtain. One might think of a scheme based upon measuring the times between cosmic ray hits in a detector, or on some physical process such as the generation of noise in an electronic circuit. Such numbers would be random in the sense that it would be impossible to predict the value of the next number from previous numbers but they are hardly convenient to use in extended calculations, and some might not have the necessary uniformity required for a Monte Carlo calculation.

In fact, it is generally preferable to use *pseudorandom numbers*, numbers generated by a computer algorithm designed to produce a sequence of apparently uncorrelated numbers that are uniformly distributed over a predefined range. In addition to the convenience of being able to generate these numbers within the Monte Carlo program itself, pseudorandom numbers have another important advantage over truly random numbers for Monte Carlo calculations. A Monte Carlo program may use a great many random numbers, and the path of the calculation through the program will depend on the numbers chosen in each run. With truly random numbers, every run of a Monte Carlo calculation would follow a different path and produce different results. Such a program would be very difficult to debug.

With pseudorandom numbers, we can repeat a calculation with the same sequence of numbers, and search for any particular problems that may be hidden in the code.

There are other advantages too. If we are studying the sensitivity of a calculation to variations in a selected parameter, we can reduce the variance of the *difference* between results calculated with two trial values of the parameter by using the same random number sequence for those parts of the calculation that are independent of the parameter in question. Finally, a pseudorandom number generator can be written to be *portable*; that is, the sequence of numbers produced by the algorithm is independent of computer hardware and language, so that a given program will produce the same results when run on different computers. In view of these advantages and the fact that we rarely, if ever, encounter situations where truly random numbers are required, we shall henceforth use the term *random numbers* to denote *pseudorandom numbers.*

In general, our random number generator must satisfy the following basic criteria:

1. The distribution of the numbers should be uniform within a specified range and should satisfy statistical tests for randomness, such as lack of predictability and of correlations among neighboring numbers.
2. The calculation should produce a large number of unique numbers before repeating the cycle.
3. The calculation should be very fast.

A simple multiplication method is often used to generate random numbers, or *uniform deviates*, as they are often called. An integer starting value or *seed* r_0 and two integer constants are chosen. Successive random numbers are derived from the recursion relation

$$r_{i+1} = (a \times r_i) \bmod m \tag{5.3}$$

where the mod operation corresponds to dividing the product in parentheses by the integer m to obtain the remainder. With appropriate choices of constants a and m, we can obtain a finite sequence of numbers that appear to be randomly selected between 1 and $m - 1$. The length of the sequence is determined by the choice of constants and is limited by the computer word size. For example, if we choose $m = 37$ and $a = 5$, Equation (5.3) gives us the cycle of 36 nicely mixed up numbers, listed in Table 5.1. Random number generators included with computer languages are often based on some variation of this multiplication technique. Careful and thorough statistical studies must be made to be sure that an untested random number generator produces an acceptable sequence of numbers.

Because the numbers generated by Equation (5.3) are not truly random, we might worry that our calculations are affected by hidden correlations in successively generated numbers. We can improve the randomness of our sample by *shuffling* the numbers. We generate two sequences of numbers with different generators a and m; one sequence is stored in an array and a number from the second sequence is used as an index to select numbers from the first sequence. For large programs that employ many random numbers, this method is limited by storage space, although *local shuffling* within a block of random numbers can be used.

TABLE 5.1
Pseudorandom numbers

i	r_i	i	r_i	i	r_i	i	r_i
1	1	10	6	19	36	28	31
2	5	11	30	20	32	29	7
3	25	12	2	21	12	30	35
4	14	13	10	22	23	31	27
5	33	14	13	23	4	32	24
6	17	15	28	24	20	33	9
7	11	16	29	25	26	34	8
8	18	17	34	26	19	35	3
9	16	18	22	27	21	36	15

Note: The generating equation is $r_{i+1} = (a \times r_i) \bmod m$, with $a = 5$ and $m = 37$. The cycle repeats $a_{37} = a_1$, $a_{38} = a_2$, and so forth.

Even a modest Monte Carlo program can require many random numbers and, to assure the statistical significance of results, we must be certain that the calculation does not use more than the maximum number generated by the algorithm before the sequence repeats. The sample generator of Equation (5.3) cannot produce more than $m - 1$ different values of r_i. The actual cycle length may be less than this range, depending on the choice of constants. The cycle length can be increased by employing two or more independent sequences such that the resulting cycle length is proportional to the product of the lengths of the component cycles.

A generator developed by Wichmann and Hill,[1] based on a simple linear combination of numbers from three independent sequences, is said to have a very long cycle ($\sim 7 \times 10^{12}$) and appears to be well tested. Because the algorithm uses three seeds, it is a little longer and slower than one- or two-seed algorithms, but its long repeat cycle, portability, and lack of correlations seem to make it a convenient, worry-free generator for most purposes. The algorithm is listed in Appendix E.

Although the fact that pseudorandom number generators always produce the same sequences of numbers from the same seeds is an advantage in program debugging, it may be a disadvantage in production running. For example, a simulation program developed for use as a science museum display could be very uninteresting if it repeated the same sequence of events every time it was run. If unpredictable seeds are required, they can easily be derived from the least counts of the computer clock. Commercial routines often include such a method of randomizing the starting seeds. On the other hand, if we wish to run a simulation program several times and to combine the results of the several different runs, the safest method to assure the statistical independence of the separate runs is to record the last values of the seeds at the end of each run and use these as starting seeds for the next run.

A thorough discussion of random number generation and of the Monte Carlo technique is given in Knuth (1981).

---

[1]The authors include a thorough and very useful discussion of the tests applied to a random number sequence, and of the development and testing of the published algorithm.

Warning

If you are using random numbers provided in commercial programs such as spread sheets or even scientific data analysis programs, you should always check the random number distributions for correlations, and make sure that the function behaves as advertised. For example, in early versions of one very popular scientific data analysis program, the choice of seed had no effect on the numbers produced by the random number routine.

5.3 RANDOM NUMBERS FROM PROBABILITY DISTRIBUTIONS

Transformation Method

Most number generators scale their output to provide real numbers uniformly distributed between 0 and 1. In general, however, we require numbers drawn from specific probability distributions. Let us define uniform deviates $p(r)$ drawn from a standard probability density distribution that is uniform between $r = 0$ and $r = 1$:

$$p(r) = \begin{cases} 1 & \text{for } 0 \leq r < 1 \\ 0 & \text{otherwise} \end{cases} \tag{5.4}$$

The distribution is *normalized* so that

$$\int_{-\infty}^{\infty} p(r)\, dr = \int_{0}^{1} 1\, dr = 1 \tag{5.5}$$

We shall refer to $p(r)$ as the *uniform distribution.*

Suppose that we require random deviates from a different normalized probability density distribution $P(r)$, which is defined to be uniform between $x = -1$ and 1; that is, the distribution

$$P(x) = \begin{cases} 1/2 & \text{for } -1 \leq x < 1 \\ 0 & \text{otherwise} \end{cases} \tag{5.6}$$

If we choose a random deviate r between 0 and 1 from the uniform distribution of Equation (5.4), it is obvious that we can calculate another random deviate x as a function of r:

$$x = f(r) = 2r - 1 \tag{5.7}$$

which will be uniformly distributed between -1 and $+1$. This is an example of a simple linear transformation.

To pick a random sample x from the distribution Equation (5.6), we started with a random deviate r drawn from the uniform distribution of Equation (5.4) and found a function $f(r)$ that gave the required relation between x and r. Let us find a general relation for obtaining a random deviate x from any probability density distribution $P(x)$, in terms of the random deviate r drawn from the uniform probability distribution $p(r)$.

Conservation of probability requires that the intervals Δr and Δx be related by the following expression

$$|p(r)\,\Delta r| = |P(x)\Delta x| \tag{5.8}$$

and, therefore, we can write

$$\int_{r=-\infty}^{r} p(r)\,dr = \int_{x=-\infty}^{x} P(x)\,dx \quad \text{or} \quad \int_{r=0}^{r} 1\,dr = \int_{x=-\infty}^{x} P(x)\,dx \tag{5.9}$$

which gives the general result

$$r = \int_{x=-\infty}^{x} P(x)\,dx \tag{5.10}$$

Thus, to find x, selected randomly from the probability distribution $P(x)$, we generate a random number r from the uniform distribution and find the value of the limit x that satisfies the integral equation (5.10).

Example 5.2. Consider the distribution described by the equation

$$p(x) = \begin{cases} A(1 + ax^2) & \text{for } -1 \le x < 1 \\ 0 & \text{otherwise} \end{cases} \tag{5.11}$$

where $P(x)$ is positive or zero everywhere within the specified range, and the normalizing constant A is chosen so that

$$\int_{-1}^{1} P(x)\,dx = 1 \tag{5.12}$$

We have

$$r = \int_{-\infty}^{x} P(x)\,dx = \int_{-1}^{x} A(1 + ax^2)\,dx \tag{5.13}$$

which gives

$$r = A(x + ax^3/3 + 1 + a/3) \tag{5.14}$$

and therefore, to find x we must solve the third-degree equation (5.14).

The procedure we have described is referred to as the *transformation method* of generating random deviates from probability distributions. In general, neither the integral equation (5.13) nor the solution of the resulting equation (5.14) can be obtained analytically, so numerical calculations are necessary.

The following steps are required to generate random deviates from a specific probability distribution by the transformation method with a numerical integration:

1. Decide on the range of x. Some probability density functions are defined in a finite range, as in Equation (5.6); others, such as the Gaussian function, extend to infinity. For numerical calculations, reasonable finite limits must be set on the range of the variable.

2. Normalize the probability function. If it is necessary to impose limits on the range of the variable x, then the function must be renormalized to assure that the integral is unity over the newly defined range. The normalization integral should be calculated by the same analytical integration or numerical integration routine that is used to find y.
3. Generate a random variable r drawn from the uniform distribution $p(r)$.
4. Integrate the normalized probability function $P(x)$ from negative infinity (or its defined lower limit) to the value $x = \mathcal{x}$, where $\mathcal{x}$ satisfies Equation (5.10).

Because the Monte Carlo method usually requires the generation of large numbers of individual events, it is essential to have available fast numerical interpolation and integration routines. To reduce computing time, it is often efficient to set up tables of repeatedly used solutions or integrals within the initializing section of a Monte Carlo program. For example, to pick a random deviate x from the distribution of Equation (5.11), we could do the integral of Equation (5.13) numerically at the beginning of our program, and set up a table of values of r versus x. Then, when we require a random number from the distribution, we generate a random number r and search the table for the corresponding value of x. In general, the search should be followed by an interpolation within the table (see Appendix A.1.) to avoid introducing excessive graininess into the resulting distribution. It would be even more convenient, but a little trickier, to produce a table of x versus r, so that the required value of x could be obtained from an index derived from r. In all cases of precalculated tables, it is important to consider the resolution required in the generated variable, because this will determine the intervals at which data must be stored, and therefore the size of the table, and the time required for a search.

Rejection Method

Although the *transformation method* is probably the most useful method for obtaining random deviates drawn from particular distributions, the *rejection method* is often the easiest to use. This is the method that we used in Example 5.1 to find the area of a circle, by generating random numbers uniformly over the surface of the circle and rejecting all except those that fell within the circumference.

Example 5.3. Suppose we wish to obtain random deviates between $x = -1$ and $x = +1$, drawn from the distribution function

$$P(x) = 1 + ax^2 \tag{5.15}$$

which is just the unnormalized distribution of Equation (5.11). To use the rejection method, we begin by generating a random deviate x' uniformly distributed between -1 and $+1$, corresponding to the allowed range of x, and a second random deviate y' uniformly distributed between 0 and $(1 + a)$, corresponding to the allowed range of $P(x)$. We can see that x' and y' must be given by

$$x' = -1 + 2r_i \quad \text{and} \quad y' = (1 + a)r_{i+1} \tag{5.16}$$

where r_i and r_{i+1} are successively generated random values of r drawn from the uniform distribution.

We count an event as a "hit" if the point (x', y') falls between the curve defined by $P(x)$ and the x axis, that is, if $y' < P(x')$, and a "miss" if it falls above the curve. In the limit of a large number of trials, the entire plot, including the area between the curve and the x axis, will be uniformly populated by this operation and our selected samples will be the x coordinates of the "hits," or the values of x', drawn randomly from the distribution $P(x)$. Note that with this method it is not necessary to normalize the distribution to form a true probability function. It is sufficient that the distribution be positive and well behaved within its allowed range.

The advantage of the rejection method over the transformation method is its simplicity. An integration is not required—only the probability function itself must be calculated. A disadvantage of the method is often its low efficiency. In a complex Monte Carlo program only a small fraction of the events may survive the complete calculation to become successful "hits" and the generation and subsequent rejection of so many random numbers may be very time consuming. To reduce this problem, it is advisable to place the strictest possible limits on the random coordinates used to map out the distribution function when using the rejection method.

5.4 SPECIFIC DISTRIBUTIONS

Gaussian Distribution

Almost any Monte Carlo calculation that simulates experimental measurements will require the generation of deviates drawn from a Gaussian distribution, or *Gaussian deviates.* A common application is simulation of measuring uncertainties by *smearing* variables. Fortunately, because of the convenient scaling properties of the Gaussian function, it is only necessary to generate Gaussian deviates from the standard distribution

$$P_G(z)\,dz = \frac{1}{\sqrt{2\pi}} \exp\left[-\frac{z^2}{2}\right]dz \tag{5.17}$$

with mean 0 and standard deviation 1, and to scale to different means μ and standard deviations σ by calculating

$$x = \sigma z + \mu \tag{5.18}$$

There are several different ways of obtaining random samples of the variable z from the distribution $P_G(z)$ of Equation (5.17). The two most obvious are the rejection and transformation methods discussed previously. Because the Gaussian function is defined between $-\infty$ and $+\infty$, these methods require that limits be placed on the range of z. For low-statistics calculations in which the Gaussian function is being used to simulate smearing of data caused by measuring errors, a range of $\pm 3\sigma$ should be satisfactory because all but ~0.3% of normally distributed events lie within this range.

Because the Gaussian function cannot be integrated analytically, numerical integrations are required for the transformation method. Decisions must be made on the order of integration and the step size as well as on the limits. A first- or second-order numerical integration (Appendix A.3.) is generally satisfactory, with a linear interpolation to find an approximation to the value of x in Equation (5.10) at the required value of the integral.

An interesting method for generating Gaussian deviates is based on the fact that if we repeatedly calculate the means of groups of numbers drawn randomly from any distribution, the distribution of those means tends to a Gaussian as the number of means increases. Thus, if we calculate many times the sums of N uniform deviates, drawn from the uniform distribution, we should expect the sums to fall into a truncated Gaussian distribution, bounded by 0 and N, with mean value $N/2$. If we generate N values of r from the distribution of Equation (5.4) and calculate

$$r_G = \sum_{i=1}^{N} r_i - N/2 \tag{5.19}$$

the variable r_G will be drawn from an approximately Gaussian distribution with mean $\mu = 0$ and standard deviation $\sigma = \sqrt{N/12}$. We should note that the maximum range of r_G will be limited to $\mu \pm N/2$ or $\mu \pm \sigma\sqrt{3N}$. For $N = 2$, the sum is a triangle function and as N increases, the distribution quickly takes on a Gaussian-like shape. Values of N as small as $N = 4$ are suitable for low statistics calculations. With $N = 4$, we have $\sigma = \sqrt{1/3} \simeq 0.058$ and the range of r_G from -2 to $+2$ corresponds to $\mu \pm \sigma\sqrt{12}$ or $\mu \pm 3.46\sigma$. If a better approximation to the Gaussian function is require and calculation time is not a problem, $N = 12$ is particularly convenient because the resulting variance and standard deviation are unity.

A particularly elegant method for obtaining random numbers drawn from the Gaussian distribution was suggested by Box and Müller (1958). This method makes use of the fact that, although the simple transformation method requires an integration of the Gaussian function, it is possible to find a function that generates the two-dimensional Gaussian distribution,

$$f(z_1, z_2) = \frac{1}{2\pi} \exp\left(-\frac{(z_1^2 + z_2^2)}{2}\right) = \frac{1}{\sqrt{2\pi}} \exp\left(-\frac{z_1^2}{2}\right) \times \frac{1}{\sqrt{2\pi}} \exp\left(-\frac{z_2^2}{2}\right) \tag{5.20}$$

From this equation, the authors obtained expressions that generate two Gaussian deviates, z_1 and z_2, from two uniform deviates, r_1 and r_2:

$$\begin{aligned} z_1 &= \sqrt{-2 \ln r_1} \cos 2\pi r_2 \\ z_2 &= \sqrt{-2 \ln r_1} \sin 2\pi r_2 \end{aligned} \tag{5.21}$$

Example 5.4. A uniform 10-cm long rod has one end held at 0°C and the other at 100°C so that the temperature along the rod is expected to vary linearly from 0° to 100°C. Let us attempt to simulate data that would be obtained by measuring the temperature at regular intervals along the rod. We shall assume that the parent population is described by the equation

$$T = a_0 + b_0 x \tag{5.22}$$

with $a_0 = 0$°C and $b_0 = 10$°C/cm, and that 10 measurements are made at 1-cm intervals from $x = 0.5$ to $x = 9.5$ cm, with negligible uncertainties in x_i and uniform measuring uncertainties in T_i of $\sigma_T = 1.0$°C.

Example 5.4 illustrates a common Monte Carlo technique: simulating the effects of measuring uncertainties by smearing data points. If a particular variable has a mean value T_i, with uncertainties σ_i and Gaussian uncertainties are assumed, then we obtain the smeared value of T_i from the relation

$$T_i' = T_i + \sigma_i r_i \tag{5.23}$$

where r_i is a random variable drawn from the standard Gaussian distribution with mean 0 and standard deviation 1. The calculation is equivalent to drawing the random variable T_i' directly from a Gaussian distribution with mean T_i and standard deviation σ_i.

Program 5.1. HOTROD (Appendix E) A simple Monte Carlo calculation to simulate the measurements described in Example 5.4. The program uses routines in the program unit MONTELIB.

Program 5.3. MONTELIB (Appendix E) Some useful Monte Carlo routines.

The data generated by the program HOTROD are shown in Table 5.2, with values of T_i for the parent population, predicted by Equation (5.22), and of T_i for the sample population, calculated from Equation (5.23) for various values of x_i. Note that, as we should expect, the modified values of T are scattered about the values calculated from Equation (5.22).

Choice of a Method

Which of these methods for generating samples from the Gaussian probability distribution is the best? The answer depends on need and circumstance. For general use it is convenient to keep a version of the Box-Müller method in your program library.

TABLE 5.2
Simulated temperature versus position data for a 10-cm rod held at $T = 0°C$ at $x = 0.0$ cm and at $T = 100°C$ at $x = 10.0$ cm

i	x_i (cm)	T_i (°C)	T_i' (°C)
1	0.5	5.00	4.71
2	1.5	15.00	15.43
3	2.5	25.00	23.24
4	3.5	35.00	35.77
5	4.5	45.00	45.39
6	5.5	55.00	52.26
7	6.5	65.00	65.71
8	7.5	75.00	76.96
9	8.5	85.00	85.97
10	9.5	95.00	93.77

Note: A uniform temperature gradient was assumed. The uncertainty in the measurement of T was assumed to be $\sigma_T = 1.0$ °C.

This routine produces a continuous range of samples limited only by the computer word size. For high-precision work, however, we should be aware that subtle correlations between adjacent uniform deviates have been shown to distort the tails of the Gaussian distribution of these numbers. If highest speed is essential, then the transformation method with a precalculated table of the integral and some pointers for quick access to the table should be the choice. This method requires making decisions on the range and resolution of the generated variable and some extra programming to create and access the integral table, but the lookup method can be very fast. Finally, if you are stranded on a desert island with only your laptop computer and have an urgent need for random selections from a Gaussian distribution, the method of summing N random numbers is sufficiently simple that you should be able to write and debug the routine in a few minutes, provided you can remember that the magic number is $N = 12$ for a variance of 1.

Poisson Distribution

Poisson statistics are important in most Monte Carlo calculations, but they are usually implied rather than calculated explicitly. Nevertheless, we sometimes wish to generate data that are distributed according to the Poisson function, and application of the transformation method to the problem is particularly simple and instructive. To find an integer x drawn from the Poisson distribution with mean μ, a *Poisson deviate*, we generate a random variable r from the uniform distribution, replace the integral of Equation (5.10) by the sum

$$r = \sum_{x=0}^{x} P_P(x; \mu) = \sum_{x=0}^{x} \frac{\mu^x}{x!} e^{-\mu} \tag{5.24}$$

and solve Equation (5.24) for x.

Although the Poisson function does not have the convenient scaling properties of the Gaussian function, and thus different calculations are required for each value of the mean μ, very few calculations are actually needed because we are interested in this distribution only at small values of μ, say $\mu \leq 16$, and only at integral values of the argument x. At larger values of μ, the Poisson distribution becomes indistinguishable from the Gaussian and it is generally more convenient to employ the Gaussian function in calculations.

Example 5.5. An instructor is preparing an exercise on Poisson statistics for his class. He plans to provide each student with a simulated data set corresponding to 200 Geiger counter measurements of cosmic ray flux recorded in 10-s intervals with an assumed mean counting rate of 8.4 counts per interval. The data will correspond to the number of counts recorded in each 10-s interval.

Students will be asked to make histograms of their individual data samples, find the means and standard deviations of the data, and compare their distributions with the predictions of Gaussian and Poisson probability functions.

For each student, a set of values of x is generated from Equation (5.24) with $\mu = 8.4$ and 200 different random numbers. The transformation method is used with a precalculated table of sums so that the value of x associated with each value of r can be selected by a simple search. To assure that each student's data set is independent, either all sets are generated in a single computer run or else the random number seeds are saved at the end of each run and used to start the next run.

Program 5.2. POISDECAY (Appendix E) Generates 200 random variables drawn from the Poisson probability distribution with mean $\mu = 8.4$ to illustrate Example 5.5. The program uses routines in the program unit MONTELIB.

The program calls the function POISSONDEVIATE with second argument INIT = TRUE to set up a table of sums of $P_P(i; \mu)$ from $i = 0$ to n indexed by n; that is, to form the array

$$S_n = \sum_{i=0}^{n} P_P(i; \mu) \quad \text{for } n = 1, 2, \ldots, n_{max} \tag{5.25}$$

so that

$$S_n = S_{n-1} + P_P(n; \mu) \quad \text{with } S_0 = P_P(0; \mu) = e^{-\mu} \tag{5.26}$$

where $n_{max} = N + 8\sqrt{\mu}$ is selected as a reasonable upper range for the Poisson curve.

For each event, the program calls POISSONDEVIATE with second argument INIT = FALSE to select a value from the table. The routine POISSONDEVIATE generates a random number r from the uniform distribution and searches the table beginning at S_0, to find the value of n for which $S_h \geq r$. The value of n at which this occurs is the desired random sample from the Poisson distribution. As the samples are generated they are entered in a histogram by calls to the routine HISTOGRAM.

A histogram of 200 variables drawn from the Poisson distribution Program 5.2 is shown in Figure 5.3 with the parent distribution represented as a solid curve (although it is, of course, not defined between integer values of the abscissa). The values of the Poisson function, calculated by the routine POISSONRECUR, and the sums, calculated by the routine POISSONDEVIATE, for $\mu = 8.4$ and for n ranging from 0 to 31, are displayed in Table 5.3.

We note that with the precalculated table it is only necessary to increment a counter a few times and compare two real numbers to obtain each random variable, whereas, without the table, it would have been necessary to calculate the Poisson function several times for each generated sample, in addition to comparing the two real numbers.

Exponential Distribution

If the Monte Carlo problem includes the generation of unstable states, random numbers drawn from an exponential distribution will be needed. Here the transformation method is clearly the method of choice because the integral equation (5.10) and resultant equation can be solved analytically.

Example 5.6. Consider an experiment to study the decay rate of a radioactive source with estimated mean life of τ seconds. The experiment involves collecting counts over successive time intervals Δt with a Geiger counter and scaler combination and plotting the number of counts in each interval against the mean interval time.

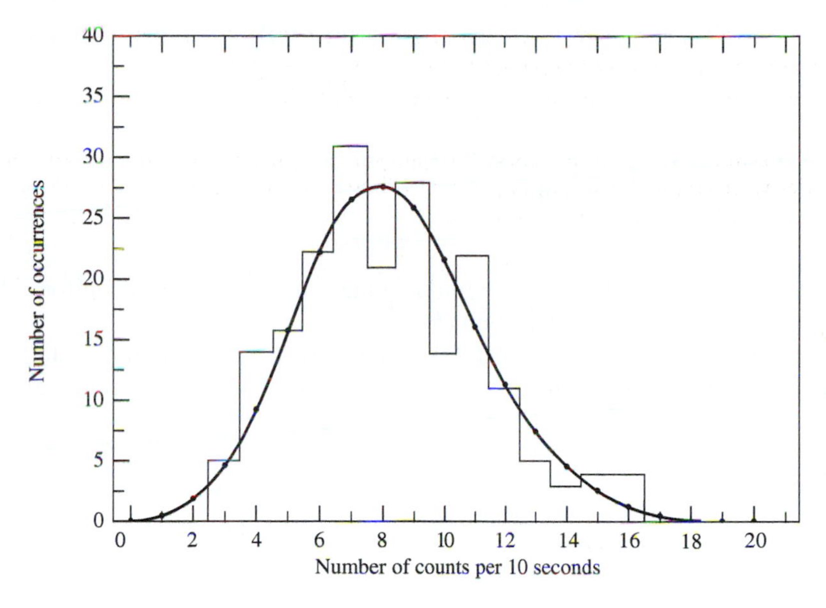

FIGURE 5.3
Histogram of 200 random variables generated by Program 5.3 from the Poisson distribution with mean $\mu = 8.4$.

TABLE 5.3
Poisson probability $P_P(i; \mu)$ and summed probability $S_i = \sum_{i=0}^{n} P_P(i; \mu)$ for $\mu = 8.4$

n	$P_P(n; \mu)$	S_n	n	$P_P(n; \mu)$	S_n
0	0.0002248673	0.0002248673	16	0.0066035175	0.9940781736
1	0.0018888855	0.0021137528	17	0.0032629145	0.9973410882
2	0.0079333192	0.0100470720	18	0.0015226935	0.9988637816
3	0.0222132938	0.0322603658	19	0.0006731908	0.9995369724
4	0.0466479169	0.0789082827	20	0.0002827401	0.9998197126
5	0.0783685004	0.1572767830	21	0.0001130961	0.9999328086
6	0.1097159005	0.2669926835	22	0.0000431821	0.9999759908
7	0.1316590806	0.3986517641	23	0.0000157709	0.9999917616
8	0.1382420346	0.5368937988	24	0.0000055198	0.9999972814
9	0.1290258990	0.6659196977	25	0.0000018547	0.9999991361
10	0.1083817551	0.7743014529	26	0.0000005992	0.9999997353
11	0.0827642494	0.8570657023	27	0.0000001864	0.9999999217
12	0.0579349746	0.9150006768	28	0.0000000559	0.9999999776
13	0.0374349066	0.9524355835	29	0.0000000162	0.9999999938
14	0.0224609440	0.9748965275	30	0.0000000045	0.9999999983
15	0.0125781286	0.9874746561	31	0.0000000012	1.0000000000

Note: The summation was terminated arbitrarily at $n \simeq \mu + 8\sqrt{\mu} \simeq 31$, and $P_P(31; \mu)$ was set to 1.

We wish to simulate this experiment with a Monte Carlo calculation. The normalized probability density function for obtaining a count at time t from an exponential distribution with mean life τ is given by

$$P_e(t;\tau) = \begin{cases} 0 & \text{for } t < 0 \\ \dfrac{e^{-t/\tau}}{\tau} & \text{for } t \geq 0 \end{cases} \tag{5.27}$$

We can obtain an expression for random samples t_i from this distribution by applying Equation (5.10) to obtain

$$t_i = -\tau \ln r_i \tag{5.28}$$

Thus, to obtain each value of t_i, we find a random number from the uniform distribution and calculate t_i from Equation (5.28).

Let us consider a second method of generating a histogram of data for this example, a method that is much more efficient, but that severely limits any later treatment of the data.

We can calculate the fraction of events that the parent distribution predicts would fall into each of the Δt wide histogram bins from the equation

$$\Delta N'(t) = \int_{t-d}^{t+d} \frac{e^{-x/\tau}}{\tau}\, dx = e^{-t/\tau}\Big|_{t-d}^{t+d} \simeq \frac{\Delta t}{\tau}\, e^{-t/\tau} \tag{5.29}$$

where we have written $d = \Delta t/2$. The effect of the statistical errors is to smear each of these calculated values in a way consistent with the Poisson distribution with mean $\mu = \Delta N_i'$. For small values of $\Delta N_i'$ we find the smeared value ΔN_i directly from Equation (5.24):

$$r = \sum_{x=0}^{\Delta N} P_P(x;\, \Delta N') \tag{5.30}$$

For larger values of $\Delta N_i'$ calculation with the Poisson equation would be too tedious, but we can use Gaussian smearing as in Example 5.4 with $\sigma_i = \sqrt{\mu}$. Note that the Poisson equation *must* be used for bins with low statistics to assure a positive number of counts in each bin. (A reminder: The overall distribution of events in this example is exponential; the expected distribution of events *in each individual bin* follows the Poisson distribution, as discussed in Section 4.3.)

Although these two methods of generating a data set or histogram produce equivalent statistical results for Example 5.6, they differ in important details. The full Monte Carlo method required generating individual "events" that can be recorded and studied. For example, we could check the statistical behavior of the data by subdividing the sample into several smaller groups. We could also investigate the effect of decreasing as well as increasing the binning intervals Δt. Finally, if we should wish to expand the study, perhaps to consider experimental geometry and detector efficiency, the full Monte Carlo method will allow that. The smearing method, on the other hand, produces only the ten numbers, representing the counts in the ten bins. Aside from merging the bins, we have no control over the data for future calculations. It is strictly a fast, "one-shot" procedure with a specific limited aim.

Example 5.7. Consider an experiment to determine the mean life of an elementary particle, the short-lived K_S^0 meson (which we shall refer to as the kaon), from measurements of the decay in flight of many such particles. In principle, we can determine the mean life τ by measuring the distribution of decay times, fitting the probability density function of Equation (5.27) to the data and solving for τ. In practice, we must make corrections for biases resulting from detection inefficiencies, including those associated with the finite sizes of our detectors. We can use a Monte Carlo calculation to estimate these biases and enable us to apply the appropriate correction.

The experimental arrangement is sketched in Figure 5.4. A high-energy charged particle p_i interacts in the target at the production vertex V_1 to produce several charged and neutral secondary particles, including a neutral kaon. The kaon travels a distance L before decaying into two pions, π_1 and π_2, at the decay vertex V_2. We determine the coordinates of the production vertex by measuring in the *production vertex detector* the trajectories of charged particles that are produced with the kaon, and tracing back these trajectories to their intersection point in the target. Similarly, we determine the coordinates of the decay vertex by measuring in the *decay vertex detector* the trajectories of the two charged pions from the kaon decay, and tracing these trajectories back to their intersection point, V_2. (The trajectories of neutral particles are much more difficult to measure than those of the charged particles.) We calculate the momentum of the neutral kaon from measurements of the momentum vectors of its two decay products, π_1 and π_2.

The geometry of the detector plays a critical role in the analysis of the data. We can make useful measurements only on events in which the trajectories of the charged particles can be measured in the vertex detectors. To assure precise measurements of the secondary tracks from the decay of the kaon, we define a *fiducial* region in which the decay must occur. The dashed rectangle on Figure 5.4 indicates the *fiducial* region with its limits d_1 and d_2 along the *x-axis*. With these limits, very short-lived and long-lived particles will be eliminated from the data sample, introducing a bias into the determination of the mean life.

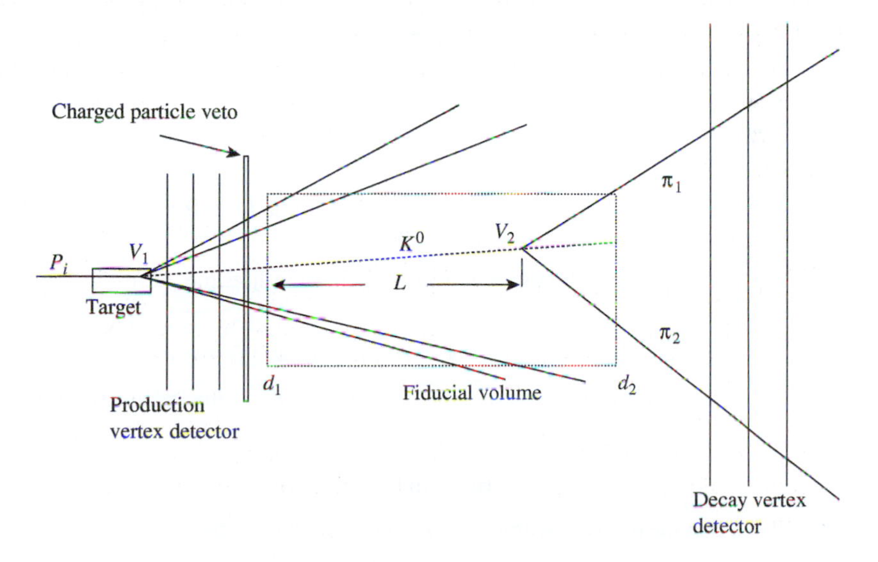

FIGURE 5.4
Experimental arrangement to measure the lifetime of an elementary particle.

In a Monte Carlo study of these biases, we could take the following steps to simulate measurements of decaying kaons:

1. Generate the production vertex coordinates and kaon momentum vector P from the known cross section for kaon production in the interaction of the incident and target particles.
2. Consider the efficiency of the production detector. If the detector successfully records charged particles produced in the initial interaction, proceed to step 3; if not, mark the event a failure and go to step 8.
3. Apply Equation (5.28) to find the time of flight (or lifetime) T of each individual kaon in its own rest frame. Use the current best-known value for the mean life τ.
4. Apply the Lorentz transformation to T to find the lifetime T' in the laboratory system.
5. Calculate the range r of the kaon in the laboratory and from this, the coordinate of the decay point.
6. Check that the kaon decays within the fiducial volume. If so, proceed to step 7; otherwise, mark the event a failure and go to step 8.
7. In the rest frame of the kaon, generate the pair of pion vectors. Transform to the laboratory system and check whether or not both particles can be detected in the decay vertex detector. If they can be detected, mark the event a success; if not, mark the event a failure.
8. Record details of the event and return to step 1 to generate a new event, or terminate if the desired number of events has been generated.

Program 5.4. KDECAY (website) Illustration of Example 5.7.

For this sample program, we simplify the problem by treating it in two dimensions and simplify or skip some of the steps as noted below.

1. Assume that each kaon is produced in the plane illustrated in Figure 5.4 and travels along the x-axis. Generate a vertex x-coordinate x_0 and the magnitude of the kaon's momentum P from suitable Gaussian distributions.
2. Skip
3. Find the lifetime T of the kaon in its own rest frame from the published value of the kaon mean life τ and Equation (5.28).
4. Apply the Lorentz transformation to T to find the lifetime T' in the laboratory system:

$$T' = \gamma T_{cm}, \text{ where } \gamma = 1/\sqrt{1-\beta^2} \text{ and } \beta = v/c$$

where v is the velocity of the kaon in the laboratory and c is the velocity of light.

5. Calculate the range r and decay point x_d:

$$r = \beta c T' \text{ and } x_d = x_0 + r$$

6. Check that the decay is within the fiducial area, that is, that

$$d_1 \leq x_d < d_2$$

If it is not, mark the event as a failure; otherwise, mark the event as a success.

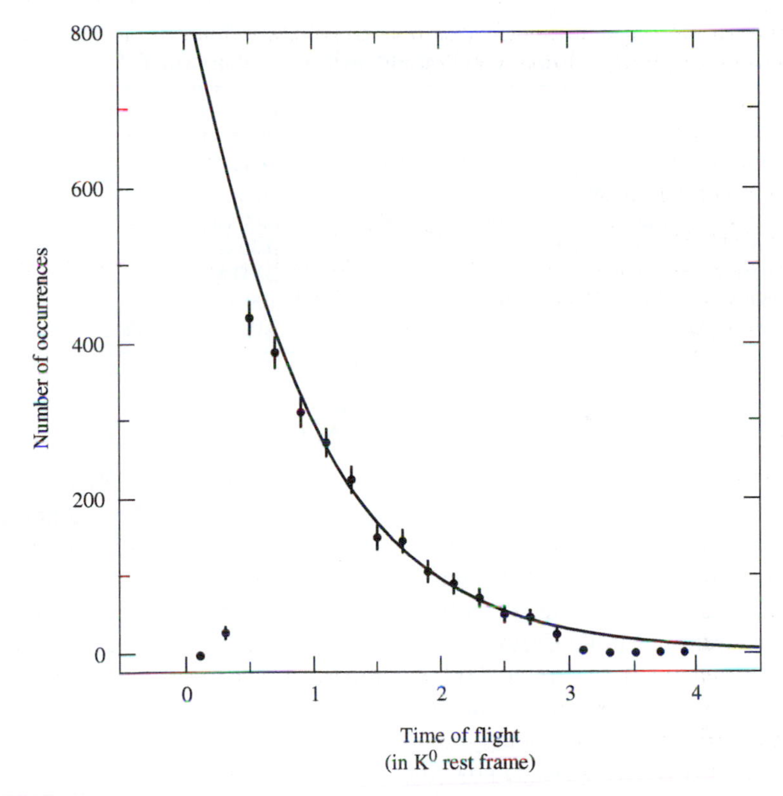

FIGURE 5.5
Distribution of times of flight (in units of 10^{-10} s) of 2355 successful K^0 decays from a total sample of 4000 generated events. The curve shows the predicted exponential distribution of the total 4000-event sample.

7. Skip this step.
8. Increment the event counters and record only successful events. If the desired number of events has been generated, terminate the calculation; otherwise, go to step 1 to begin generating the next event.

The properties of the two Gaussians and the other constants of the calculation are listed in Table 5.4. Note that we must use as input to our program a reasonable value of the kaon mean life, the quantity that we are attempting to measure. If the quantity had been only poorly measured previously, or perhaps not at all, it might be necessary to run the Monte Carlo program with several different trial values of τ, bracketing the expected value.

For this example, we generated 4000 events of which 2355 passed the fiducial cut. Figure 5.5 shows the distribution of the times of flight T (or lifetimes) in the rest frame of the kaon for successful events. The curve shows the expected distribution of the times of flight if no events had been rejected. We obtain the efficiency of the system as a function of the time of flight T by calculating the ratio of the number $N'(T)$ of successful events to the total number $N(T)$ generated

TABLE 5.4
Constants used in the Monte Carlo generation of Program 5.3

TauKaon (K_0 mean life)	0.894×10^{-10} s
MassKaon (K_0 mass)	497.7 Mev/c^2
*d*1 (Lower limit of fiducial range)	10 m
*d*2 (Upper limit of fiducial range)	40 m
xMean (mean coordinate of the production vertex, V_1)	5.00 cm
xSig (Standard deviation of production vertex)	0.50 cm
pMean (mean K_0 momentum)	2000 MeV/c
pSig (Standard deviation of K_0 momentum)	100 MeV/c
c (velocity of light)	3.00×10^{10} cm/s

$$\epsilon(T) = N'(T)/N(T) \tag{5.31}$$

We note that there are large losses of events at short times, below about $T = 0.5 \times 10^{-10}$ s, caused by the gap between the production vertex V_1 and the beginning of the fiducial region d_1, and smaller, but significant losses at long times of events that decayed beyond the end of the fiducial region, d_2.

To correct data obtained in an actual experiment and distributed as $N_{\text{exp}}(T)$, we should first run the Monte Carlo to generate sufficient numbers of events so that the uncertainties in the $N'(T)$ are negligible compared to the uncertainties in the experimental data sample. We should then select a continuous region of our data sample where the efficiency is reasonably good (and definitely not zero!) and correct the measurements by scaling $N_{\text{exp}}(T)$ by $1/\epsilon(T)$. Note that the statistical uncertainties in the measured data must also be scaled, so there is little point in including data from very low-efficiency regions of the sample. We can then obtain our estimate of the mean life of the kaon from a least-squares fit of Equation (5.27) to the corrected data. (A reminder: Although the overall distribution of events in this example is exponential, the expected distribution of events *in each individual bin* follows the Poisson distribution, as discussed in Section 4.4.)

A more detailed discussion of analysis techniques for this experiment is in Chapter 10.

5.5 EFFICIENT MONTE CARLO GENERATION

Because the relative error in a result calculated by the Monte Carlo method is inversely proportional to the square root of the number of *successful* events generated, it is important, especially for a long calculation, to have the highest possible program efficiency. Rejected events do not improve the statistical accuracy and every effort should be made to reduce the time spent on calculations that lead to "misses" rather than "hits." There are several ways to improve generation efficiency:

1. Don't be a purist. The Monte Carlo method is basically a way of doing complicated multidimensional integrals. If you can save time by doing part of the problem by analytic methods, do so.

2. Program carefully. Do not repeat calculations if the results can be saved for later use.
3. If possible, test the low-yield sections of the simulation early and cut out as soon as a "miss" occurs. Except for particular loss studies, it is usually not profitable to follow the calculation of an event that is known to end in failure.
4. Try to reduce the variance of the results by limiting ranges wherever possible. One application of this technique can be illustrated in Example 5.1, where the area of a circle of radius r_c is calculated by inscribing it within a square. Making the side of the square larger than the diameter of the circle would be wasteful and would increase the variance of the area determination.
5. When repeating a calculation to find the effects of varying a parameter, consider setting up the program in such a way that the identical sequence of random numbers is repeated throughout the calculation, except for calculations specifically associated with the change. This technique will not improve the variance of the *overall* calculation, but will reduce the variance of the *difference* of results from two calculations.
6. Inspect each probability function carefully before beginning a calculation and estimate the resolution and detail that will be required in the calculation. If a distribution has fine structure, try to determine whether or not such structure is of interest and must be preserved. If necessary, consider breaking the calculations into separate regions and varying the sampling sensitivity as appropriate for each region.
7. Be critical. Examine your generated variables to see that they fall within the expected ranges and follow expected distributions. In a large program, errors that affect the results in subtle ways may be buried within the program and be very difficult to detect. The only way to prevent problems is to make detailed checks at every stage of the program.

SUMMARY

Pseudorandom numbers: Numbers created by a computer algorithm such that successive numbers appear to be uncorrelated with previous numbers. They are referred to as *random numbers* or *random deviates.*

Uniform deviates: Pseudorandom numbers that are uniformly distributed between 0 and 1:

$$p(r) = \begin{cases} 1 & \text{for } 0 \le r < 1 \\ 0 & \text{otherwise} \end{cases}$$

Normalized distribution: A distribution that is scaled so that its integral over a specified range is equal to unity.

Transformation integral: Transforms the variable r drawn randomly from the uniform distribution into a variable x drawn randomly from the distribution $P(x)$:

$$\int_{r=0}^{r} 1\, dr = \int_{x=-\infty}^{x} P(x)\, dx$$

Rejection method: A method of generating random numbers drawn from particular distributions by rejecting those that fall outside the geometrical limits of the specified distribution.
Gaussian deviate: Random number drawn from a Gaussian distribution.
Quick Gaussian deviate: The sum of N random numbers is approximately Gaussian distributed with $\mu = N/2$ and $\sigma = \sqrt{N/12}$. Choose $N = 12$ and calculate $r_G = \Sigma r_i - N/2$ to obtain r_G drawn from
the standard Gaussian distribution with $\mu = 0$ and $\sigma = 1$.
Box-Müller method for Gaussian deviates: Select r_1 and r_2 from the uniform distribution and calculate

$$z_1 = \sqrt{-2 \ln r_1} \cos 2\pi r_2 \quad \text{and} \quad z_2 = \sqrt{-2 \ln r_1} \sin 2\pi r_2$$

to obtain z_1 and z_2 drawn from the standard Gaussian distribution.
Data smearing: Method for adding random variations to calculations to simulate the effects of finite measuring errors, $T_i' = T_i + \sigma_i r_i$.
Random numbers from the exponential distribution: To obtain a random number t_i drawn from the exponential distribution, calculate $t_i = -\tau \ln r_i$ from a random deviate r_i.

EXERCISES

5.1. Write a computer program that incorporates the Wichmann and Hill pseudorandom number generator and use it to generate 100 random numbers beginning with seeds $s_1 = 13$, $s_2 = 117$, and $s_3 = 2019$. Make a histogram of the numbers and draw a line representing the expected number of events in each bin. Calculate χ^2 for the agreement between the expected and generated number of events and find the associated probability.

5.2. (*a*) Generate 1000 random numbers uniformly distributed between $-\pi$ and $+\pi$.
(*b*) Generate 1000 random numbers between $x = 0$ and 1, distributed according to the distribution function $P(x) = (5x + 3)$. Use the transformation method with an analytic integration.
(*c*) Find the mean and standard deviation of each distribution and compare them to the predicted values.
(*d*) Make a 20-bin histogram of each distribution and plot on each the predicted distribution.
(*e*) Calculate χ^2 to compare each generated distribution to its parent distribution.

5.3. Write a general routine to generate random integers drawn from the binomial distribution by the transformation method. Use the routine to generate 1000 events corresponding to the distribution of heads or tails when a coin is tossed 50 times. Plot your results and compare them to the direct prediction of Equation (2.4).

5.4. Write a Monte Carlo routine to simulate 200 rolls of a pair of dice and find the frequency of occurrences of each possible sum. Plot a histogram of the occurrences with statistical error bars and plot the prediction of the binomial distribution. Calculate χ^2 for the agreement between the prediction and the data, and find the χ^2 probability. Compare your results to the exact probability calculation of Exercise 2.4.

5.5. Make a histogram of 200 random numbers that follow the Gaussian distribution by finding the distribution of the sums of groups of 12 random variates drawn from the uniform distribution. Calculate the mean and standard deviation of the generated numbers and the uncertainty in the mean.

5.6. Generate 1000 random numbers between $x = -3$ and $+3$, distributed according to the Lorentzian distribution with mean $\mu = 0$ and half-width $\Gamma = 1.0$. Use the transformation method with a numerical integration and interpolation. (See Appendix A.1 and A.3.) Make a 20-bin histogram of the generated numbers and plot Lorentzian the curve on the distribution. Calculate χ^2 to compare the generated distribution to the parent distribution.

5.7. Use the transformation method to produce a sequence of 200 random numbers x drawn from the distribution

$$P(x) = \sin x \quad \text{for } 0 \leq x < \pi$$
$$= 0 \quad \text{elsewhere}$$

Make a histogram of the events and compare it to the expected distribution. Note that the calculation can be done analytically and requires an inverse trigonometric function.

5.8. Use the rejection method to generate 500 random deviates between $x = 0$ and $x = 1$, drawn from the distribution $y(x) = a_1 + a_2x^2$, with $a_1 = 3.4$ and $a_2 = 12.1$. Find the mean and standard deviation of the generated numbers and compare them to the expected values.

5.9. Write a Monte Carlo program to generate 200 cubes with sides $a = 2.0 \pm 0.1$ cm, $b = 3.0 \pm 0.1$ cm, and $c = 4.0 \pm 0.2$ cm. Plot the distribution of the volumes of the cubes and find the mean volume, the standard deviation of the distribution, and the uncertainty in the mean. Compare the standard deviation of the distribution to the value predicted by the error propagation equation.

5.10. A *Pascal triangle* provides an interesting illustration of the relation between the binomial and Gaussian probability distributions. Assume an arrangement of pins in the form of a triangle as illustrated.

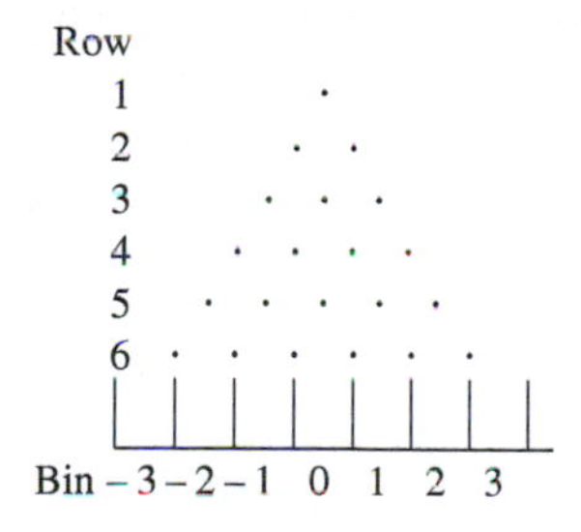

A ball, dropped into the device strikes the top pin and has a 50% probability of striking either of the two pins below it in the next row. The ball bounces down until it reaches the bottom where it is collected in one of the vertical bins.

(*a*) Find a general expression for the probability that a ball will land in a given bin after dropping through *N* rows of pins.

(*b*) Assume that 512 balls are dropped onto the top pin. Find the number of balls in each bottom bin for a device with three rows of pins above the bins. Repeat for devices with four, five, and six rows of pins.

(*c*) Find the standard deviation of the distribution of balls for each example; that is, assume that the bin number is the independent variable so that $\bar{x} = 0$.

(*d*) Plot histograms of the distribution of the balls with Gaussian curves with the means and standard deviations determined in (*c*).

5.11. Write a Monte Carlo program to simulate the Pascal triangle device described in the previous exercise. Compare the results obtained by the two methods.

CHAPTER 6

LEAST-SQUARES FIT TO A STRAIGHT LINE

6.1 DEPENDENT AND INDEPENDENT VARIABLES

We often wish to determine one characteristic y of an experiment as a function of some other quantity x. That is, instead of making a number of measurements of a single quantity x, we make a series of N measurements of the pair (x_i, y_i), one for each of several values of the index i, which runs from 1 to N. Our object is to find a function $y = y(x)$ that describes the relation between these two measured variables. In this chapter we consider the problem of pairs of variables (x_i, y_i) that are linearly related to one another, and refer to data from two undergraduate laboratory experiments as examples. In the following chapters, we shall discuss methods of finding relationships that are not linear.

Example 6.1. A student is studying electrical currents and potential differences. He has been provided with a 1-m nickel-silver wire mounted on a board, a lead-acid battery, and an analog voltmeter. He connects cells of the battery across the wire and measures the potential difference or voltage between the negative end and various positions along the wire. From examination of the meter, he estimates the uncertainty in each potential measurement to be 0.05 V. The uncertainty in the position of the probe is less than 1 mm and is considered to be negligible.

The data are listed in Table 6.1 and are plotted in Figure 6.1 to show the potential difference as a function of wire length x. The estimated common uncertainty in each measured potential difference is indicated on the graph by the vertical error bars. From these measurements, we wish to find the linear function $y(x)$ (shown as a solid line) that describes the way in which the voltage V varies as a function of position x along the wire.

Example 6.2. In another experiment, a student is provided with a radioactive source enclosed in a small 8-mm-diameter plastic disk and a Geiger counter with a 1-cm-diameter end window. Her object is to investigate the $1/r^2$ law by recording Geiger counter measurements over a fixed period of time at various distances from the source between 20 and 100 cm. Because the counting rate is not expected to vary from measurement to measurement, except for statistical fluctuations, the student can record data long enough to obtain good statistics over the entire range of the experiment. She uses an automatic recording system and records counts for thirty 15-s intervals at each position. For analysis in this experiment, she sums the counts from each set of 30 measurements to obtain the number of counts in 7.5 m intervals. The separate 15-s interval measurements at each position can be used in other statistical studies.

The data are listed in Table 6.2 and plotted against $x = 1/r^2$ in Figure 6.2. The vertical error bars on the data points represent the statistical uncertainties in the measured numbers of counts and are equal to the square roots of the numbers of counts. The uncertainties in the measurements of the distances from the source to the counter were assumed to be negligible.

Linear Approximation

In both of these examples, the functional relationship between the dependent and independent variables can be approximated by a straight line of the form

$$y(x) = a + bx \tag{6.1}$$

We shall consider in this chapter a method for determining the most probable values for the coefficients a and b.

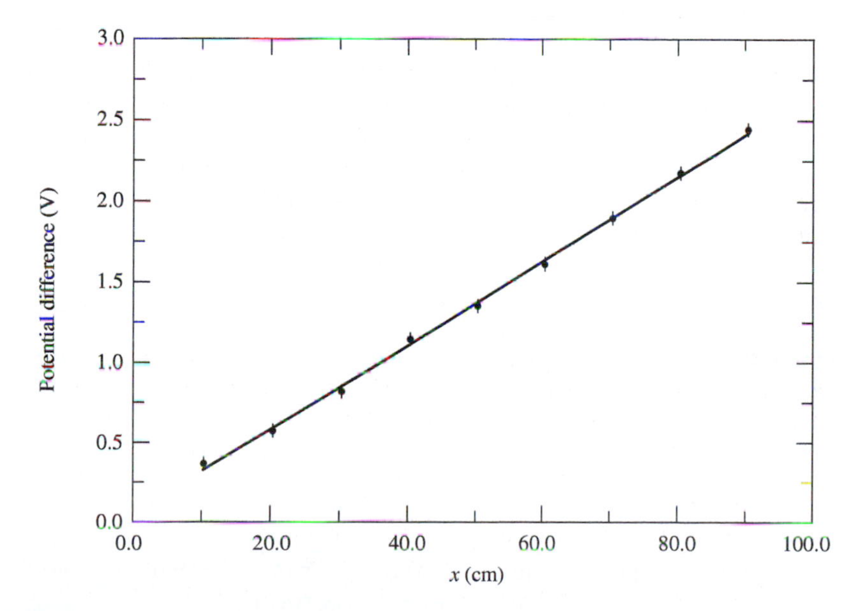

FIGURE 6.1
Potential difference as a function of position along a conducting wire (Example 6.1). The uniform uncertainties in the potential measurements are indicated by the vertical error bars. The straight line is the result of a least-squares fit to the data.

TABLE 6.1
Potential difference V as a function of position along a current-carrying nickel-silver wire

Point number	Postition x_i (cm)	Potential difference V_i (V)	x_i^2	x_iV_i	Fitted potential difference $a + bx$
1	10.0	0.37	100	3.70	0.33
2	20.0	0.58	400	11.60	0.60
3	30.0	0.83	900	24.90	0.86
4	40.0	1.15	1,600	46.00	1.12
5	50.0	1.36	2,500	68.00	1.38
6	60.0	1.62	3,600	97.20	1.64
7	70.0	1.90	4,900	133.00	1.91
8	80.0	2.18	6,400	174.40	2.17
9	90.0	2.45	8,100	220.50	2.43
Sums	450.0	12.44	28,500	779.30	

$\Delta = N\Sigma x_i^2 - (\Sigma x_i)^2 = (9 \times 28{,}500) - (450)^2 = 54{,}000$
$a = (\Sigma x_i^2 \Sigma V_i - \Sigma x_i \Sigma x_i V_i)/\Delta = (28{,}500 \times 12.44 - 450.0 \times 779.30)/54{,}000 = 0.0714$
$b = (N\Sigma x_i V_i - \Sigma x_i \Sigma V_i)/\Delta = (9 \times 779.30 - 450.0 \times 12.44)/54{,}000 = 0.0262$
$\sigma_a^2 \simeq \sigma_V^2 \Sigma x_i^2/\Delta = 0.05^2 \times 28{,}500 / 54{,}000 = 0.001319$ $\quad \sigma_a \simeq 0.036\ \sigma_a' = 0.019$
$\sigma_b^2 \simeq N\sigma_V^2/\Delta = 9 \times 0.05^2 / 54{,}000 = 0.417 \times 10^{-6}$ $\quad \sigma_b \simeq 0.00065\ \sigma_b' = 0.00034$

Note: A uniform uncertainty in V of 0.05 V is assumed. A linear fit to the data, calculated by the method of determinants, gives $a = 0.07 \pm 0.04$ V and $b = 0.0262 \pm 0.0006$ V/cm, with $\chi^2 = 1.95$ for 7 degrees of freedom. The χ^2 probability for the fit is approximately 96%.

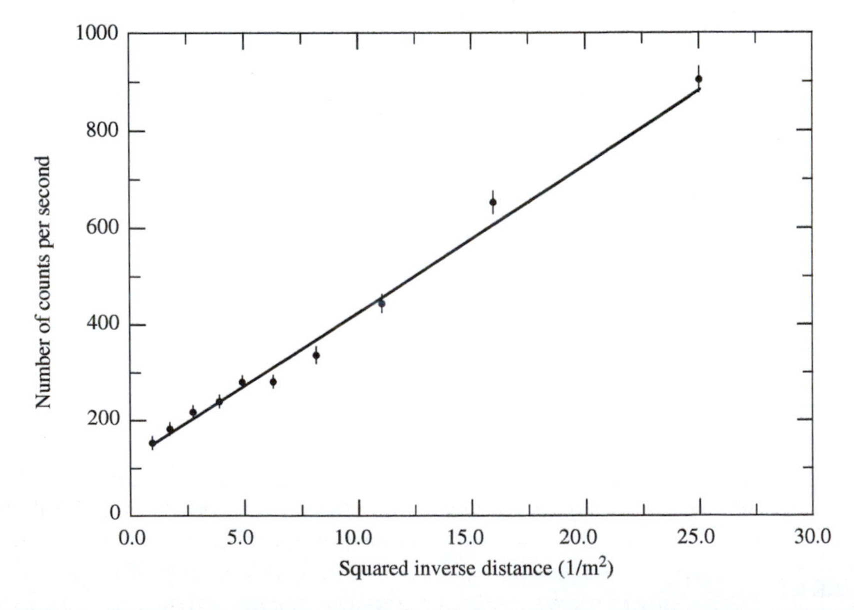

FIGURE 6.2
Number of counts in constant time intervals from a radioactive source as a function of the inverse distance from source to Geiger counter (Example 6.2). The vertical error bars indicate the statistical uncertainties in the counts. The straight line is the result of a least-squares fit to the data.

TABLE 6.2
Number of counts detected in 7½-min intervals as a function of distance from the source

i	Distance d_i (m)	$x_i = 1/d_i^2$ (m^{-2})	Counts C_i	σ_{C_i}	Weight ($1/C_i^2$) w_i	$w_i x_i$	$w_i C_i$	$w_i x_i^2$	$w_i x_i C_i$	Fitted counts $a + bx_i$
1	0.20	25.00	901	30.0	0.00111	0.0278	1	0.694	25.0	887
2	0.25	16.00	652	25.5	0.00153	0.0254	1	0.393	16.0	610
3	0.30	11.11	443	21.0	0.00226	0.0251	1	0.279	11.1	461
4	0.35	8.16	339	18.4	0.00295	0.0241	1	0.197	8.2	370
5	0.40	6.25	283	16.8	0.00353	0.0221	1	0.138	6.3	311
6	0.45	4.94	281	16.8	0.00356	0.0176	1	0.087	4.9	271
7	0.50	4.00	240	15.5	0.00417	0.0167	1	0.067	4.0	242
8	0.60	2.78	220	14.8	0.00455	0.0126	1	0.035	2.8	205
9	0.75	1.78	180	13.4	0.00556	0.0099	1	0.018	1.8	174
10	1.00	1.00	154	12.4	0.00649	0.0065	1	0.007	1.0	150
Sums					0.03570	0.1868	10	1.912	81.0	

$\sigma_i = \sqrt{y_i} \quad w_i = 1/\sigma_i^2 = 1/y_i$

$\Delta = \Sigma w_i \Sigma w_i x_i^2 - (\Sigma w_i x_i)^2 = 0.03570 \times 1.912 - (0.1868)^2 = 0.0334$

$a = [\Sigma w_i C_i \Sigma w_i x_i^2 - \Sigma w_i x_i \Sigma w_i x_i C_i]/\Delta = [10 \times 1.912 - 0.1868 \times 81.0]/\Delta = 119.5$

$b = [\Sigma w_i \Sigma w_i x_i C_i - \Sigma w_i x_i \Sigma w_i C_i]/\Delta = [0.03570 \times 81.0 - 0.1868 \times 10]/\Delta = 30.7$

$\sigma_a^2 \simeq \Sigma w_i x_i^2/\Delta = 1.912 / 0.0334 = 57.3 \qquad \sigma_a \simeq 7.6$

$\sigma_b^2 \simeq \Sigma w_i/\Delta = 0.03570 / 0.0334 = 1.07 \qquad \sigma_b \simeq 1.1$

Note: A linear fit to the data of the function $C = a + bx$ by the method of determinants gives $a = 119 \pm 8$ and $b = 31 \pm 1$, with $\chi^2 = 11.1$ for 8 degrees of freedom. The χ^2 probability for the fit is about 20%.

We cannot fit a straight line to the data exactly in either example because it is impossible to draw a straight line through all the points. For a set of N arbitrary points, it is always possible to fit a polynomial of degree $N - 1$ exactly, but for our experiments, the coefficients of the higher-order terms would have questionable significance. We assume that the fluctuations of the individual points above and below the solid curves are caused by experimental uncertainties in the individual measurements. In Chapter 11 we shall develop a method for testing whether higher-order terms are significant.

Measuring Uncertainties

If we were to make a series of measurements of the dependent quantity y_i for one particular value x_i of the independent quantity, we would find that the measured values were distributed about a mean in the manner discussed in Chapter 5 with a probability of ~68% that any single measurement of y_i be within 1 standard deviation of the mean. By making a number of measurements for each value of the independent quantity x_i, we could determine mean values $\bar{y}_i$ with any desired precision. Usually, however, we can make only one measurement y_i for each value of $x = x_i$, so that we must determine the value of y corresponding to that value of x with an uncertainty that is characterized by the standard deviation σ_i of the distribution of data for that point.

We shall assume for simplicity in all the following discussions that we can ascribe all the uncertainty in each measurement to the dependent variable. This is equivalent to assuming that the precision of the determination of x is considerably higher than that of y. This difference is illustrated in Figures 6.1 and 6.2 by the fact that the uncertainties are indicated by error bars for the dependent variables but not for the independent variables.

Our condition, that we neglect uncertainties in x and consider just the uncertainties in y, will be valid only if the uncertainties in y that would be produced by variations in x corresponding to the uncertainties in the measurement of x are much smaller than the uncertainties in the measurement of y. This is equivalent, in first order, to the requirement at each measured point that

$$\sigma_x \frac{dy}{dx} \ll \sigma_y$$

where dy/dx is the slope of the function $y = y(x)$.

We are not always justified in ascribing all uncertainties to the dependent parameter. Sometimes the uncertainties in the determination of both quantities x and y are nearly equal. But our fitting procedure will still be fairly accurate if we estimate the indirect contribution σ_{yI} from the uncertainty σ_x in x to the total uncertainty in y by the first-order relation

$$\sigma_{yI} = \sigma_x \frac{dy}{dx} \tag{6.2}$$

and combine this with the direct contribution σ_{yD}, which is the measuring uncertainty in y, to get

$$\sigma_y^2 = \sigma_{yI}^2 + \sigma_{yD}^2 \tag{6.3}$$

For both Examples 6.1 and 6.2 the condition would be reasonable because we predict a linear dependence of y with x. With the linear assumption, we treat the uncertainties in our data as if they were in the dependent variable only, while realizing that the corresponding fluctuations may have been originally derived from uncertainties in the determinations of both dependent and independent variables.

In those cases where the uncertainties in the determination of the independent quantity are considerably greater than those in the dependent quantity, it might be wise to interchange the definition of the two quantities.

6.2 METHOD OF LEAST SQUARES

Our data consist of pairs of measurements (x_i, y_i) of an independent variable x and a dependent variable y. We wish to find values of the parameters a and b that minimize the discrepancy between the measured values y_i and calculated values $y(x)$. We cannot determine the parameters exactly with only a finite number of observations, but can hope to extract the most probable estimates for the coefficients in the same way that we extracted the most probable estimate of the mean in Chapter 4.

Before proceeding, we must define our criteria for minimizing the discrepancy between the measured and predicted values y_i. For any arbitrary values of a

and b, we can calculate the deviations Δy_i between each of the observed values y_i and the corresponding calculated or fitted values

$$\Delta y_i = y_i - y(x_i) = y_i - a - bx_i \tag{6.4}$$

With well chosen parameters, these deviations should be relatively small. However, the sum of these deviations is not a good measure of how well our calculated straight line approximates the data because large positive deviations can be balanced by negative ones to yield a small sum even when the fit of the function $y(x)$ to the data is bad. We might consider instead summing the absolute values of the deviations, but this leads to difficulties in obtaining an analytical solution. Instead we sum the squares of the deviations.

There in no correct unique method for optimizing the parameters valid for all problems. There exists, however, a method that can be fairly well justified, that is simple and straightforward, and that is well established experimentally. This is the *method of least squares,* similar to the method discussed in Chapter 4, but extended to include more than one variable. It may be considered as a special case of the more general *method of maximum likelihood.*

Method of Maximum Likelihood

Our data consist of a sample of observations drawn from a parent distribution that determines the probability of making any particular observation. For the particular problem of an expected linear relationship between dependent and independent variables, we define parent parameters a_0 and b_0 such that the actual relationship between y and x is given by

$$y_0(x) = a_0 + b_0 x \tag{6.5}$$

We shall assume that each individual measured value of y_i is itself drawn from a Gaussian distribution with mean $y_0(x_i)$ and standard deviation σ_i. We should be aware that the Gaussian assumption may not always be exactly true. In Example 6.2 the $y_i = C_i$ were obtained in a counting experiment and therefore follow a Poisson distribution. However, for a sufficiently large number of counts y_i the distribution may be considered to be Gaussian. We shall discuss fitting with Poisson statistics in Section 6.6.

With the Gaussian assumption, the probability P_i for making the observed measurement y_i with standard deviation σ_i for the observations about the actual value $y_0(x_i)$ is

$$P_i = \frac{1}{\sigma_i\sqrt{2\pi}} \exp\left\{-\frac{1}{2}\left[\frac{y_i - y_0(x_i)}{\sigma_i}\right]^2\right\} \tag{6.6}$$

The probability for making the observed set of measurements of the N values of y_i is the product of the probabilities for each observation:

$$P(a_0, b_0) = \Pi P_i = \prod\left(\frac{1}{\sigma_i\sqrt{2\pi}}\right) \exp\left\{-\frac{1}{2}\sum\left[\frac{y_i - y_0(x_i)}{\sigma_i}\right]^2\right\} \tag{6.7}$$

where the product Π is taken with i ranging from 1 to N and the product of the exponentials has been expressed as the exponential of the sum of the arguments. In these products and sums, the quantities $1/\sigma_i^2$ act as weighting factors.

Similarly, for any *estimated* values of the parameters a and b, we can calculate the probability of obtaining the observed set of measurements

$$P(a, b) = \prod\left(\frac{1}{\sigma_i\sqrt{2\pi}}\right)\exp\left\{-\frac{1}{2}\sum\left[\frac{y_i - y(x_i)}{\sigma_i}\right]^2\right\} \tag{6.8}$$

with $y(x)$ defined by Equation (6.1) and evaluated at each of the values x_i.

We assume that the observed set of measurements is more likely to have come from the parent distribution of Equation (6.5) than from any other similar distribution with different coefficients and, therefore, the probability of Equation (6.7) is the maximum probability attainable with Equation (6.8). Thus, the maximum-likelihood estimates for a and b are those values that maximize the probability of Equation (6.8).

Because the first factor in the product of Equation (6.8) is a constant, independent of the values of a and b, maximizing the probability $P(a, b)$ is equivalent to minimizing the sum in the exponential. We define this sum to be our goodness-of-fit parameter χ^2:

$$\chi^2 = \sum\left[\frac{y_i - y(x_i)}{\sigma_i}\right]^2 = \sum\left[\frac{1}{\sigma_i}(y_i - a - bx_i)\right]^2 \tag{6.9}$$

We use the same symbol χ^2, defined earlier in Equation (4.32), because this is essentially the same definition in a different context.

Our method for finding the optimum fit to the data will be to find values of a and b that minimize this weighted sum of the squares of the deviations χ^2 and hence, to find the fit that produces the smallest sum of the squares or the *least-squares fit.* The magnitude of χ^2 is determined by four factors:

1. Fluctuations in the measured values of the variables y_i, which are random samples from a parent population with expectation values $y_0(x_i)$.
2. The values assigned to the uncertainties σ_i in the measured variables y_i. Incorrect assignment of the uncertainties σ_i will lead to incorrect values of χ^2.
3. The selection of the analytical function $y(x)$ as an approximation to the "true" function $y_0(x)$. It might be necessary to fit several different functions in order to find the appropriate function for a particular set of data.
4. The values of the parameters of the function $y(x)$. Our objective is to find the "best values" of these parameters.

6.3 MINIMIZING χ^2

To find the values of the parameters a and b that yield the minimum value for χ^2, we set to zero the partial derivatives of χ^2 with respect to each of the parameters

$$
\begin{aligned}
\frac{\partial}{\partial a}\chi^2 &= \frac{\partial}{\partial a}\sum\left[\frac{1}{\sigma_i^2}(y_i - a - bx)^2\right] \\
&= -2\sum\left[\frac{1}{\sigma_i^2}(y_i - a - bx_i)\right] = 0 \\
\frac{\partial}{\partial b}\chi^2 &= \frac{\partial}{\partial b}\sum\left[\frac{1}{\sigma_i^2}(y_i - a - bx_i)^2\right] \\
&= -2\sum\left[\frac{1}{\sigma_i^2}(y_i - a - bx_i)\right] = 0
\end{aligned}
\tag{6.10}
$$

These equations can be rearranged as a pair of linear simultaneous equations in the unknown parameters a and b:

$$
\begin{aligned}
\sum\frac{y_i}{\sigma_i^2} &= a\sum\frac{1}{\sigma_i^2} + b\sum\frac{x_i}{\sigma_i^2} \\
\sum\frac{x_i y_i}{\sigma_i^2} &= a\sum\frac{x_i}{\sigma_i^2} + b\sum\frac{x_i^2}{\sigma_i^2}
\end{aligned}
\tag{6.11}
$$

The solutions can be found in any one of a number of different ways, but, for generality we shall use the method of determinants. (See Appendix B.) The solutions are

$$
\begin{aligned}
a &= \frac{1}{\Delta}\begin{vmatrix}\sum\frac{y_i}{\sigma_i^2} & \sum\frac{x_i}{\sigma_i^2} \\ \sum\frac{x_i y_i}{\sigma_i^2} & \sum\frac{x_i^2}{\sigma_i^2}\end{vmatrix} = \frac{1}{\Delta}\left(\sum\frac{x_i^2}{\sigma_i^2}\sum\frac{y_i}{\sigma_i^2} - \sum\frac{x_i}{\sigma_i^2}\sum\frac{x_i y_i}{\sigma_i^2}\right) \\
b &= \frac{1}{\Delta}\begin{vmatrix}\sum\frac{1}{\sigma_i^2} & \sum\frac{y_i}{\sigma_i^2} \\ \sum\frac{x_i}{\sigma_i^2} & \sum\frac{x_i y_i}{\sigma_i^2}\end{vmatrix} = \frac{1}{\Delta}\left(\sum\frac{1}{\sigma_i^2}\sum\frac{x_i y_i}{\sigma_i^2} - \sum\frac{x_i}{\sigma_i^2}\sum\frac{y_i}{\sigma_i^2}\right) \\
\Delta &= \begin{vmatrix}\sum\frac{1}{\sigma_i^2} & \sum\frac{x_i}{\sigma_i^2} \\ \sum\frac{x_i}{\sigma_i^2} & \sum\frac{x_i^2}{\sigma_i^2}\end{vmatrix} = \sum\frac{1}{\sigma_i^2}\sum\frac{x_i^2}{\sigma_i^2} - \left(\sum\frac{x_i}{\sigma_i^2}\right)^2
\end{aligned}
\tag{6.12}
$$

For the special case in which all the uncertainties are equal ($\sigma = \sigma_i$), they cancel and the solutions may be written

$$
\begin{aligned}
a &= \frac{1}{\Delta'}\begin{vmatrix}\Sigma y_i & \Sigma x_i \\ \Sigma x_i y_i & \Sigma x_i^2\end{vmatrix} = \frac{1}{\Delta'}(\Sigma x_i^2 \Sigma y_i - \Sigma x_i \Sigma x_i y_i) \\
b &= \frac{1}{\Delta'}\begin{vmatrix}N & \Sigma y_i \\ \Sigma x_i & \Sigma x_i y_i\end{vmatrix} = \frac{1}{\Delta'}(N\Sigma x_i y_i - \Sigma x_i \Sigma y_i) \\
\Delta' &= \begin{vmatrix}N & \Sigma x_i \\ \Sigma x_i & \Sigma x_i^2\end{vmatrix} = N\Sigma x_i^2 - (\Sigma x_i)^2
\end{aligned}
\tag{6.13}
$$

Examples

For the data of Example 6.1 (Table 6.1), we assume that the uncertainties in the measured voltages V are all equal and that the uncertainties in x_i are negligible. We can therefore use Equation (6.13). We accumulate four sums Σx_i, $\Sigma y_i = \Sigma V_i$, Σx_i^2, and $\Sigma x_i y_i = \Sigma x_i V_i$ and combine them according to Equation (6.13) to find numerical values for a and b. The steps of the calculation are illustrated in Table 6.1, and the resulting fit is shown as a solid line on Figure 6.1.

Determination of the parameters a and b from Equation (6.12) is somewhat more tedious, because the uncertainties σ_i must be included. Table 6.2 shows steps in the calculation of the data of Example 6.2 with the uncertainties σ_i in the numbers of counts C_i determined by Poisson statistics so that $\sigma_i^2 = C_i$. The values of a and b found in this calculation were used to calculate the straight line through the data points in Figure 6.2.

It is important to note that the value of C_i to be used in determining the uncertainty σ_i must be the actual number of events observed. If, for example, the student had decided to improve her statistics by collecting data at the larger distances over longer time periods Δt_i and to normalize all her data to a common time interval Δt_c,

$$C_i' = C_i \times \Delta t_c / \Delta t_i$$

then the statistical uncertainty in C' would be given by

$$\sigma_i' = \sqrt{C_i} \times \Delta t_c / \Delta t_i$$

Program 6.1. FITLINE (Appendix E) Solution of Equations (6.11) by the determinant method of Equation (6.12).

The program uses routines in the programs units FITVARS, FITUTIL, and GENUTIL, which are also used by other fitting programs. The sample programs use single precision variables for simplicity, although double, or higher, precision is highly recommended.

Program 6.1 uses Equation (6.12) to solve both Examples 6.1 and 6.2, although separate routines written for each problem would be slightly more efficient. Because the measurements of Example 6.1 have common errors, we could, for example, increase the fitting speed by using Equations (6.13) rather than Equations (6.12). Similarly, for Example 6.2, we could simplify the fitting routine by replacing the statistical errors SIGY[I] by the explicit expression for $\sqrt{y_i}$. However, in most calculations that involve statistical errors, there are also other errors to be considered, such as those arising from background subtractions, so the loss of generality would more than compensate for any increased efficiency in the calculations.

Program 6.2. FITVARS (website) Include file of constants, variables, and arrays for least-squares fits.

Program 6.3. FITUTIL (website) Utility routines for fitting programs Input/output routine, χ^2 calculation, χ^2-density, and χ^2-integral probability.

Program 6.4. GENUTIL (website) General Utility Routines
Includes approximate gamma function, Simpson's rule integration.

6.4 ERROR ESTIMATION

Common Uncertainties

If the standard deviations σ_i for the data points y_i are unknown but we can assume that they are all equal, $\sigma_i^2 = \sigma^2$, then we can estimate them from the data and the results of our fit. The requirement of equal errors may be satisfied if the uncertainties are instrumental and all the data are recorded with the same instrument and on the same scale, as was assumed in Example 6.1.

In Chapter 2 we obtained, for our best estimate of the variance of the data sample,

$$\sigma^2 \simeq s^2 \equiv \frac{1}{N-m}\sum (y_i - \bar{y})^2 \tag{6.14}$$

where $N - m$ is the number of degrees of freedom and is equal to the number of measurements minus the number of parameters determined from the fit. In Equation (6.14) we identify y_i with the measured value of the dependent variable, and for $\bar{y}$, the expected mean value of y_i, we use the value calculated from Equation (6.1) for each data point with the fitted parameters a and b. Thus, our estimate $\sigma_i = \sigma$ for the standard deviation of an individual measurement is

$$\sigma^2 \simeq s^2 = \frac{1}{N-2}\sum (y_i - a - bx_i)^2 \tag{6.15}$$

By comparing Equation (6.15) with Equation (6.9), we see that it is just this common uncertainty that we have minimized in the least-squares fitting procedure. Thus, we can obtain the common error in our measurements of y from the fit, although at the expense of any information about the quality of the fit.

Variable Uncertainties

In general the uncertainties σ_i in the dependent variables y_i will not all be the same. If, for example, the quantity y represents the number of counts in a detector per unit time interval (as in Example 6.2), then the errors are statistical and the uncertainty in each measurement y_i is directly related to the magnitude of y (as discussed in Section 4.2), and the standard deviations σ_i associated with these measurements is

$$\sigma_i^2 = C_i \tag{6.16}$$

In principle, the value of y_i, which should be used in calculating the standard deviations σ_i by Equation (6.16), is the value $y_0(x_i)$ of the parent population. In practice we use the measured values that are only samples from that population. In the limit of an infinite number of determinations, the average of all the measurements would very closely approximate the parent value, but generally we cannot make more than one measurement of each value of x, much less an infinite number. We could approximate the parent value $y_0(x_i)$ by using the calculated value $y(x)$ from our fit, but that would complicate the fitting procedure. We shall discuss this possibility further in the following section.

Contributions from instrumental and other uncertainties may modify the simple square root form of the statistical errors. For example, uncertainties in measuring the time interval during which the events of Example 6.2 were recorded might contribute, although statistical fluctuations generally dominate in counting experiments. Background subtractions are another source of uncertainty. In many counting experiments, there is a background under the data that may be removed by subtraction, or may be included in the fit. In Example 6.2, cosmic rays and other backgrounds contribute to a counting rate even when the source is moved far away from the detector, as indicated by the nonzero intercept of the fitted line of Figure 6.2 on the C axis. If the student had chosen to record the radiation background counts C_b in a separate measurement and to subtract C_b from each of her measurements C_i to obtain

$$C_i' = C_i - C_b$$

then the uncertainty in C' would have been given by combining in quadrature the uncertainties in the two measurements:

$$\sigma_i'^2 = \sigma_i^2 + \sigma_b^2$$

χ^2 Probability

For those data for which we know the uncertainties σ_i in the measured values y_i we can calculate the value of χ^2 from Equation (6.9) and test the goodness of our fit. For our two-parameter fit to a straight line, the number of degrees of freedom will be $N - 2$. Then, for the data of Example 6.2, we should hope to obtain $\chi^2 \simeq 10 - 2 = 8$. The actual value, $\chi^2 = 11.1$, is listed in Table 6.2, along with the probability ($p = 20\%$). (See Table C.4.) We interpret this probability in the following way. Suppose that we have obtained a χ^2 probability of $p\%$ for a certain set of data. Then, we should expect that, if we were to repeat the experiment many times, approximately $p\%$ of the experiments would yield χ^2 values as high as the one that we obtained or higher. This subject will be discussed further in Chapter 11.

In Example 6.1, we obtained a value of $\chi^2 = 1.95$ for 7 degrees of freedom, corresponding to a probability of about 96%. Although this probability may seem to be gratifyingly high, the very low value of χ^2 gives a strong indication that the common uncertainty in the data may have been overestimated and it might be wise to use the value of χ^2 to obtain a better estimate of the common uncertainty. From Equations (6.15) and (6.9), we obtain an expression for the revised common uncertainty σ_c' in terms of χ^2 and the original estimate, σ_c:

$$\sigma_c'^2 \simeq \sigma_i^2 \times \chi^2/(N - 2) \tag{6.17}$$

or, more generally

$$\sigma_c'^2 \simeq \sigma_i^2 \times \chi_\nu^2 \tag{6.18}$$

where $\chi_\nu^2 = \chi^2/\nu$ and ν is the number of degrees of freedom in the fit. Thus, for Example 6.1, we find $\sigma_c'^2 = 0.05^2 \times 1.95/(9 - 2) = 0.0007$, or $\sigma_c' = {\sim}0.03$ V.

Uncertainties in the Parameters

In order to find the uncertainty in the estimation of the parameters a and b in our fitting procedure, we use the error propagation method discussed in Chapter 3. Each of our data points y_i has been used in the determination of the parameters and each has contributed some fraction of its own uncertainty to the uncertainty in our final determination. Ignoring systematic errors, which would introduce correlations between uncertainties, the variance σ_z^2 of the parameter z is given by Equation (3.14) as the sum of the squares of the products of the standard deviations σ_i of the data points with the effects that the data points have on the determination of z:

$$\sigma_z^2 = \sum\left[\sigma_i^2\left(\frac{\partial z}{\partial y_i}\right)^2\right] \tag{6.19}$$

Thus, to determine the uncertainties in the parameters a and b, we take the partial derivatives of Equation (6.12):

$$\begin{aligned}\frac{\partial a}{\partial y_j} &= \frac{1}{\Delta}\left(\frac{1}{\sigma_j^2}\sum\frac{x_i^2}{\sigma_i^2} - \frac{x_j}{\sigma_j^2}\sum\frac{x_i}{\sigma_i^2}\right)\\ \frac{\partial b}{\partial y_j} &= \frac{1}{\Delta}\left(\frac{x_j}{\sigma_j^2}\sum\frac{1}{\sigma_i^2} - \frac{1}{\sigma_j^2}\sum\frac{x_i}{\sigma_i^2}\right)\end{aligned} \tag{6.20}$$

We note that the derivatives are functions only of the variances and of the independent variables x_i. Combining these equations with the general expression of Equation (6.19) and squaring, we obtain for σ^2,

$$\begin{aligned}\sigma_a^2 &\simeq \sum_{j=1}^{N}\frac{\sigma_j^2}{\Delta^2}\left[\frac{1}{\sigma_j^4}\left(\sum\frac{x_i^2}{\sigma_i^2}\right)^2 - \frac{2x_j}{\sigma_j^4}\sum\frac{x_i^2}{\sigma_i^2}\sum\frac{x_i}{\sigma_i^2} + \frac{x_j^2}{\sigma_j^4}\left(\sum\frac{x_i}{\sigma_i^2}\right)^2\right]\\ &= \frac{1}{\Delta^2}\left[\sum\frac{1}{\sigma_j^2}\left(\sum\frac{x_i^2}{\sigma_i^2}\right)^2 - 2\sum\frac{x_j}{\sigma_j^2}\sum\frac{x_i^2}{\sigma_i^2}\sum\frac{x_i}{\sigma_i^2} + \sum\frac{x_j^2}{\sigma_j^2}\left(\sum\frac{x_i}{\sigma_i^2}\right)^2\right]\\ &= \frac{1}{\Delta^2}\left(\sum\frac{x_i^2}{\sigma_i^2}\right)\left[\sum\frac{1}{\sigma_j^2}\sum\frac{x_i^2}{\sigma_i^2} - \left(\sum\frac{x_i}{\sigma_i^2}\right)^2\right]\\ &= \frac{1}{\Delta}\sum\frac{x_i^2}{\sigma_i^2}\end{aligned} \tag{6.21}$$

and for σ_b^2,

$$\begin{aligned}\sigma_b^2 &\simeq \sum_{j=1}^{N}\frac{\sigma_j^2}{\Delta^2}\left[\frac{x_j^2}{\sigma_j^4}\left(\sum\frac{1}{\sigma_i^2}\right)^2 - \frac{2x_j}{\sigma_j^4}\sum\frac{1}{\sigma_i^2}\sum\frac{x_i}{\sigma_i^2} + \frac{1}{\sigma_j^4}\left(\sum\frac{x_i}{\sigma_i^2}\right)^2\right]\\ &= \frac{1}{\Delta^2}\left[\sum\frac{x_j^2}{\sigma_j^2}\left(\sum\frac{1}{\sigma_i^2}\right)^2 - 2\sum\frac{x_j}{\sigma_j^2}\sum\frac{1}{\sigma_i^2}\sum\frac{x_i}{\sigma_i^2} + \sum\frac{1}{\sigma_j^2}\left(\sum\frac{x_i}{\sigma_i^2}\right)^2\right]\\ &= \frac{1}{\Delta^2}\left(\sum\frac{x_j^2}{\sigma_i^2}\right)\left[\sum\frac{1}{\sigma_j^2}\sum\frac{1}{\sigma_i^2} - \left(\sum\frac{x_i}{\sigma_i^2}\right)^2\right]\\ &= \frac{1}{\Delta}\sum\frac{1}{\sigma_i^2}\end{aligned} \tag{6.22}$$

For the special case of common uncertainties in y_i, $\sigma_i = \sigma$, these equations reduce to

$$\sigma_a^2 = \frac{\sigma^2}{\Delta'}\Sigma x_i^2 \qquad \text{and} \qquad \sigma_b^2 = N\frac{\sigma^2}{\Delta'} \tag{6.23}$$

with σ given by Equation (6.15) and Δ' given by Equation (6.13).

The uncertainties in the parameters σ_a and σ_b, calculated from the original error estimates, are listed in Tables 6.1 and 6.2. For Example 6.1, revised uncertainties σ_a' and σ_b', based on the revised common data uncertainty calculated from Equation (6.18), are also listed.

6.5 SOME LIMITATIONS OF THE LEAST-SQUARES METHOD

When a curve is fitted by the least-squares method to a collection of statistical counting data, the data must first be *histogrammed*; that is, a histogram must be formed of the corrected data, either during or after data collection. In Example 6.2, the data were collected over intervals of time Δt, with the size of the interval chosen to assure that a reasonable number of counts would be collected in each time interval. For data that vary linearly with the independent variable, this treatment poses no special problems, but one could imagine a more complex problem in which fine details of the variation of the dependent variable y with the independent variable x are important. Such details might well be lost if the binning were too coarse. On the other hand, if the binning interval were too fine, there might not be enough counts in each bin to justify the Gaussian probability hypothesis. How does one choose the appropriate bin size for the data?

A handy rule of thumb when considering the Poisson distribution is to assume that *large enough* = 10. A comparison of the Gaussian and Poisson distributions for mean $\mu \simeq 10$ and standard deviation $\sigma = \sqrt{\mu}$ (see Figures 2.4 and 2.5) shows very little difference between the two distributions. We might expect this because the mean is more than 3 standard deviations away from the origin. Thus, we may be reasonably confident about the results of a fit if no histogram contains less than ten counts and if we are not placing excessive reliance on the actual value of χ^2 obtained from the fit. If a bin does have fewer than the allowed minimum number of counts, it may be possible to merge that bin with an adjacent one. Note that there is no requirement that intervals on the abscissa be equal, although we must be careful in our choice of the appropriate value of x_i for the merged bin. We should also be aware that such mergers necessarily reduce the resolution of our data and may, when fitting functions more complicated than a straight line, obscure some interesting features.

In general, the choice of bin width will be a compromise between the need for sufficient statistics to maintain a small relative error in the values of y_i and thus in the fitted parameters, and the need to preserve interesting structure in the data. When full details of any structure in the data must be preserved, it might be advisable to apply the maximum-likelihood method directly to the data, event by event, rather than to use the least-squares method with its necessary binning of the data. We return to this subject in Chapter 10.

There is also a question about our use of the experimental errors in the fitting process, rather than the errors predicted by our estimate of the parent distribution. For Example 6.2, this corresponds to our choosing $\sigma_i^2 = y_i$ rather than $\sigma_i^2 = y(x_i) = a + bx_i$. We shall consider the possibility of using errors from our estimate of the parent distribution, as well as the direct application of the Poisson probability function, in the following section.

Another important point to consider when fitting curves to data is the possibility of rounding errors, which can reduce the accuracy of the results. With manual calculations, it is important to avoid rounding the numbers until the very end of the calculation. With computers, problems may arise because of finite computer word length. This problem can be especially severe with matrix and determinant calculations, which often involve taking small differences between large numbers. Depending on the computer and the software, it may be necessary to use double-precision variables in the fitting routine.

We discuss in Chapter 7 the interaction of parameters in a multiparameter fit. For now, it is worth noting that, for a nominally "flat" distribution of data, the intercept obtained from a fit to a straight line may not be identical to the mean value of the data points on the ordinate. See Exercise 6.7 for an example of this effect.

6.6 ALTERNATE FITTING METHODS

In this section we attempt to solve the problem of fitting a straight line to a collection of data points by using errors determined from the estimated parent distribution rather than from the measurements, and by directly applying Poisson statistics, rather than Gaussian statistics. Because it is not possible to derive a set of independent linear equations for the parameters with these conditions, explicit expressions for the parameters a and b cannot be obtained. However, with fast computers, solving coupled, nonlinear equations is not difficult, although the clarity and elegance of the straightforward least-squares method can be lost.

Poisson Uncertainties

Let us consider a collection of purely statistical data that obey Poisson statistics (as in Example 6.2) so that the uncertainties can be expressed by Equation (6.16). We begin by substituting the approximation $\sigma_i^2 = y(x_i) = a + bx_i$ into the definition of χ^2 in Equation (6.9), which is based on Gaussian probability, and minimizing the value of χ^2 as in Equations (6.10). The result is a pair of simultaneous equations that can be solved for a and b:

$$N = \sum \frac{y_i^2}{(a + bx_i)^2}$$
$$\Sigma x_i = \sum \frac{x_i y_i^2}{(a + bx_i)^2} \qquad (6.24)$$

Poisson Probability

Next, let us replace the Gaussian probability $P(a, b)$ of Equation (6.8) by the corresponding probability for observing y_i counts from a Poisson distribution with mean $\mu_i = y(x_i)$,

$$P(a, b) = \prod \left(\frac{[y(x_i)]^{y_i}}{y_i!} e^{-y(x_i)} \right) \tag{6.25}$$

and apply the method of maximum likelihood to this probability. It is easier and equivalent to maximize the natural logarithm of the probability with respect to each of the parameters a and b:

$$\ln P(a, b) = \Sigma[y_i \ln y(x_i)] - \Sigma y(x_i) + \text{constant} \tag{6.26}$$

where the constant term is independent of the parameters a and b. The result of taking partial derivatives of Equation (6.26) is a pair of simultaneous equations similar to those of Equation (6.24),

$$\begin{aligned} N &= \sum \frac{y_i}{a + bx_i} \\ \Sigma x_i &= \sum \frac{x_i y_i}{a + bx_i} \end{aligned} \tag{6.27}$$

but with less emphasis on fitting the larger values of y_i.

Neither the coupled simultaneous Equations (6.24) nor the Equations (6.27) can be solved directly for a and b, but each pair can be solved by an iterative method in which values of a and b are chosen and then adjusted until the two simultaneous equations are satisfied. (See Appendix A.5.)

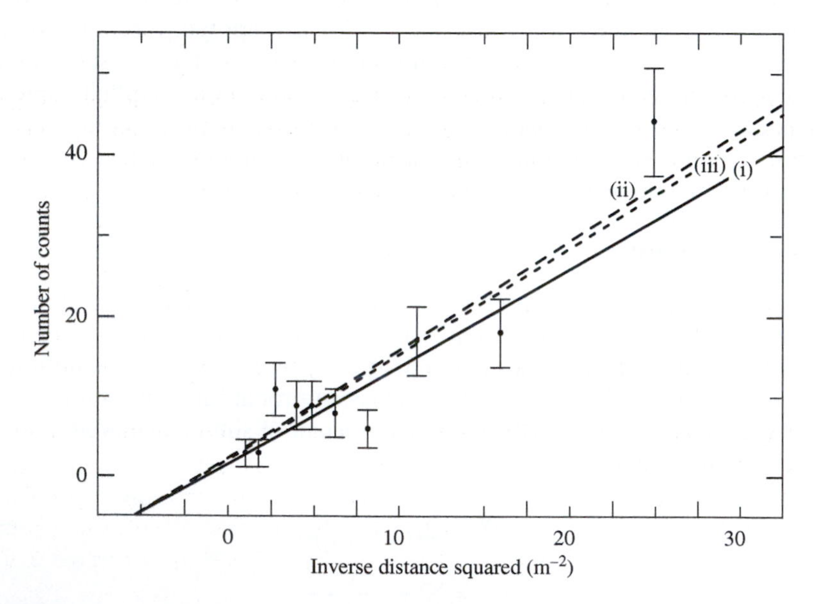

FIGURE 6.3

Least-squares fit of a straight line to the data by three different methods. (i) Standard least-squares method with Gaussian statistics and experimental uncertainties; (ii) Gaussian statistics and analytic uncertainties; (iii) Poisson statistics and analytic uncertainties. The analytic errors are expressed as $\sigma_i^2 = a + bx_i$.

TABLE 6.3

Comparison of fits to a selection of statistical data from Example 6.2 for three different fitting methods

i	Inverse distance squared x_i (m^{-2})	Number of counts C_i	(1) Standard	(2) Gaussian $\sigma^2 = y(x_i)$	(3) Poisson $\sigma^2 = y(x_i)$
1	25.00	44	32.0	36.3	35.1
2	16.00	18	21.0	24.1	23.2
3	11.11	17	15.1	17.5	16.8
4	8.16	6	11.5	13.5	12.9
5	6.25	8	9.1	10.9	10.4
6	4.94	9	7.5	9.2	8.6
7	4.00	9	6.4	7.9	7.4
8	2.78	11	4.9	6.3	5.8
9	1.78	3	3.7	4.9	4.5
10	1.00	3	2.7	3.9	3.4
Sums		128	114.0	134.4	128.0
		a	1.52	2.50	2.11
		b	1.22	1.35	1.32
		χ^2	13.7	17.6	15.5

Note: (1) Standard least-squares method with Gaussian statistics and experimental uncertainties; (2) Gaussian statistics and analytic uncertainties; (3) Poisson statistics and analytic uncertainties. The analytic uncertainties are expressed as $\sigma^2 = a + bx_i$.

Example 6.3. Because we expect the methods discussed here to be equivalent to the standard method for large data samples, we selected a low statistics sample to emphasize the differences. We chose from the measurements of Example 6.2 only those events collected at each detector position during the first 15-s interval, a total of 128 events at ten different positions. The results of (i) calculations by the standard method, (ii) calculations with Gaussian statistics and with errors given by $\sigma_i = y(x_i) = a + bx_i$, and (iii) calculations with Poisson statistics with errors as in method (ii) are listed in Table 6.3 and illustrated in Figure 6.3. We note that method (i) appears to underestimate the number of events in the sample, whereas method (ii) overestimates the number. Method (iii) with Poisson statistics and errors calculated as in method (ii) finds the exact number.

We can avoid questions of finite binning and the choice of statistics by making direct use of the maximum-likelihood method, treating the fitting function as a probability distribution. This method also allows detailed handling of problems in which the probability associated with individual measurements varies in a complex way from observation to observation. We shall pursue this subject further in Chapter 10.

In general, however, the simplicity of the least-squares method and the difficulty of solving the equations that result from other methods, particularly with more complicated fitting functions, leads us to choose the standard method of least squares for most problems. We make the following two assumptions to simplify the calculation:

1. The shapes of the individual Poisson distributions governing the fluctuations in the observed y_i are nearly Gaussian.
2. The uncertainties σ_i in the observations y_i may be obtained from the uncertainties in the data and may be approximated by $\sigma_i^2 \simeq y_i$ for statistical uncertainties.

SUMMARY

Linear function: $y(x) = a + bx$.
Chi-square:

$$\chi^2 = \sum \left[\frac{1}{\sigma_i}(y_i - a - bx_i) \right]^2$$

Least-squares fitting procedure: Minimize χ^2 with respect to each of the parameters simultaneously.
Solutions for least-squares fit of a straight line:

$$a = \frac{1}{\Delta} \begin{vmatrix} \sum \frac{y_i}{\sigma_i^2} & \sum \frac{x_i}{\sigma_i^2} \\ \sum \frac{x_i y_i}{\sigma_i^2} & \sum \frac{x_i^2}{\sigma_i^2} \end{vmatrix} = \frac{1}{\Delta} \left(\sum \frac{x_i^2}{\sigma_i^2} \sum \frac{y_i}{\sigma_i^2} - \sum \frac{x_i}{\sigma_i^2} \sum \frac{x_i y_i}{\sigma_i^2} \right)$$

$$b = \frac{1}{\Delta} \begin{vmatrix} \sum \frac{1}{\sigma_i^2} & \sum \frac{y_i}{\sigma_i^2} \\ \sum \frac{x_i}{\sigma_i^2} & \sum \frac{x_i y_i}{\sigma_i^2} \end{vmatrix} = \frac{1}{\Delta} \left(\sum \frac{1}{\sigma_i^2} \sum \frac{x_i y_i}{\sigma_i^2} - \sum \frac{x_i}{\sigma_i^2} \sum \frac{y_i}{\sigma_i^2} \right)$$

$$\Delta = \begin{vmatrix} \sum \frac{1}{\sigma_i^2} & \sum \frac{x_i}{\sigma_i^2} \\ \sum \frac{x_i}{\sigma_i^2} & \sum \frac{x_i^2}{\sigma_i^2} \end{vmatrix} = \sum \frac{1}{\sigma_i^2} \sum \frac{x_i^2}{\sigma_i^2} - \left(\sum \frac{x_i}{\sigma_i^2} \right)^2$$

Estimated uniform variance s^2:

$$\sigma^2 \simeq s^2 = \frac{1}{N-2} \sum (y_i - \bar{y})^2$$

Statistical fluctuations:

$$\sigma_i^2 \simeq y_i \qquad \text{(raw data counts)}$$

Uncertainties in coefficients:

$$\sigma_a^2 = \frac{1}{\Delta} \sum \frac{x_i^2}{\sigma_i^2} \qquad \sigma_b^2 = \frac{1}{\Delta} \sum \frac{1}{\sigma_i^2}$$

EXERCISES

6.1. Fit the data of Example 6.2 as if all the data had equal uncertainties $\sigma_i = \bar{\sigma} = 18.5$, where $\bar{\sigma}$ is the average of the given values of σ. Note that the fitted parameters are independent of the value of $\bar{\sigma}$, but the values of χ^2, σ_a, and σ_b are not.

6.2. Derive Equation (6.23) from Equations (6.21) and (6.22).

6.3. Show that Equation (6.12) reduces to Equation (6.13) if $\sigma_i = \sigma$.

6.4. Derive a formula for making a linear fit to data with an intercept at the origin so that $y = bx$. Apply your method to fit a straight line through the origin to the following coordinate pairs. Assume uniform uncertainties $\sigma_i = 1.5$ in y_i. Find χ^2 for the fit and the uncertainty in b.

x_i	2	4	6	8	10	12	14	16	18	20	22	24
y_i	5.3	14.4	20.7	30.1	35.0	41.3	52.7	55.7	63.0	72.1	80.5	87.9

6.5. A student hangs masses on a spring and measures the spring's extension as a function of the applied force in order to find the spring constant k. Her measurements are:

Mass (kg)	200	300	400	500	600	700	800	900
Extension (cm)	5.1	5.5	5.9	6.8	7.4	7.5	8.6	9.4

There is an uncertainty of 0.2 in each measurement of the extension. The uncertainty in the masses is negligible. For a perfect spring, the extension ΔL of the spring will be related to the applied force by the relation $k\Delta L = F$, where $F = mg$, and $\Delta L = L - L_0$, and L_0 is the unstretched length of the spring. Use these data and the method of least squares to find the spring constant k, the unstretched length of the spring L_0, and their uncertainties. Find χ^2 for the fit and the associated probability.

6.6. Outline a procedure for solving the simultaneous Equations (6.27). Refer to Appendix A.

6.7. A student measures the temperature (T) of water in an insulated flask at times (t) separated by 1 minute and obtains the following values:

t(s)	0	1	2	3	4	5	6	7	8
T(°C)	98.51	98.50	98.50	98.49	98.52	98.49	98.52	98.45	98.47

(*a*) Calculate the mean temperature and its standard error.

(*b*) To test whether or not the water is cooling, plot a graph of the temperatures versus the time and make a least-squares fit of a straight line to the data. Is there a statistically significant slope to the graph?

(*c*) Note that the intercept is not identical to the mean value of the temperature you calculated in part (*a*). Now, shift the time coordinates by 4 s so that the mean time is 0. Refit the data with the new values of T. Is the intercept now identical to the mean value of T?

(*d*) Clearly, the results of this experiment cannot depend upon the time at which the measurements were made. Show that, if the mean value of x is equal to zero, then the intercept b calculated from Equation (6.13) is identically equal to the mean value of y.

CHAPTER 7

LEAST-SQUARES FIT TO A POLYNOMIAL

7.1 DETERMINANT SOLUTION

So far we have discussed fitting a straight line to a group of data points. However, suppose our data (x_i, y_i) were not consistent with a straight line fit. We might construct a more complex function with extra parameters and try varying the parameters of this function to fit the data more closely. A very useful function for such a fit is a power-series polynomial

$$y(x) = a_1 + a_2 x + a_3 x^2 + a_4 x^3 + \cdots + a_m x^{m-1} \tag{7.1}$$

where the dependent variable y is expressed as a sum of power series of the independent variable x with coefficients a_1, a_2, a_3, a_4, and so forth.

For problems in which the fitting function is linear in the parameters, the method of least squares is readily extended to any number of terms m, limited only by our ability to solve m linear equations in m unknowns and by the precision with which calculations can be made. We can rewrite Equation (7.1) as

$$y(x) = \sum_{k=1}^{m} a_k x^{k-1} \tag{7.2}$$

where the index k runs from 1 to m. In fact, we can generalize the method even further by writing Equation (7.2) as

$$y(x) = \sum_{k=1}^{m} a_k f_k(x) \tag{7.3}$$

where the functions $f_k(x)$ could be the powers of x as in Equation (7.2), $f_1(x) = 1$, $f_2(x) = x$, $f_3(x) = x^2$, and so forth, or they could be other functions of x as long as they *do not involve the parameters* a_1, a_2, a_3, and so forth.

With this definition, the probability function of Equation (6.8) can be written as

$$P(a_1, a_2, \ldots, a_m) = \prod \left(\frac{1}{\sigma_i\sqrt{2\pi}}\right) \exp\left\{-\frac{1}{2}\sum \frac{1}{\sigma_i^2}\left[y_i - \sum_{k=1}^{m} a_k f_k(x_i)\right]^2\right\} \tag{7.4}$$

and Equation (6.9) for χ^2 becomes

$$\chi^2 = \sum \left[\frac{1}{\sigma_i}\left[y_i - \sum_{k=1}^{m} a_k f_k(x_i)\right]\right]^2 \tag{7.5}$$

The method of least squares requires that we minimize χ^2, our measure of the goodness of fit to the data, with respect to the parameters a_1, a_2, a_3, and so forth. The minimum is determined by taking partial derivatives with respect to each parameter in the expression for χ^2 of Equation (7.5), and setting them to zero:

$$\frac{\partial}{\partial a_l}\chi^2 = \frac{\partial}{\partial a_l}\sum \left[\frac{1}{\sigma_i}\left[y_i - \sum_{k=1}^{m} a_k f_k(x_i)\right]\right]^2$$

$$= -2\sum \left\{\frac{f_l(x_i)}{\sigma_i^2}\left[y_i - \sum_{k=1}^{m} a_k f_k(x_i)\right]\right\} = 0 \tag{7.6}$$

Thus, we obtain a set of m coupled linear equations for the m parameters a_l, with the index l running from 1 to m:

$$\sum y_i \frac{f_l(x_i)}{\sigma_i^2} = \sum_{k=1}^{m}\left\{a_k \sum\left[\frac{1}{\sigma_i^2} f_l(x_i) f_k(x_i)\right]\right\}$$

or

$$\begin{aligned}
\sum y_i \frac{f_1(x_i)}{\sigma_i^2} &= \sum \frac{f_1(x_i)}{\sigma_i^2}[a_1 f_1(x_i) + a_2 f_2(x_i) + a_3 f_3(x_i) \cdots] \\
\sum y_i \frac{f_2(x_i)}{\sigma_i^2} &= \sum \frac{f_2(x_i)}{\sigma_i^2}[a_1 f_1(x_i) + a_2 f_2(x_i) + a_3 f_3(x_i) \cdots] \\
\sum y_i \frac{f_3(x_i)}{\sigma_i^2} &= \sum \frac{f_3(x_i)}{\sigma_i^2}[a_1 f_1(x_i) + a_2 f_2(x_i) + a_3 f_3(x_i) \cdots]
\end{aligned} \tag{7.7}$$

and so forth.

The solutions can be found by the method of determinants, as in Chapter 6. We shall display the full solution for the particular case of $m = 3$:

$$a_1 = \frac{1}{\Delta}\begin{vmatrix} \sum y_i \frac{f_1(x_i)}{\sigma_i^2} & \sum \frac{f_1(x_i)f_2(x_i)}{\sigma_i^2} & \sum \frac{f_1(x_i)f_3(x_i)}{\sigma_i^2} \\ \sum y_i \frac{f_2(x_i)}{\sigma_i^2} & \sum \frac{f_2(x_i)f_2(x_i)}{\sigma_i^2} & \sum \frac{f_2(x_i)f_3(x_i)}{\sigma_i^2} \\ \sum y_i \frac{f_3(x_i)}{\sigma_i^2} & \sum \frac{f_3(x_i)f_2(x_i)}{\sigma_i^2} & \sum \frac{f_3(x_i)f_3(x_i)}{\sigma_i^2} \end{vmatrix}$$

$$a_2 = \frac{1}{\Delta}\begin{vmatrix} \sum \frac{f_1(x_i)f_1(x_i)}{\sigma_i^2} & \sum y_i \frac{f_1(x_i)}{\sigma_i^2} & \sum \frac{f_1(x_i)f_3(x_i)}{\sigma_i^2} \\ \sum \frac{f_2(x_i)f_1(x_i)}{\sigma_i^2} & \sum y_i \frac{f_2(x_i)}{\sigma_i^2} & \sum \frac{f_2(x_i)f_3(x_i)}{\sigma_i^2} \\ \sum \frac{f_3(x_i)f_1(x_i)}{\sigma_i^2} & \sum y_i \frac{f_3(x_i)}{\sigma_i^2} & \sum \frac{f_3(x_i)f_3(x_i)}{\sigma_i^2} \end{vmatrix} \qquad (7.8)$$

$$a_3 = \frac{1}{\Delta}\begin{vmatrix} \sum \frac{f_1(x_i)f_1(x_i)}{\sigma_i^2} & \sum \frac{f_1(x_i)f_2(x_i)}{\sigma_i^2} & \sum y_i \frac{f_1(x_i)}{\sigma_i^2} \\ \sum \frac{f_2(x_i)f_1(x_i)}{\sigma_i^2} & \sum \frac{f_2(x_i)f_2(x_i)}{\sigma_i^2} & \sum y_i \frac{f_2(x_i)}{\sigma_i^2} \\ \sum \frac{f_3(x_i)f_1(x_i)}{\sigma_i^2} & \sum \frac{f_3(x_i)f_2(x_i)}{\sigma_i^2} & \sum y_i \frac{f_3(x_i)}{\sigma_i^2} \end{vmatrix}$$

with

$$\Delta = \begin{vmatrix} \sum \frac{f_1(x_i)f_1(x_i)}{\sigma_i^2} & \sum \frac{f_1(x_i)f_2(x_i)}{\sigma_i^2} & \sum \frac{f_1(x_i)f_3(x_i)}{\sigma_i^2} \\ \sum \frac{f_2(x_i)f_1(x_i)}{\sigma_i^2} & \sum \frac{f_2(x_i)f_2(x_i)}{\sigma_i^2} & \sum \frac{f_2(x_i)f_3(x_i)}{\sigma_i^2} \\ \sum \frac{f_3(x_i)f_1(x_i)}{\sigma_i^2} & \sum \frac{f_3(x_i)f_2(x_i)}{\sigma_i^2} & \sum \frac{f_3(x_i)f_3(x_i)}{\sigma_i^2} \end{vmatrix}$$

We note that, as in the straight-line fits in Chapter 6, the denominator Δ is a function only of the independent variable x and the uncertainties σ_i in the dependent variable, and is not a function of the dependent variable y_i itself. For the special case of a quadratic power series in x, $y(x_i) = a_1 + a_2 x_i + a_3 x_i^2$, we have $f_1(x_i) = 1$, $f_2(x_i) = x_i$, and $f_3(x_i) = x^2$, so that Equations (7.8) become

$$a_1 = \frac{1}{\Delta}\begin{vmatrix} \sum y_i \frac{1}{\sigma_i^2} & \sum \frac{x_i}{\sigma_i^2} & \sum \frac{x_i^2}{\sigma_i^2} \\ \sum y_i \frac{x_i}{\sigma_i^2} & \sum \frac{x_i^2}{\sigma_i^2} & \sum \frac{x_i^3}{\sigma_i^2} \\ \sum y_i \frac{x_i^2}{\sigma_i^2} & \sum \frac{x_i^3}{\sigma_i^2} & \sum \frac{x_i^4}{\sigma_i^2} \end{vmatrix}$$

$$a_2 = \frac{1}{\Delta}\begin{vmatrix} \sum \frac{1}{\sigma_i^2} & \sum y_i \frac{1}{\sigma_i^2} & \sum \frac{x_i^2}{\sigma_i^2} \\ \sum \frac{x_i}{\sigma_i^2} & \sum y_i \frac{x_i}{\sigma_i^2} & \sum \frac{x_i^3}{\sigma_i^2} \\ \sum \frac{x_i^2}{\sigma_i^2} & \sum y_i \frac{x_i^2}{\sigma_i^2} & \sum \frac{x_i^4}{\sigma_i^2} \end{vmatrix} \qquad (7.9)$$

$$a_3 = \frac{1}{\Delta}\begin{vmatrix} \sum \frac{1}{\sigma_i^2} & \sum \frac{x_i}{\sigma_i^2} & \sum y_i \frac{1}{\sigma_i^2} \\ \sum \frac{x_i}{\sigma_i^2} & \sum \frac{x_i^2}{\sigma_i^2} & \sum y_i \frac{x_i}{\sigma_i^2} \\ \sum \frac{x_i^2}{\sigma_i^2} & \sum \frac{x_i^3}{\sigma_i^2} & \sum y_i \frac{x_i^2}{\sigma_i^2} \end{vmatrix}$$

with

$$\Delta = \begin{vmatrix} \sum \frac{1}{\sigma_i^2} & \sum \frac{x_i}{\sigma_i^2} & \sum \frac{x_i^2}{\sigma_i^2} \\ \sum \frac{x_i}{\sigma_i^2} & \sum \frac{x_i^2}{\sigma_i^2} & \sum \frac{x_i^3}{\sigma_i^2} \\ \sum \frac{x_i^2}{\sigma_i^2} & \sum \frac{x_i^3}{\sigma_i^2} & \sum \frac{x_i^4}{\sigma_i^2} \end{vmatrix}$$

Example 7.1. A student plans to use a thermocouple to monitor temperatures and must first calibrate it against a thermometer. The thermocouple consists of a junction of a copper wire and a constantan wire. In order to measure the junction voltage with high precision, she connects the sample junction in series with a reference junction that is held at 0°C in an ice water bath. The data, therefore, will be valid only for calibrating the relative variation of the junction voltage with temperature. The absolute voltage must be determined in a separate experiment by measuring it at one specific temperature.

The student measures the difference in output voltage between the two junctions for a temperature variation in the sample junction from 0 to 100°C in steps of 5°C. The measurements are made on the 3-mV scale of the voltmeter, and fluctuations of the

TABLE 7.1
Experimental data for the determination of the relative output voltage V of a thermocouple junction as a function of temperature T of the junction

Trial i	Temperature T (°C)	Measured voltage V (mV)	Calculated voltage $V(T)$ (mV)
1	0.	−0.849	−0.918
2	5.	−0.738	−0.728
3	10.	−0.537	−0.536
4	15.	−0.354	−0.341
5	20.	−0.196	−0.143
6	25.	−0.019	0.058
7	30.	0.262	0.261
8	35.	0.413	0.467
9	40.	0.734	0.676
10	45.	0.882	0.888
11	50.	1.258	1.102
12	55.	1.305	1.319
13	60.	1.541	1.539
14	65.	1.768	1.761
15	70.	1.935	1.987
16	75.	2.147	2.215
17	80.	2.456	2.446
18	85.	2.676	2.679
19	90.	2.994	2.915
20	95.	3.200	3.155
21	100.	3.318	3.396

$a_1 = -0.918 \pm 0.030$
$a_2 = 0.0377 \pm 0.0013$
$a_3 = 0.000055 \pm 0.000013$

Note: The common uncertainty in the voltage measurement is assumed to be 0.05 V. The value of χ^2 for the fit was $\chi^2 = 26.6$ for 18 degrees of freedom, with a probability of 8.8%. Parameters obtained from the fit are listed at the bottom of the table.

needle indicate that the uncertainties in the measurements are approximately 0.05 mV for all readings.

Data from the experiment are listed in Table 7.1 and are plotted in Figure 7.1. To a first approximation, the variation of V with T is linear, but close inspection of the graph reveals a slight curvature. Theoretically, we expect a good fit to these data with a quadratic curve of the form $V = a_1 + a_2T + a_3T^2$.

The parameters for the fit to the data of Example 7.1 have been obtained by evaluating the sums and determinants of Equations (7.9). For a second-degree polynomial with 21 data points, Equation (7.5) becomes

$$\chi^2 \equiv \sum_{i=1}^{21} \frac{1}{\sigma_i^2}[y_i - a_1 - a_2x_i - a_3x_i^2]^2 \tag{7.10}$$

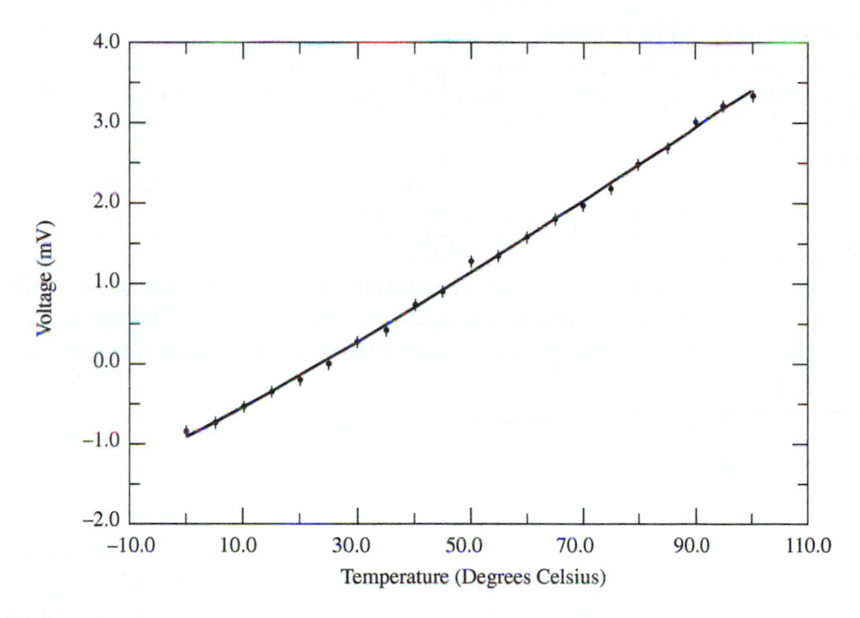

FIGURE 7.1
Thermocouple voltage versus temperature (Example 7.1). The curved line was calculated by fitting to the data second-degree polynomial $V = a_1 + a_2T + a_3T^2$ by the least-squares method. Uniform uncertainties were assumed.

The values of χ^2 and the parameters a_1, a_2, and a_3 determined from the fit are listed in Table 7.1, as are the calculated values of $V(T_i) = y(x_i)$. The calculated values of V are also represented by the solid line on the graph of Figure 7.1. We obtain $\chi^2 = 26.6$ for this fit, or $\chi^2_\nu = \chi^2/\nu = 1.5$, where the number of degrees of freedom ν is related to the number of events N and the number of free parameters m by $\nu = N - m$. The probability for obtaining χ^2 this high or higher can be determined from the χ^2-probability distribution (see Table C.4) and is about 8.8%, indicating a reasonable fit to the data.

As an alternative to calculating χ^2 from the fit, we could extend Equation (6.15) to three parameters and calculate the average uncertainty in the temperature readings to obtain

$$\sigma^2 \simeq s^2 = \frac{1}{N-m}\sum_{i=1}^{21}[[y_i - (a_1 + a_2x_i + a_3x_i^2)]]^2 = \frac{\chi^2}{N-m} \tag{7.11}$$

which is just the value of the uncertainty that would make $\chi^2_\nu = 1$. For Example 7.1, we obtain for an estimate of the variance,

$$\sigma'^2 = \sigma^2 \times \chi^2/(N-n) = 0.05 \times 26.6/18 = 0.06°\text{C}$$

suggesting, perhaps, that the student slightly underestimated the uncertainty in her measurements of V.

7.2 MATRIX SOLUTION

The techniques of least-squares fitting fall under the general name of regression analysis. Because we have been considering only problems in which the fitting function

$$y(x_i) = \sum_{k=1}^{m} a_k f_k(x_i) \tag{7.12}$$

is linear in the *parameters* a_k, we are considering only linear regression or multiple linear regression, usually shortened to multiple regression. In Chapter 8 we deal with techniques for handling problems with fitting functions that are not linear in the parameters.

Matrix Equations

We have not yet determined the uncertainties in the three parameters we obtained when we fitted the second-order equation to the data of Example 7.1. We could find the uncertainties by extending the method used for the linear fits of Examples 6.1 and 6.2. However, the algebra becomes even more tedious as the number of terms in the fitted equation increases, and in fact, our method only yielded estimates of the variances σ_k^2 and not of the covariances σ_{kl}^2, which are often important for fitted parameters. Rather than pursue the determinant method, we shall discuss immediately the more elegant and general matrix method for solving the multiple regression problem. Some of the properties of matrices are discussed in Appendix B.

Equations (7.7) can be expressed in matrix form as the equivalence between a row matrix $\boldsymbol{\beta}$ and the product of a row matrix $\mathbf{a}$ with a symmetric matrix $\boldsymbol{\alpha}$, all of order m:

$$\boldsymbol{\beta} = \mathbf{a}\boldsymbol{\alpha} \tag{7.13}$$

The elements of the row matrix $\boldsymbol{\beta}$ are defined by

$$\beta_k \equiv \sum \left[\frac{1}{\sigma_i^2} y_i f_k(x_i) \right] \tag{7.14}$$

those of the symmetric matrix $\boldsymbol{\alpha}$ by

$$\alpha_{lk} \equiv \sum \left[\frac{1}{\sigma_i^2} f_l(x_i) f_k(x_i) \right] \tag{7.15}$$

and the elements of the row matrix $\mathbf{a}$ are the parameters of the fit. For $m = 3$, the matrices may be written as

$$\boldsymbol{\beta} = [\beta_1 \ \beta_2 \ \beta_3] \qquad \mathbf{a} = [a_1 \ a_2 \ a_3] \tag{7.16}$$

and

$$\boldsymbol{\alpha} = \begin{bmatrix} \alpha_{11} & \alpha_{12} & \alpha_{13} \\ \alpha_{21} & \alpha_{22} & \alpha_{23} \\ \alpha_{31} & \alpha_{32} & \alpha_{33} \end{bmatrix} \tag{7.17}$$

To solve for the parameter matrix $\mathbf{a}$ we multiply both sides of Equation (7.13) on the right by the inverse $\boldsymbol{\epsilon}$ of the matrix $\boldsymbol{\alpha}$, defined such that $\boldsymbol{\alpha\epsilon} = \boldsymbol{\alpha\alpha}^{-1} = \mathbf{1}$, the unity matrix. We obtain

$$\boldsymbol{\beta\epsilon} = \mathbf{a}\boldsymbol{\alpha\epsilon} = \mathbf{a} \tag{7.18}$$

which gives

$$\mathbf{a} = \boldsymbol{\beta\epsilon} = \boldsymbol{\beta\alpha}^{-1} \tag{7.19}$$

Equation (7.19) can also be expressed as

$$a_l = \sum_{k=1}^{m}(\beta_k \epsilon_{kl}) = \sum_{k=1}^{m}\left\{\epsilon_{kl}\sum\left[\frac{1}{\sigma_i^2}y_i f_k(x_i)\right]\right\} \tag{7.20}$$

where the β_k is given by Equation (7.14).

The solution of Equation (7.19) requires that the matrix $\boldsymbol{\alpha}$ be inverted. This generally is not a simple procedure, except for matrices of very low order, but computer routines are readily available. The inversion of a matrix is discussed in Appendix B.

The symmetric matrix $\boldsymbol{\alpha}$ is called the *curvature matrix* because of its relationship to the curvature of the χ^2 function in parameter space. The relationship becomes apparent when we take the second derivatives of χ^2 with respect to the parameters. From Equation (7.6), we have for the partial derivative of χ^2 with respect to any arbitrary parameter a_l,

$$\frac{\partial\chi^2}{\partial a_l} = -2\sum\left\{\frac{f_l(x_i)}{\sigma_i^2}\left[y_i - \sum_{k=1}^{m}a_k f_k(x_i)\right]\right\} \tag{7.21}$$

and the second cross-partial derivative with respect to two such parameters is

$$\frac{\partial^2\chi^2}{\partial a_l\,\partial a_k} = 2\sum\left[\frac{1}{\sigma_i^2}f_l(x_i)f_k(x_i)\right] = 2\alpha_{lk} \tag{7.22}$$

Estimation of Errors

The variance $\sigma_{a_l}^2$ of any parameter a_l is the sum of the variances of each of the data points σ_i multiplied by the square of the effect that each data point has on the determination of the parameter a_l [see Equation (6.19)]. Similarly, the covariance of two parameters a_j and a_l is given by

$$\sigma_{a_j a_l}^2 = \sum\left[\sigma_i^2\frac{\partial a_j}{\partial y_i}\frac{\partial a_l}{\partial y_i}\right] \tag{7.23}$$

(which also gives the variance for $j = l$), where we have assumed that there are no correlations between uncertainties in the measured variables y_i. Taking the derivatives in Equation (7.23) of a_l with respect to y_i we obtain

$$\frac{\partial a_l}{\partial y_i} = \sum_{k=1}^{m}\left[\epsilon_{lk}\frac{1}{\sigma_i^2}f_k(x_i)\right] \tag{7.24}$$

and, substituting into Equation (7.23), we obtain for the weighted sum of the squares of the derivatives,

$$\begin{aligned}\sigma_{a_j a_l}^2 &= \sum\left\{\sigma_i^2 \sum_{k=1}^{m}\left[\epsilon_{jk}\frac{1}{\sigma_i}f_k(x_i)\right]\sum_{p=1}^{m}\left[\epsilon_{lp}\frac{1}{\sigma_i}f_p(x_i)\right]\right\}\\ &= \sum_{k=1}^{m}\left\{\epsilon_{jk}\sum_{p=1}^{m}\left[\epsilon_{lp}\sum\left(\frac{1}{\sigma_i^2}f_p(x_i)f_k(x_i)\right)\right]\right\}\\ &= \sum_{k=1}^{m}\left\{\epsilon_{jk}\sum_{p=1}^{m}\left[\epsilon_{lp}\cdot\alpha_{pk}\right]\right\}\\ &= \sum_{k=1}^{m}\left[\epsilon_{kj}\cdot \mathbf{1}_{lk}\right] = \epsilon_{jl}\end{aligned} \tag{7.25}$$

where we have switched the order of the sums over the dummy indices i, k, and l and have used the fact that because the curvature matrix $\boldsymbol{\alpha}$ is symmetric, its inverse $\boldsymbol{\epsilon}$ must also be symmetric, so that $\epsilon_{kj} = \epsilon_{jk}$. The elements of the unity matrix, which result from the summed products of the elements of $\boldsymbol{\alpha}$ with its inverse $\boldsymbol{\epsilon}$, are represented by $\mathbf{1}_{lk}$.

The inverse matrix $\boldsymbol{\epsilon} \equiv \boldsymbol{\alpha}^{-1}$ is called the error matrix or the covariance matrix because its elements are the variances and covariances of the fitted parameters $\sigma_{a_j a_l}^2 = \epsilon_{jl}$.

Example 7.2. The matrix method is illustrated by a straight-line fit $V = a_1 + a_2 T$ to a selection of data from Example 7.1. To show clearly each step of the calculation, we have selected just six points spaced at 25° intervals between 0 and 100° and have assumed a common uncertainty in the dependent variable $\sigma_v = 0.05$ mV. The data are listed in the columns 2 and 3 of Table 7.2*a*.

We begin by calculating each of the fitting functions $f_1 = 1$ and $f_2 = x$ at each value of the independent variable T. These are listed in columns 4 and 5 of Table 7.2*a*. For each measured value of x, the values of β_k, the elements of the column matrix $\boldsymbol{\beta}$, and of α_{lk}, the elements of the symmetric matrix $\boldsymbol{\alpha}$, are calculated according to Equations (7.14) and (7.15). The individual terms in the calculation of β_1 and β_2 are listed in columns 6 and 7 of Table 7.2*a* and the individual terms in the calculation of α_{lk} are listed in columns 8 through 10. (We assume symmetry in $\boldsymbol{\alpha}$.) The resulting matrices are displayed in Table 7.2*b*.

The symmetric matrix $\boldsymbol{\alpha}$ is inverted to obtain the variance matrix $\boldsymbol{\epsilon}$ with elements ϵ_{kl}, shown in Table 7.2*b*, and the product matrix of the fitted parameters $\boldsymbol{\alpha} = \boldsymbol{\beta\epsilon}$ is calculated and displayed in Table 7.2*b*. The calculated values of the fitted variable V for each value of the independent variable T are listed in the last column of Table 7.2*a*.

Program 7.1. MULTREGR (Appendix E) Least-squares fitting with matrices.

Multiple regression problems are usually solved by computer. The program MULTREGR calls a set of routines for fitting any function that is linear in the parameters $a_1, a_2, \ldots, a_m$ to a set of N data points. Branches in the program on the character variable PAE permit selection of the fitting function for each example in this chapter, with PAE = 'P' for the power series in x, PAE = 'A' for all terms of a fourth-order Legendre polynomial, or PAE = 'E' for only the even terms in the Legendre polynomial. The program uses several program units in addition to those referred to in Chapter 6.

TABLE 7.2
Matrix solution for linear fit to data of Example 7.2

(*a*) Data and components of matrix elements

i	T	V	$f_1(x_i)$	$f_2(x_i)$	β_1	β_2	α_{11}	α_{12}	α_{22}	V_{fit}
1	0	−0.849	1	0	−339.6	0	400	0	0	−0.947
2	20	−0.196	1	20	−78.4	−1,458	400	8,000	160,000	−0.101
3	40	0.734	1	40	293.6	11,744	400	16,000	640,000	0.745
4	60	1.541	1	60	616.4	36,984	400	24,000	1,440,000	1.590
5	80	2.456	1	80	982.4	78,592	400	32,000	2,560,000	2.436
6	100	3.318	1	100	1327.6	132,720	400	40,000	4,000,000	3.281
					2802.0	258,472	2,400	120,000	8,800,000	

(*b*) Matrices

$$\alpha = \begin{bmatrix} 2{,}400 & 120{,}000 \\ 120{,}000 & 8{,}800{,}00 \end{bmatrix} \qquad \epsilon = \begin{bmatrix} 1.310 \times 10^{-03} & -1.786 \times 10^{-05} \\ -1.786 \times 10^{-05} & 3.571 \times 10^{-07} \end{bmatrix}$$

$$\beta = [2{,}802 \quad 258{,}472] \qquad \mathbf{a} = [-0.947 \quad 0.0423]$$

Note: The uniform uncertainty in V was assumed to be 0.05 mV as in Example 7.1. The columns labeled β_1 and α_{11}, etc. correspond to the individual contributions by each measured coordinate pair to the summed values of β and α. The value of χ^2 for the fit was 9.1 for 4 degrees of freedom corresponding to a probability of 5.5%.

Program 7.2. FITFUNC7 (Appendix E) Fitting functions and χ^2 calculation. In general, every fitting problem requires such a routine. The function POWERFUNC calculates the individual terms in a power function of any order in x for Example 7.2, or Legendre polynomials for Example 7.3.

Program 7.3. MAKEAB7 (Appendix E) Form the arrays for the matrices $\boldsymbol{\alpha}$ and $\boldsymbol{\beta}$.

Program B.1. MATRIX (Appendix E) Matrix products and inversion.

When we use the matrix method to fit a polynomial function to a data sample, the resulting parameters must be identical to those calculated by the determinant method, but we also obtain the full error matrix. The error matrix obtained by fitting a second-degree polynomial to the complete data sample of Example 7.1 is listed in Table 7.3.

The error matrix can be used to estimate the uncertainty in a calculated result, including the effects of the correlations of the errors. As an example, let us suppose that we wish to find the predicted value of the voltage V and its uncertainty for a temperature of exactly 80°C. We should calculate

$$V = a_1 + a_2 T + a_3 T^2 \tag{7.26}$$

using the parameters determined by the fit to the data. The uncertainty in the calculated value of V, which results from the uncertainty in the parameters, is given by Equation (3.13),

TABLE 7.3
Error matrix from a fit by the matrix method to the data of Table 7.1

$$\begin{bmatrix} 8.907 \times 10^{-04} & -3.473 \times 10^{-05} & 2.823 \times 10^{-07} \\ -3.473 \times 10^{-05} & 1.913 \times 10^{-06} & -1.783 \times 10^{-08} \\ 2.823 \times 10^{-07} & -1.783 \times 10^{-08} & 1.783 \times 10^{-10} \end{bmatrix}$$

Note: The table gives the variances and covariances of the fitted parameters. The values of the parameters and of χ^2 are listed in Table 7.1.

$$\begin{aligned} s^2 &= \left(\frac{\partial V}{\partial a_1}\right)^2 \sigma_1^2 + \left(\frac{\partial V}{\partial a_2}\right)^2 \sigma_2^2 + \left(\frac{\partial V}{\partial a_3}\right)^2 \sigma_3^2 \\ &\quad + 2\left(\frac{\partial V}{\partial a_1}\frac{\partial V}{\partial a_2}\right)\sigma_{12}^2 + 2\left(\frac{\partial V}{\partial a_1}\frac{\partial V}{\partial a_3}\right)\sigma_{13}^2 + 2\left(\frac{\partial V}{\partial a_2}\frac{\partial V}{\partial a_3}\right)\sigma_{23}^2 \\ &= 1 \cdot \epsilon_{11} + T^2 \cdot \epsilon_{22} + T^4 \cdot \epsilon_{33} + 2(T \cdot \epsilon_{12} + T^2 \cdot \epsilon_{13} + T^3 \cdot \epsilon_{23}) \end{aligned} \tag{7.27}$$

where ϵ_{12} and so on are the covariant terms in the symmetric error matrix. If we used only the diagonal terms in the error matrix, our result would be $V = (2.45 \pm 0.14)$ V. However, the off-diagonal terms are mainly negative, and including them reduces the uncertainty by almost a factor of 10 to 0.015, so that we should quote $V = (2.45 \pm 0.02)$ V.

Linear Least-Squares Fitting with a Spreadsheet

Table 7.4 illustrates the use of a spreadsheet (without taking advantage of the spreadsheet's built-in least-squares fitting routine) to fit a straight line to the data of Example 7.2 by the matrix method. We entered the data in columns labeled T, V, and σ_v and calculated component terms to be summed for $\boldsymbol{\beta}_1$, $\boldsymbol{\beta}_2$, and $\boldsymbol{\alpha}_{11}$, $\boldsymbol{\alpha}_{12}$, and $\boldsymbol{\alpha}_{22}$ in the labeled columns using the indicated equations. We summed each $\boldsymbol{\alpha}$ column to form the elements of the square matrix $\boldsymbol{\alpha}$, and the $\boldsymbol{\beta}$ columns to form the linear matrix $\boldsymbol{\beta}$. The spreadsheet's matrix-handling routines were applied to invert the $\boldsymbol{\alpha}$-matrix to form the ϵ-matrix, and to multiply $\boldsymbol{\epsilon}$ by $\boldsymbol{\beta}$ to find the parameter matrix **a**. Uncertainties in the parameters were calculated from the square roots of the diagonal terms in the ϵ-matrix. Although we used absolute cell addresses to illustrate the procedure, we could have simplified the calculation by naming the arrays of cells and using the array-handling capabilities of the spreadsheet.

It may seem inefficient to write a program to solve such a simple problem, which most spreadsheets can handle with ease. However, there are advantages. First, it would be relatively easy to expand the program to fit more parameters, or to fit a series of functions more complicated than simple powers of the independent variable. A second advantage is that the solution provides the full error matrix. While most fitting programs should provide the uncertainties in the fitted parameters, the covariances may not be available. In some problems, they are essential.

We used *Quatro Pro* for this example, but the procedure with *Excel* is similar.

TABLE 7.4
Matrix solution by spreadsheet calculation for linear fit to data of Example 7.2

***(a)* Data and components of the matrix elements and sums**

T(°C)	V (mV)	σ_v (mV)	β_1	β_2	α_{11}	$\alpha_{12} = \alpha_{21}$	α_{22}	Y_{calc}	χ^2
Column Equations			$V*1/\sigma^2$	$V*T/\sigma^2$	$1*1/\sigma^2$	$1*T/\sigma^2$	$T*T/\sigma^2$	$a_1 + a_2*T$	$[(Y - Y_{calc})/\sigma]^2$
0	−0.849	0.050	−339.6	0	400	0	0	−0.947	3.81
20	−0.196	0.050	−78.4	−1,568	400	8,000	160,000	−0.101	3.61
40	0.734	0.050	293.6	11,744	400	16,000	640,000	0.745	0.04
60	1.541	0.050	616.4	36,984	400	24,000	1,440,000	1.590	0.97
80	2.456	0.050	982.4	78,592	400	32,000	2,560,000	2.436	0.16
100	3.318	0.050	1327.2	132,720	400	40,000	4,000,000	3.281	0.54
SUMS			2801.6	258,472	2400	120,000	8,800,000		9.13

***(b)* Matrices and fitted coefficients with uncertainties (*Quatro Pro* matrix algebra used to calculate ϵ and a)**

$$\alpha = \begin{bmatrix} 2400 & 120000 \\ 120000 & 8800000 \end{bmatrix} \qquad \epsilon = \begin{bmatrix} 1.310\text{E-}03 & -1.786\text{E-}05 \\ -1.786\text{E-}05 & 3.571\text{E-}07 \end{bmatrix}$$

$$\beta = |2801.6 \quad 258472| \qquad \mathbf{a} = [-0.947 \quad 0.0423]$$

$$\chi^2 = 9.13 \qquad \sigma_a = \boxed{0.036 \quad 0.006}$$

7.3 INDEPENDENT PARAMETERS

Suppose we take the data of Example 6.1 or Example 6.2 and fit to them the quadratic polynomial function $y = a_1 + a_2x + a_3x^2$ as we did for Example 7.1. We should expect to find a rather small and possibly meaningless result for the coefficient a_3 of the quadratic term, but, because a_3 was not set equal to zero by definition, as in the analysis of Chapter 6, we might also find that the values of a_1 and a_2 have changed, sometimes considerably, from the values obtained in the linear fit. In general, the polynomial fitting procedure that we have considered will yield values for the coefficients that depend on the degree of the polynomial fitted to the data.

This interdependence arises from the fact that we have specified our coordinate system without regard to the region of parameter space from which our data points are extracted. The value of a_1 represents the intercept on the ordinal axis, the coefficient a_2 represents the slope at this same point, and other coefficients represent higher orders of curvature at this same intercept point. If the data are not clustered about this intercept point, its location might be highly dependent on the polynomial used to fit the data.

We might be able to extract more meaningful information about the data if we were to determine instead coefficients a_1', a_2', a_3', and so forth, which represent the average value, the average slope, the average curvature, and so forth, of the data. Such coefficients would be independent of our choice of coordinate system and would represent physical characteristics of the data that are independent of the degree of the fitted polynomial.

Orthogonal Polynomials

We want to fit the data to a function that is similar to that of Equation (7.1) but that yields the desired independence of the coefficients. The appropriate function to use is the sum of orthogonal polynomials,[1] which has the form

$$\begin{aligned} y(x) = a_1 + a_2(x - \beta) + a_3(x - \gamma_1)(x - \gamma_2) \\ + a_4(x - \delta_1)(x - \delta_2)(x - \delta_3) + \cdots \end{aligned} \tag{7.28}$$

Following the development of Section 7.1, we must minimize χ^2 to determine the coefficients a_1, a_2, a_3, a_4, and so on, with the further criterion that the addition of higher-order terms to the polynomial will not affect the evaluation of lower-order terms. This criterion will be used to determine the extra parameters β, γ_1, γ_2, and so on.

The goodness-of-fit parameter χ^2 is defined as

$$\chi^2 \equiv \sum \left[\frac{\Delta y_i}{\sigma_i}\right]^2 = \sum \left[\frac{1}{\sigma_i^2}[y_i - y(x_i)]^2\right] \tag{7.29}$$

Setting the derivatives of χ^2 with respect to each of the m coefficients a_1, a_2, and so forth to 0 yields m simultaneous equations

$$\begin{aligned} \sum y_i = Na_1 + a_2\Sigma(x_i - \beta) + a_3\Sigma(x_i - \gamma_1)(x_i - \gamma_2) \\ + a_4\Sigma(x_i - \delta_1)(x_i - \delta_2)(x_i - \delta_3) + \cdots \end{aligned} \tag{7.30}$$

$$\begin{aligned} \sum x_i y_i = a_1\Sigma x_i + a_2\Sigma x_i(x_i - \beta) + a_3\Sigma x_i(x_i - \gamma_1)(x_i - \gamma_2) \\ + a_3\Sigma x_i(x_i - \delta_1)(x_i - \delta_2)(x_i - \delta_3) + \cdots \end{aligned} \tag{7.31}$$

$$\begin{aligned} \sum x_i^2 y_i = a_1\Sigma x_i^2 + a_2\Sigma x_i^2(x_i - \beta) + a_3\Sigma x_i^2(x_i - \gamma_1)(x_i - \gamma_2) \\ + a_4\Sigma x_i^2(x_i - \delta_1)(x_i - \delta_2)(x_i - \delta_3) + \cdots \end{aligned} \tag{7.32}$$

$$\begin{aligned} \sum x_i^3 y_i = a_1\Sigma x_i^3 + a_2\Sigma x_i^3(x_i - \beta) + a_3\Sigma x_i^3(x_i - \gamma_1)(x_i - \gamma_2) \\ + a_4\Sigma x_i^3(x_i - \delta_1)(x_i - \delta_2)(x_i - \delta_3) + \cdots \end{aligned} \tag{7.33}$$

where we have omitted a factor of σ_i^2 in the denominator for clarity.

---

[1]Any polynomial such as that of Equation (7.1) can be rewritten as a sum of orthogonal polynomials

$$y = a + \sum_{j=1}^{n} [b_j f_j(x_i)]$$

with the orthogonal property that $\Sigma[f_j(x_i)f_k(x_i)] = 0$ for $j \neq k$.

Additional Parameters

Let us examine Equation (7.30). If we restrict ourselves to a zeroth-degree polynomial, that is, to only one coefficient a_1, all the other coefficients are equal to 0 by definition. The coefficient a_1, therefore, is specified completely by the first term on the right-hand side of Equation (7.30):

$$a_1 = \frac{1}{N}\sum y_i = \bar{y} \tag{7.34}$$

If we restrict ourselves to a first-degree polynomial, the coefficient a_2 of the second term of Equation (7.30) is not 0. However, if a_1 is to be independent of the value of a_2, the second term itself must be 0. Hence, the requirement that

$$\sum (x_i - \beta) = 0$$

leads to the value for β,

$$\beta = \frac{1}{N}\sum x_i = \bar{x} \tag{7.35}$$

and a_2 can be determined directly from Equation (7.31) by substituting the values of a_1 and β with higher-order coefficients (a_3, a_4, etc.) set to 0.

Similarly, if we consider a quadratic function, the third term of Equation (7.30) must be 0 even when the coefficient a_3 is not 0. This constraint leads to a quadratic equation in γ_1 and γ_2 that is not sufficient to specify either parameter. We have the additional constraint, however, that the coefficient a_2 must be specified completely by Equations (7.30) and (7.31). Thus, the third term in both Equations (7.30) and (7.31) must be 0 regardless of the value of the coefficient a_3, and we have two simultaneous quadratic equations for the parameters γ_1 and γ_2,

$$\sum (x_i - \gamma_1)(x_i - \gamma_2) = 0 \qquad \text{and} \qquad \sum x_i(x_i - \gamma_1)(x_i - \gamma_2) = 0 \tag{7.36}$$

Similarly, the coefficient a_3 must be determined completely by Equation (7.32) (and the predetermined values of a_1 and a_2), and this constraint yields three simultaneous equations for the parameters δ_1, δ_2, and δ_3:

$$\begin{aligned}
\sum (x_i - \delta_1)(x_i - \delta_2)(x_i - \delta_3) &= 0 \\
\sum x_i(x_i - \delta_1)(x_i - \delta_2)(x_i - \delta_3) &= 0 \\
\sum x_i^2(x_i - \delta_1)(x_i - \delta_2)(x_i - \delta_3) &= 0
\end{aligned} \tag{7.37}$$

The extrapolation to higher order is straightforward. (Note that these *additional parameters* are functions only of the independent variable x_i.)

Estimates of the Coefficients

Once the parameters β, γ, δ, and so on have been determined by the constraint equations, estimates of the coefficients a_1, a_2, and so on can be found from the resulting

$n + 1$ simultaneous equations. The value for the first coefficient a_1 is specified completely by minimizing χ^2 with respect to a_1 in Equation (7.30) and is given in Equation (7.34). The value of the second coefficient a_2 is determined by minimizing χ^2 with respect to both a_1 and a_2 in Equations (7.30) and (7.31) and substituting the value of a_1 from Equation (7.34) into Equation (7.31). Similarly, the value of a_3 can be determined from Equation (7.32) after substituting the values of a_1 and a_2 determined from Equations (7.30) and (7.31). Each succeeding equation yields a value for the next higher-order coefficient.

Note that the value determined for any coefficient is thus independent of the value specified for any higher-order coefficient, but is not independent of the value of lower-order coefficients. The parameters, representing our best estimates of the coefficients, are given by

$$\begin{aligned}
a_1 &= \frac{1}{N}\sum y_i = \bar{y} \\
a_2 &= \frac{\sum y_i(x_i - \beta)}{\sum (x_i - \beta)^2} \\
a_3 &= \frac{\sum y_i(x_i - \gamma_1)(x_i - \gamma_2)}{\sum [(x_i - \gamma_1)(x_i - \gamma_2)]^2} \\
a_4 &= \frac{\sum y_i(x_i - \delta_1)(x_i - \delta_2)(x_i - \delta_3)}{\sum [(x_i - \delta_1)(x_i - \delta_2)(x_i - \delta_3)]^2}
\end{aligned} \tag{7.38}$$

and so forth.

Simplification

For the general case of arbitrarily chosen data points (x_i, y_i), this procedure is cumbersome even with computer techniques because it requires the solution of coupled, nonlinear equations. There is, however, a special type of data for which the calculations can be considerably simplified, namely, data that meet the following two criteria: (1) the independent variables x_i are equally spaced, and (2) the uncertainties are constant, $\sigma_i = \sigma$, and can therefore be ignored.

Consider the experiments of Examples 6.1 (measurement of temperature versus position) and 7.1 (voltage versus temperature). Those data satisfy the required conditions and, therefore, we could use a simplified method of independent parameters to obtain a fit. The resulting values of the coefficients for these particular experiments might not have any great physical significance (that is, $a_1 = \bar{T}$ the average temperature of the data points in Example 6.1 is not a particularly useful number), but by using this technique of fitting orthogonal polynomials we could try fitting higher-degree polynomials without changing the values of the coefficients already calculated for a straight-line or quadratic fit. The experiment of Example 6.2 (the decay of a radioactive state) fulfills only the first of the two criteria, because the x data points are equally spaced but the uncertainties are statistical, so that we cannot ignore the factor of σ_i^2 that belongs in the denominators of the fitting Equations (7.30) through (7.33).

For an experiment similar to that of Example 7.1, where we have made N measurements of equally spaced values of the independent variable x ranging from x_1 to x_N in steps of Δ,

$$\Delta = x_{i+1} + x_i$$

and the uncertainties are due to instrumental errors with a common standard deviation $\sigma_i = \sigma$, Equations (7.35) through (7.37) reduce to

$$\begin{aligned}
\beta &= \frac{1}{N}\sum x_i = \bar{x} = \frac{1}{2}(x_i - x_n) \\
\gamma &= \beta \pm \sqrt{\frac{1}{N}\sum (x_i - \beta)^2} = \beta \pm \Delta\sqrt{\frac{1}{12}(N^2 - 1)} \qquad (7.39) \\
\delta &= \beta, \beta \pm \sqrt{\frac{\sum [x_i(x_i - \beta)^3]}{\sum [x_i(x_i - \beta)]}} = \beta, \beta \pm \Delta\sqrt{\frac{1}{20}(3N^2 - 7)}
\end{aligned}$$

A more comprehensive list of parameters for orthogonal polynomials can be found in Anderson and Houseman (1942).

Table 7.5 shows coefficients a_1, a_2, a_3, and a_4 as well as the values of χ^2 and the χ^2-probability obtained when we fit the data of Example 7.1, by the standard least-squares method and by the independent parameter method of Equation (7.39). We have made separate fits with first-, second-, and third-degree polynomials ($m = 2$, 3, and 4). As expected, adding extra terms does not change the values of the lower-order coefficients obtained by the independent parameter method and therefore we display them only once in Table 7.5. There is a marked improvement in χ^2 in going from the two-parameter (linear) fits to three-parameter (quadratic) fits. Unless a theoretical reason dictates that our data should follow a cubic distribution, there is no justification in making a four-parameter (cubic) fit to these data, because the value of χ^2 for $m = 3$ is satisfactory (26.6 for 18 degrees of freedom,

TABLE 7.5

Values of χ^2 and parameters obtained by fitting the data of Example 7.1 by the standard least-squares method and by the method of independent parameters, as a function of the number of parameters m of the fit

	Standard least squares			
m	2	3	4	Independent parameters
χ^2	43.5 (0.12%)	26.6(8.8%)	24.9(9.4%)	
a_1	-1.01 ± 0.02	(-0.92 ± 0.03)	(-0.89 ± 0.03)	1.15
a_2	$(4.31 \pm 0.04)10^{-2}$	$(3.8 \pm 0.1)10^{-2}$	$(3.4 \pm 0.3)10^{-2}$	4.31×10^{-2}
a_3		$(5.5 \pm 1.3)10^{-5}$	$(1.5 \pm 0.8)10^{-4}$	5.49×10^{-5}
a_4			$(-6.5 \pm 5.1)10^{-7}$	6.51×10^{-7}

Note: The values of χ^2 are the same for both methods. The numbers in parentheses correspond to the χ^2 probability for the fit with 21-m degrees of freedom.

corresponding to $P = 8.8\%$), and adding more terms does not improve the fit. If a cubic function had been predicted by theoretical considerations, we should be obligated to say that our data are not sensitive to the presence of a cubic term.

Legendre Polynomials

Although the method of fitting to orthogonal polynomials outlined in the previous section can be tedious, there are predefined sets of orthogonal polynomials that are often useful in fitting data. One important set is the *Legendre polynomials*

$$y(x) = a_0P_0(x) + a_1P_1(x) + \cdots = \sum_{L=0}^{m-1}[a_LP_L(x)] \tag{7.40}$$

where $x = \cos\theta$ and the terms $P_L(x)$ in the function are given by

$$\begin{aligned} P_0(x) &= 1 \qquad & P_2(x) &= \tfrac{1}{2}(3x^2 - 1) \\ P_1(x) &= x \qquad & P_3(x) &= \tfrac{1}{2}(5x^3 - 3x) \end{aligned} \tag{7.41}$$

and higher-order terms can be determined from the recursion relation

$$P_L(x) = \frac{1}{L}[(2L-1)xP_{L-1}(x) - (L-1)P_{L-2}(x)] \tag{7.42}$$

Legendre polynomials are orthogonal when averaged over all values of $x = \cos\theta$:

$$\int_{-1}^{1}[P_L(x)P_M(x)\,dx = \begin{cases} 2/(2L+1) & \text{for } L = M \\ 0 & \text{for } L \neq M \end{cases} \tag{7.43}$$

Example 7.3. Let us consider an experiment in which ^{13}C is bombarded by 4.5-MeV protons. In the subsequent reaction, some of the protons are captured by the ^{13}C nucleus, which then decays by gamma emission, producing gamma rays with energies up to 11 MeV. A measurement of the angular distribution of the emitted gamma rays gives information about the angular momentum states of the energy levels in the residual nucleus ^{14}N.

Table 7.6 lists simulated data for this experiment. Gamma ray counts were recorded at 17 angles from 0 to 160°. Columns 1 through 4 list the angles at which the measurements were made, the cosine of the angle ($x = \cos\theta$), the measured number of counts (C_i), and the uncertainties σ_{C_i} in the counts. The uncertainties are assumed to be purely statistical. These data are plotted in Figure 7.2 as a function of the angle θ. There appears to be symmetry around $\theta = 90°$, and consideration of the reaction process predicts that the data should be described by a fourth-order Legendre polynomial with only even terms:

$$C = a_0P_0(x) + a_2P_2(x) + a_4P_4(x) \quad \text{with } x = \cos\theta \tag{7.44}$$

Let us apply the matrix method of least squares of Section 7.2 to this problem to fit the series of Legendre polynomials of Equation (7.41) to these data. We shall first fit a fourth-order Legendre polynomial that includes both odd and even terms. The fitting function is of the form

$$y(x) = a_0f_0(x) + a_1f_1(x) + \cdots + a_{m-1}f_{m-1}(x) \tag{7.45}$$

which is linear in the fitting parameters, the coefficients a_i.

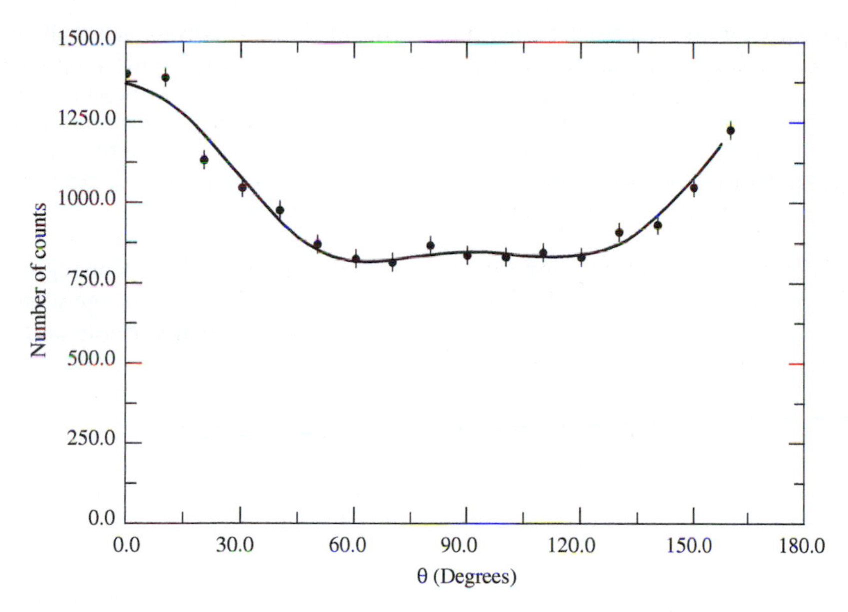

FIGURE 7.2
Angular distribution of gamma rays emitted from the simulated reaction $^{13}C(p, \gamma)^{14}N$ produced by incident protons at $E_p = 4.5$ MeV (Example 7.3). The calculated curve represents a fit to the data of a series of even Legendre polynomials up to $L = 4$. Statistical uncertainties were assumed.

TABLE 7.6

Angular distribution of gamma rays emitted in the reaction $^{13}C(p, \gamma)^{14}N$ produced by incident protons at $E_P = 4.5$ MeV

θ (degrees)	$x = \cos\theta$	C_i counts	σ_{C_i}	Y_i all terms	Y_i even term
0	1.000	1400	37.4	1365.8	1361.3
10	0.985	1386	37.2	1325.2	1321.1
20	0.940	1130	33.6	1217.0	1213.9
30	0.866	1045	32.3	1075.8	1074.5
40	0.766	971	31.2	943.5	944.4
50	0.643	862	29.4	852.5	855.6
60	0.500	819	28.6	813.9	818.6
70	0.342	808	28.4	816.9	821.9
80	0.174	862	29.4	836.5	840.2
90	0.000	829	28.2	848.6	849.6
100	−0.174	824	28.7	842.8	840.2
110	−0.342	839	29.0	827.5	821.9
120	−0.500	819	28.6	825.4	818.6
130	−0.643	901	30.0	861.0	855.6
140	−0.766	925	30.4	945.7	944.4
150	−0.866	1044	32.3	1069.8	1074.5
160	−0.940	1224	35.0	1202.9	1213.9

Note: The calculated numbers of counts were obtained from least-squares Legendre polynomials fits to the data of the form $Y_i(x) = \sum_{L=1}^{5} a_L P_{L-1}(x_i)$, for separate fits with all terms and with even terms only.

Computer fits Routines used for fitting a series of Legendre polynomials to these data are included in Program 7.1. The procedure LEGPOLY in the program unit FITFUNC7 calculates the terms of the Legendre polynomials through tenth order. The procedure is selected through a branch on the variable PAE in the function Funct with PAE = 'A' for all terms to order $n = m - 1$, or PAE = 'E' to fit with just the even terms. Note that the index k of the term in the fitting function, in general, does not correspond to the order L of the Legendre polynomial.

The efficiency of the calculation (and therefore the speed of the linear regression calculation) could be improved in a number of ways. The simplest change would be to calculate the functions once at each value of the independent variable and store the calculated values in an array.

Parameters obtained by fitting a series in Legendre polynomials for terms up to $L = 4$ are listed in Table 7.7. Separate fits were made with all terms and with only the even terms in the series. As expected, the coefficients of terms involving odd orders are comparable to their uncertainties and negligible compared to those involving even poynomials. The full error matrix for the fit with even terms is listed in Table 7.8.

In view of the strong theoretical argument that only even Legendre polynomials are required for this reaction, it would be appropriate to fit a series that includes only the even terms. The parameters obtained in this fit are also displayed in Table 7.7, and the numbers of counts calculated from these parameters are listed. The function calculated with even terms is illustrated as a curve on the data of Figure 7.2.

Because we are fitting with orthogonal functions, we might have expected to obtain identical values for the coefficient a_0 from both fits. (We expect the higher-order even coefficients to change because the presence or absence of lower-order coefficients must affect the higher coefficients.) The fact that there is some dependence of a_0 on higher-order terms is a result of the fact that a given experiment does not sample uniformly the entire range of the Legendre polynomial, so the orthogonality relation Equation (7.43) is not satisfied by a finite data set. This is in contrast to the situation in the previous section, where we set up orthogonal functions based on the data themselves. Nevertheless, it is generally good practice to use orthogonal

TABLE 7.7
Coefficients and χ^2 from least-squares fit to Legendre polynomial series

	χ^2	a_0	a_1	a_2	a_3	a_4
All terms	17.2(14%)	937.4 ± 7.6	0.7 ± 12.8	259 ± 14	10 ± 17	158 ± 18
Even terms	17.6(22%)	938.1 ± 7.5		261 ± 14		161 ± 16

TABLE 7.8
Error matrix for a least-squares fit to even Legendre polynomials

$$\begin{bmatrix} 56.24 & -5.256 & -6.272 \\ -5.256 & 186.5 & -26.90 \\ -6.272 & -26.90 & 279.8 \end{bmatrix}$$

fitting functions whenever possible to minimize both the correlations between coefficients and the dependence of higher coefficients on the presence of lower ones.

The values of χ^2 and the χ^2-probability for the two fits are also given in Table 7.7. We note that χ^2 for the three-parameter fit is necessarily higher than that for the five-parameter fit, but χ^2 per degree of freedom is smaller and the χ^2-probability is higher.

7.4 NONLINEAR FUNCTIONS

In all the procedures developed so far we have assumed that the fitting function was linear in the coefficients. By that we mean that the function can be expressed as a sum of separate terms each multiplied by a single coefficient. How can we fit data with a function that is not linear in the coefficients? For example, suppose we have measured the distribution of decay times of an unstable state and that the distribution can be represented by the normalized function $P(t) = (1/\tau)e^{-t/\tau}$, where τ is the mean lifetime of the state. Can we find the parameter τ by the least-squares method? The method of least squares does not yield a straightforward analytical solution for such functions. In Chapter 8 we investigate methods of searching parameter space for values of the coefficients that will minimize the goodness-of-fit criterion χ^2. Here we consider approximate solutions to such problems using linear-regression techniques.

Linearization

It is possible to transform some functions into linear functions. For example, if we were to fit an exponential decay problem of the form

$$y = ae^{-bx} \tag{7.46}$$

where a and b are the unknown parameters, it would seem reasonable to take logarithms of both sides and to fit the resulting straight line equation

$$\ln y = \ln a - bx \tag{7.47}$$

The method of least squares minimizes the value of χ^2 with respect to each of the coefficients $\ln a$ and $\ln b$ where χ^2 is given by

$$\chi^2 = \sum \left\{ \frac{1}{\sigma_i'^2} [\ln y_i + \ln a - bx_i]^2 \right\} \tag{7.48}$$

where we must use weighted uncertainties σ_i' instead of σ_i to account for the transformation of the dependent variable:

$$\sigma_i' = \frac{d(\ln y_i)}{dy} \sigma_i = \frac{1}{y_i} \sigma_i \tag{7.49}$$

The importance of weighting the uncertainties is illustrated in Figure 7.3, which shows the function of Equation (7.46) graphed both on a linear and on a logarithmic scale. (For plotting, we use logarithms to base 10 rather than natural logarithms.) The uncertainties are given by $\sigma_i = \sqrt{y_i}$ and therefore increase with increasing y_i.

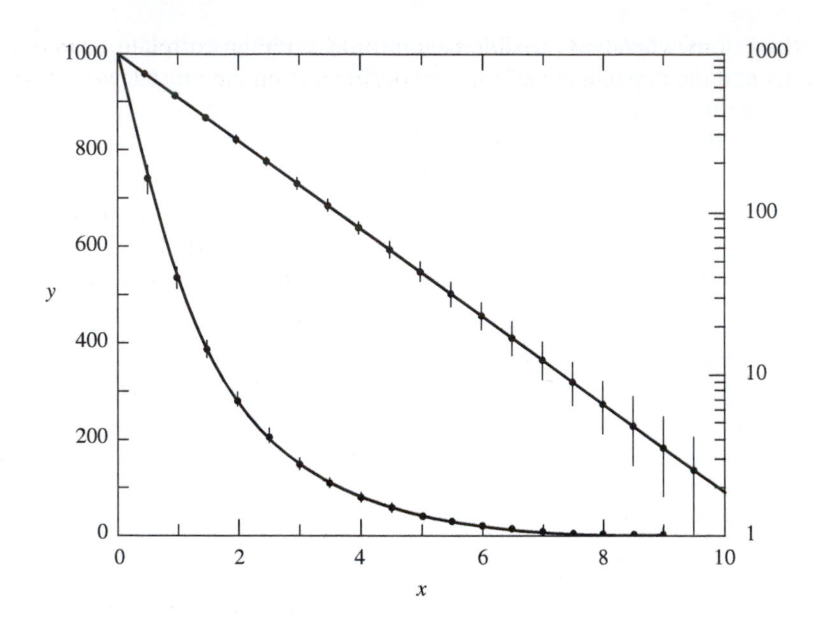

FIGURE 7.3
Graph of the function $y = ae^{-bx}$ calculated on a linear and a logarithmic scale. The error bars are given by $\sigma_i = \sqrt{y_i}$. The curved line corresponds to the linear scale on the left, and the straight line to the logarithmic scale on the right.

However, on the logarithmic scale, they appear to decrease with increasing y_i and are very large for very small $\ln y_i$. If we were to ignore this effect in fitting Equation (7.47), we would overemphasize the uncertainties for small values of y_i.

In general, if we fit the function $f(y)$ rather than y, the uncertainties σ_i in the measured quantities must be modified by

$$\sigma_i' = \frac{df(y)}{dy_i}\sigma_i \tag{7.50}$$

Errors in the Parameters

If we modify the fitting function so that instead of fitting the data points y_i with the coefficient $a, b, \ldots,$ we fit modified data points $y_i' = f(y_i)$ with coefficients a', $b', \ldots,$ then our estimates of the errors in the coefficients will pertain to the uncertainties in the modified coefficients $a', b', \ldots,$ rather than to the desired coefficients $a, b, \ldots.$ If the relationship between the two sets of coefficients is defined to be

$$a' = f_a(a) \qquad b' = f_b(b) \tag{7.51}$$

then the correspondence between the uncertainties $\sigma_a', \sigma_b', \ldots$ in the modified coefficients and the uncertainties $\sigma_a, \sigma_b, \ldots$ in the desired coefficients is obtained in a manner similar to that for σ_i' and σ_i in Equation (7.50):

$$\sigma_a' = \frac{df_a(a)}{da}\sigma_a \qquad \sigma_b' = \frac{df_b(b)}{db}\sigma_b \tag{7.52}$$

Thus, if the modified coefficient is $a' = \ln a$, the estimated error in a is determined from the estimated error in a', according to Equation (7.52) with $f_a = \ln a$:

$$\sigma'_a = \frac{d(\ln a)}{da}\sigma_a = \frac{\sigma_a}{a} \tag{7.53}$$

Values of χ^2 for testing the goodness of fit should be determined from the original uncertainties of the data σ_i and from the unmodified equation, although Equation (7.48) should give approximately equivalent results when weighted with the modified uncertainties σ'_i.

In Example 6.2, we considered an experiment to check the decrease in the number of counts C as a function of distance r from a radiative source. We expected a relation of the form

$$C(r) = b/r^2 \tag{7.54}$$

and therefore changed the independent variable to $x = 1/r^2$ and fitted a straight line to the C versus x data. Because uncertainties were assigned only to the dependent variable C, the fit was not distorted by that transformation.

Suppose, instead, that our objective had been to determine the exponent a in the expression for C:

$$C(r) = br^{-a} \tag{7.55}$$

Taking logarithms of both sides, we obtain the linear equation

$$\ln(C) = \ln(b) - a \ln r$$

or

$$C_i = b' - ar' \tag{7.56}$$

with $C' = \ln C$, $r' = \ln r$, and $b' = \ln b$. The uncertainties σ' in C' would be given by Equation (7.49) as

$$\sigma' = \sigma/C$$

and we could find the exponent a by fitting a straight line to Equation (7.56) using these weighted uncertainties.

Although the method of taking logarithms of an exponential or a power function to produce a function that is linear in the parameters may be convenient for quick estimates, with fast computers it is generally better to solve such problems by one of the approximation methods developed for fitting nonlinear functions. These methods will be explored in Chapter 8.

SUMMARY

Linear function: Function that is linear in its parameter a_k:

$$y(x) = \sum_{k=1}^{m} a_k f_k(x)$$

Least-squares fit to a function that is linear in its parameters:

$$\Delta = \begin{vmatrix} \sum \frac{f_1(x_i)f_1(x_i)}{\sigma_i^2} & \sum \frac{f_1(x_i)f_2(x_i)}{\sigma_i^2} & \sum \frac{f_1(x_i)f_3(x_i)}{\sigma_i^2} & \cdots \\ \sum \frac{f_2(x_i)f_1(x_i)}{\sigma_i^2} & \sum \frac{f_2(x_i)f_2(x_i)}{\sigma_i^2} & \sum \frac{f_2(x_i)f_3(x_i)}{\sigma_i^2} & \cdots \\ \sum \frac{f_3(x_i)f_1(x_i)}{\sigma_i^2} & \sum \frac{f_3(x_i)f_2(x_i)}{\sigma_i^2} & \sum \frac{f_3(x_i)f_3(x_i)}{\sigma_i^2} & \cdots \\ \vdots & \vdots & \vdots & \vdots \end{vmatrix}$$

$$a_1 = \frac{1}{\Delta}\begin{vmatrix} \sum y_i \frac{f_1(x_i)}{\sigma_i^2} & \sum \frac{f_1(x_i)f_2(x_i)}{\sigma_i^2} & \sum \frac{f_1(x_i)f_3(x_i)}{\sigma_i^2} & \cdots \\ \sum y_i \frac{f_2(x_i)}{\sigma_i^2} & \sum \frac{f_2(x_i)f_2(x_i)}{\sigma_i^2} & \sum \frac{f_2(x_i)f_3(x_i)}{\sigma_i^2} & \cdots \\ \sum y_i \frac{f_3(x_i)}{\sigma_i^2} & \sum \frac{f_3(x_i)f_2(x_i)}{\sigma_i^2} & \sum \frac{f_3(x_i)f_3(x_i)}{\sigma_i^2} & \cdots \\ \vdots & \vdots & \vdots & \vdots \end{vmatrix}$$

For the jth coefficient, a_j is found by replacing the jth column in the expression for Δ with the first column in the expression for a_1.
Chi square:

$$\chi^2 = \sum_{i=1}^{N}\left[\frac{1}{\sigma}[y_i - y(x_i)]\right]^2 = \sum_{i=1}^{N}\left[\frac{1}{\sigma}\left[y_i - \sum_{k=1}^{m} a_k f_k\right]\right]^2$$

Sample variance σ^2:

$$\sigma^2 \simeq s^2 = \frac{1}{N-m}\sum_{i=1}^{N}[y_i - y(x_i)]^2$$

Matrix solution: $\mathbf{a} = \boldsymbol{\beta\epsilon} = \boldsymbol{\beta\alpha}^{-1}$ where $\mathbf{a}$ is a linear matrix of the coefficients and

$$\beta_k \equiv \sum\left[\frac{1}{\sigma_i^2} y_i f_k(x_i)\right]$$

$$\alpha_{lk} \equiv \sum\left[\frac{1}{\sigma_i^2} f_l(x_i) f_k(x_i)\right]$$

Error or variance matrix: The diagonal elements of the square matrix $\boldsymbol{\epsilon} = \boldsymbol{\alpha}^{-1}$ are the variances of the parameters a_k and the off-diagonal elements are the covariances:

$$\sigma_{a_l}^2 = \epsilon_{ll} \qquad \sigma_{a_{lk}}^2 = \epsilon_{lk}$$

Orthogonal polynomials:

$$y(x) = a_1 + a_2(x-\beta) + a_3(x-\gamma_1)(x-\gamma_2) + a_4(x-\delta_1)(x-\delta_2)(x-\delta_3) + \cdots$$

$$a_1 = \bar{y} \qquad a_2 = \frac{\Sigma y_i(x_i - \beta)}{\Sigma(x_i - \beta)^2}$$

$$a_3 = \frac{\Sigma y_i(x_i - \gamma_1)(x_i - \gamma_2)}{\Sigma[(x_i - \gamma_1)(x_i - \gamma_2)]^2} \qquad a_4 = \frac{\Sigma y_i(x_i - \delta_1)(x_i - \delta_2)(x_i - \delta_3)}{\Sigma[(x_i - \delta_1)(x_i - \delta_2)(x_i - \delta_3)]^2}$$

For equally spaced values of x, $x_{i+1} - x_i = \Delta$,

$$\beta = \tfrac{1}{2}(x_i + x_N) \qquad \gamma = \beta \pm \Delta\sqrt{\tfrac{1}{12}(N^2 - 1)}$$

$$\delta = \beta,\ \beta \pm \Delta\sqrt{\tfrac{1}{20}(3N^2 - 7)}$$

Legendre polynomials:

$$y(x) = \sum_{L=1}^{m-1}[a_L P_L(x)]$$

$$P_0(x) = 1 \qquad P_1(x) = x$$

$$P_L(x) = \frac{1}{L}[(2L - 1)xP_{L-1}(x) - (L - 1)P_{L-2}(x)] \quad \text{(recursion relation)}$$

Nonlinear functions:

If $y_i' = f(y_i)$, then

$$\sigma_i' = \frac{df(y)}{dy_i}\sigma_i$$

and if $a' = f_a(a)$ and $b' = f_b(b)$, then

$$\sigma_a' = \frac{df_a(a)}{da}\sigma_a \qquad \sigma_b' = \frac{df_b(b)}{db}\sigma_b$$

EXERCISES

7.1. Show by direct calculation using the data of Example 7.2 listed in Table 7.2 that $\boldsymbol{\alpha\epsilon} = \mathbf{1}$ where $\mathbf{1}$ is the unity matrix.

7.2. The tabulated data represent the lower bin limit x and the bin contents y of a histogram of data that fall into two peaks.

i	1	2	3	4	5	6	7	8	9	10
x_i	50	60	70	80	90	100	110	120	130	140
y_i	5	7	11	13	21	43	30	16	15	10

i	11	12	13	14	15	16	17	18	19	20
x_i	150	160	170	180	190	200	210	220	230	240
y_i	13	42	90	75	29	13	8	4	6	3

Use the method of least squares to find the amplitudes a_1 and a_2 and their uncertainties by fitting to the data the function

$$y(x) = a_1 L(x; \mu_1, \Gamma_1) + a_2 L(x; \mu_2, \Gamma_2)$$

with $\mu_1 = 102.1$, $\Gamma_1 = 30$, $\mu_2 = 177.9$, and $\Gamma_2 = 20$. The function $L(x; \mu, \Gamma)$ is the Lorentzian function of Equation (2.32). Assume statistical uncertainties ($\sigma_i = \sqrt{y_i}$). Find χ^2 for the fit and the full error matrix.

7.3. From the parameters listed in Table 7.7 for the fit of even terms to the data of Example 7.3, determine the predicted value of the cross section for $\theta = 90°$ and its uncertainty. Calculate the uncertainty from the diagonal errors, listed in Table 7.7 and from the full error matrix listed in Table 7.8 and compare the two results.

7.4. Fit fourth-degree power series polynomials instead of Legendre polynomials to the data of Example 7.3. Let $x = \cos\theta$ and fit a polynomial with all terms to x^4 and another polynomial with only the even terms. Compare your results to those obtained from the fit to Legendre polynomials displayed in Table 7.7.

7.5. Derive the expression for γ_1 and γ_2 of Equation (7.36).

7.6. Derive an expression for $P_4(\cos\theta)$. [See Equation (7.42).]

7.7. Show by direct integration that $P_0(x)$, $P_1(x)$, and $P_2(x)$ are orthogonal and obey Equation (7.43).

7.8. In an experiment to measure the angular distribution of elastically scattered particles, a beam of particles strikes a liquid hydrogen target and counts are recorded at selected angles to the direction of the incident beam. Measurements are made both with the target filled with liquid hydrogen *(full target)* and with an empty target *(empty target)*. The empty-target measurements were made with one-half the number of incident particles used for the full-target signal. By subtracting the suitably scaled empty-target signal from the full-target signal, the angular distribution of scattering on pure hydrogen can be determined.

Assume that the following data were obtained in such an experiment. Uncertainties in the numbers of counts are statistical.

$\cos\theta$ (lower limit)	−1.0	−0.8	−0.6	−0.4	−0.2	0.0	0.2	0.4	0.6	0.8
$\cos\theta$ (upper limit)	−0.8	−0.6	−0.4	−0.2	0.0	0.2	0.4	0.6	0.8	1.0
Counts, full target	184	128	99	49	53	55	70	81	136	216
Counts, empty target	5	4	4	1	3	1	4	9	8	7

(*a*) Scale the empty-target data to the same number of incident antiprotons used in recording the full-target data and make a subtraction to obtain the number of interactions on the hydrogen. Pay particular attention to the uncertainties in the difference.

(*b*) Use the least-squares method to fit the function

$$y(x) = a_1 P_0(x) + a_2 P_1(x) + a_3 P_2(x)$$

to the subtracted data, to obtain the coefficients a_1, a_2, and a_3, where the functions $P_L(x)$ are the Legendre polynomials defined in Equation (7.41).

7.9. Follow the procedure outlined in Section 7.4 to find the exponent a in Equation (7.55), using the data of Example 6.2 (Table 6.2).

7.10. A 1-m-long plastic plate with rulings at 10-cm intervals is dropped through a photogate to measure the acceleration of gravity g in an undergraduate laboratory experiment. The time is recorded as each ruling passes through the gate. The passage of the first ruling starts the timer. Data from such an experiment are tabulated. The recorded time is related to the distance that the ruler has fallen by $y = y_0 - v_0 t - 1/2 g t^2$. Note that neither the initial height y_0 nor the initial speed v_0 are known.

Ruling #	0	1	2	3	4	5	6	7	8	9	10
Time(s)	0.000	0.079	0.132	0.174	0.212	0.244	0.271	0.301	0.325	0.349	0.373

Use the least-squares method with a second-degree polynomial to find g and its uncertainty. Measure y from the photogate so that you can set $y = 0$ when ruling #0 passes the gate, $y = 1$ when ruling #1 passes, and so forth. Choose t as the independent and y as the dependent variable. Assume a uniform uncertainty in t of 0.001 s and a negligible uncertainty in y. Because the uncertainty is in the independent variable, it must be transformed to the dependent variable by the method discussed in Section 6.1. This will require initial estimates of g and v_0. After the fit has been made you may wish to repeat the fit using estimates of g and v_0 from the previous fit to improve the results.

CHAPTER

8

LEAST-SQUARES FIT TO AN ARBITRARY FUNCTION

8.1 NONLINEAR FITTING

The methods of least squares and multiple regression developed in the previous chapters are restricted to fitting functions that are linear in the parameters as in Equation (7.3):

$$y(x) = \sum_{j=1}^{m} [a_j f_j(x)] \tag{8.1}$$

This limitation is imposed by the fact that, in general, minimizing χ^2 can yield a set of coupled equations that are linear in the m unknown parameters only if the fitting functions $y(x)$ are themselves linear in the parameters. We shall distinguish between the two types of problems by referring to *linear fitting* for problems that involve equations that are linear in the parameters, such as those discussed in Chapters 6 and 7, and *nonlinear fitting* for those problems that are nonlinear in the parameters.

Example 8.1. In a popular undergraduate physics laboratory experiment, a real silver quarter is irradiated with thermal neutrons to create two short-lived isotopes of silver, ${}_{47}Ag^{108}$ and ${}_{47}Ag^{110}$, that subsequently decay by beta emission. Students count the emitted beta particles in 15-s intervals for about 4 min to obtain a decay curve. Data collected from such an experiment are listed in Table 8.1 and plotted on a semilogarithmic graph in Figure 8.1. The data are reported at the end of each 15-s interval, just as they were recorded by a scaler. The data points do not fall on a straight

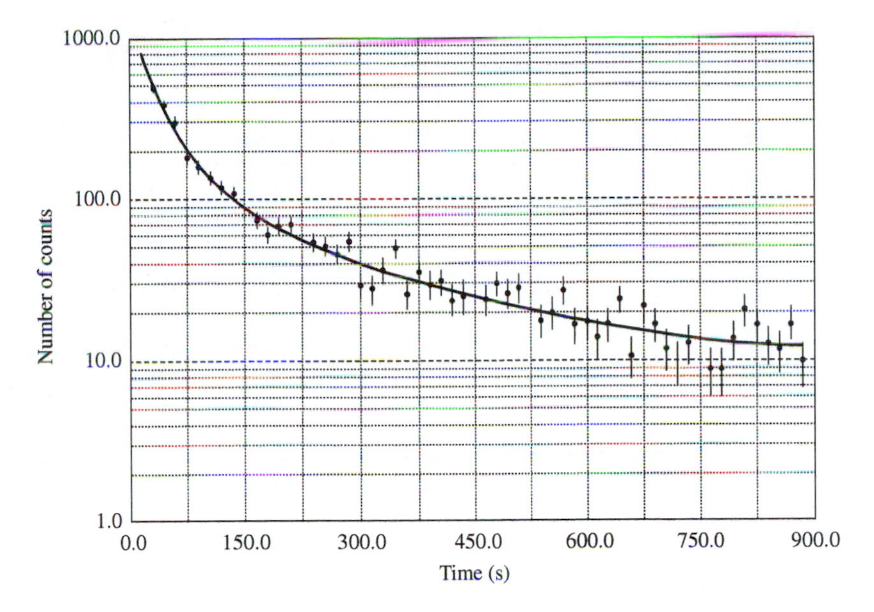

FIGURE 8.1
Number of counts detected from the decay of two excited states of silver as a function of time (Example 8.1). Time is reported at the end of each interval. Statistical uncertainties are assumed. The curve was obtained by a nonlinear least-squares fit of Equation (8.2) to the data.

line because the probability function that describes the process is the sum of two exponential functions plus a constant background. We can represent the decay by the fitting function

$$y(x_i) = a_1 + a_2 e^{-t/a_4} + a_3 e^{-t/a_5} \tag{8.2}$$

where the parameter a_1 corresponds to the background radiation and a_2 and a_3 correspond to the amplitudes of the two excited states with mean lives a_4 and a_5, respectively. Clearly, Equation (8.2) is not linear in the parameters a_4 and a_5, although it is linear in the parameters a_1, a_2, and a_3.

We can use a graphical analysis method to find the two mean lifetimes by plotting the data on semilogarithmic paper after first subtracting from each data point the constant background contribution, which has been measured separately. (Note that the background counts have not been subtracted in Figure 8.1.) We then consider two regions of the plot: region *a*, at small values of T (e.g., $T < 120$ s) in which the short-lived state dominates the plot, and region *b*, at large values of T (e.g., $T > 200$ s) in which only the long-lived state contributes to the data. We can estimate the mean lifetime of the long-lived state by finding the slope of our best estimate of the straight line that passes through the data points in region *b*. From this result we can estimate the contribution of the long-lived component to region *a* and subtract that contribution from each of the data points, and thus make a new plot of the number of counts in region *a*, which we attribute to the short-lived state alone. The slope of the line through the corrected points gives us the mean lifetime of the short-lived state. Linear regression techniques discussed in Section 7.4 could be used to find the slope of the graph in each region.

TABLE 8.1
Geiger counter data from an irradiated silver piece, recorded in 15-s intervals

Point number	Time	Measured counts	Calculated counts	Point number	Time	Measured counts	Calculated counts
1	15	775	748.3	31	465	24	24.0
2	30	479	519.8	32	480	30	23.0
3	45	380	370.4	33	495	26	22.1
4	60	302	272.0	34	510	28	21.3
5	75	185	206.7	35	525	21	20.5
6	90	157	162.7	36	540	18	19.8
7	105	137	132.5	37	555	20	19.2
8	120	119	111.5	38	570	27	18.5
9	135	110	96.3	39	585	17	18.0
10	150	89	85.0	40	600	17	17.4
11	165	74	76.5	41	615	14	16.9
12	180	61	69.7	42	630	17	16.5
13	195	66	64.2	43	645	24	16.0
14	210	68	59.5	44	660	11	15.6
15	225	48	55.5	45	675	22	15.2
16	240	54	51.9	46	690	17	14.9
17	255	51	48.8	47	705	12	14.6
18	270	46	45.9	48	720	10	14.3
19	285	55	43.3	49	735	13	14.0
20	300	29	40.9	50	750	16	13.8
21	315	28	38.7	51	765	9	13.5
22	330	37	36.7	52	780	9	13.3
23	345	49	34.8	53	795	14	13.1
24	360	26	33.1	54	810	21	12.9
25	375	35	31.5	55	825	17	12.7
26	390	29	30.0	56	840	13	12.6
27	405	31	28.6	57	855	12	12.4
28	420	24	27.3	58	870	18	12.3
29	435	25	26.1	59	885	10	12.1
30	450	35	25.0				

Note: The time is reported at the end of each interval. The calculated number of counts was found by method 4.

Because analytic methods of least-squares fitting cannot be used for nonlinear fitting problems, we must consider approximation methods and make searches of parameter space. In the following sections we discuss four nonlinear fitting methods: a simple grid-search method in which we simply calculate χ^2 at trial values of the parameters, and search for those values of the parameters that yield a minimum value of χ^2, a gradient-search method that uses the slope of the function to improve the efficiency of the search, and two semianalytic methods that make use of the matrix method developed in Chapter 7, with a linear approximation to the nonlinear functions. As examples, we shall determine the parameters ($a_1 \ldots a_5$) by fitting Equation (8.2) to the data of Example 8.1 using each of the four methods. The curve on Figure 8.1 is the result of such a fit.

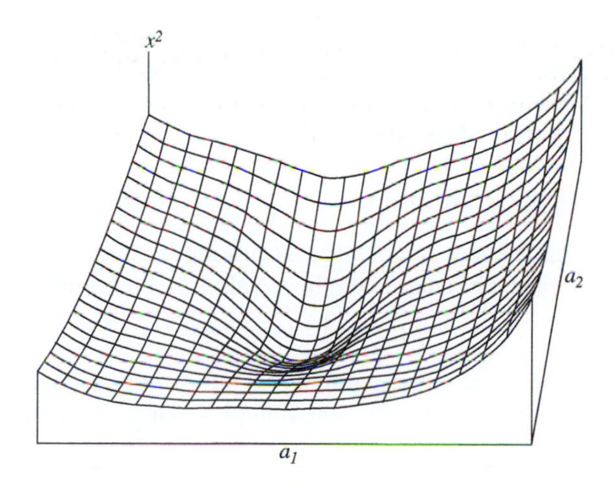

FIGURE 8.2
Chi-square hypersurface as a function of two parameters.

Method of Least Squares

We can generalize the probability function, or *likelihood function,* of Equation (6.7) to any number of parameters,

$$P(a_1, a_2, \ldots, a_m) = \prod \left[\frac{1}{\sigma_i\sqrt{2\pi}}\right] \exp\left\{-\frac{1}{2}\sum\left[\frac{y_i - y(x_i)}{\sigma_i}\right]^2\right\} \tag{8.3}$$

and, as in the previous chapters, maximize the likelihood with respect to the parameters by minimizing the exponent, or the goodness-of-fit parameter χ^2:

$$\chi^2 \equiv \sum\left\{\frac{1}{\sigma_i^2}[y_i - y(x_i)]^2\right\} \tag{8.4}$$

where x_i and y_i are the measured variables, σ_i is the uncertainty in y_i, and $y(x_i)$ are values of the function calculated at x_i. According to the method of least squares, the optimum values of the parameters a_j are obtained by minimizing χ^2 simultaneously with respect to each parameter,

$$\begin{aligned}\frac{\partial\chi^2}{\partial a_j} &= \frac{\partial}{\partial a_j}\sum\left\{\frac{1}{\sigma_i^2}[y_i - y(x_i)]^2\right\} = 0 \\ &= -2\sum\left\{\frac{1}{\sigma_i^2}[y_i - y(x_i)]\frac{\partial y(x_i)}{\partial a_j}\right\}\end{aligned} \tag{8.5}$$

Taking partial derivatives of χ^2 with respect to each of the m parameters a_j will yield m coupled equations in the m unknown parameters a_j as in Section 7.1. If these equations are not linear in all the parameters, we must, in general, treat χ^2 as a continuous function of the m parameters, describing a hypersurface in an m-dimensional space, as expressed by Equation (8.4), and search that space for the appropriate minimum value of χ^2. Figure 8.2 illustrates such a hyperspace for a function of two parameters. Alternatively, we may apply to the m equations obtained from Equations (8.5) approximation methods developed for finding roots of coupled, nonlinear equations. A combination of both methods is often used.

Variation of χ^2 Near a Minimum

For a sufficiently large event sample, the likelihood function becomes a Gaussian function of each parameter centered on those values a_j' that minimize χ^2:

$$P(a_j) = Ae^{-(a_j - a_j')^2/2\sigma_i^2} \tag{8.6}$$

where A is a function of the other parameters, but not of a_j. Comparing Equation (8.3) for the likelihood function with Equation (8.4) for χ^2, we observe that we can express χ^2 as

$$\chi^2 = -2\ln[P(a_1, a_2, \ldots, a_m)] + 2\sum \ln(\sigma_i\sqrt{2\pi}) \tag{8.7}$$

Then, from Equation (8.6), we can write

$$\chi^2 = \frac{(a_j - a_j')^2}{\sigma_j^2} + C \tag{8.8}$$

to show the variation of χ^2 with any single parameter a_j in the vicinity of a minimum with respect to that parameter. The constant C is a function of the uncertainties σ_i and the parameters a_k for $k \neq j$. Thus χ^2 varies as the square of distance from a minimum, and an increase of 1 standard deviation (σ) in the parameter from the value a_j' at the minimum increases χ^2 by 1. For a more general proof, see Arndt and MacGregor (1966), appendix II.

We can see that this result is consistent with that obtained from a second-order Taylor expansion of χ^2 about the values a_j', where the values of χ^2 and its derivatives at $a = a'$ are written as χ_0^2, $\partial\chi_0^2/\partial a_j$, and so forth:

$$\chi^2 \simeq \chi_0^2 + \sum_{j=1}^{m}\left\{\frac{\partial\chi_0^2}{\partial a_j}(a_j - a_j')\right\} + \frac{1}{2}\sum_{k=1}^{m}\sum_{j=1}^{m}\left\{\frac{\partial^2\chi_0^2}{\partial a_k \partial a_j}(a_k - a_k')(a_j - a_j')\right\} \tag{8.9}$$

Because the condition for minimizing χ^2 is that the first partial derivative with respect to each parameter vanish (i.e., $\partial\chi^2/\partial a_j = 0$), we can expect that near a local minimum in any parameter a_j, χ^2 will be a quadratic function of that parameter.

We can obtain another useful relation from Equation (8.8) by taking the second derivative of χ^2 with respect to the parameter a_j to obtain

$$\frac{\partial^2\chi^2}{\partial a_j^2} = \frac{2}{\sigma_j^2} \tag{8.10}$$

We obtain the following expression for the uncertainty in the parameter in terms of the curvature of the χ^2 function in the region of the minimum:

$$\sigma_j^2 = 2\left(\frac{\partial^2\chi^2}{\partial a_j^2}\right)^{-1} \tag{8.11}$$

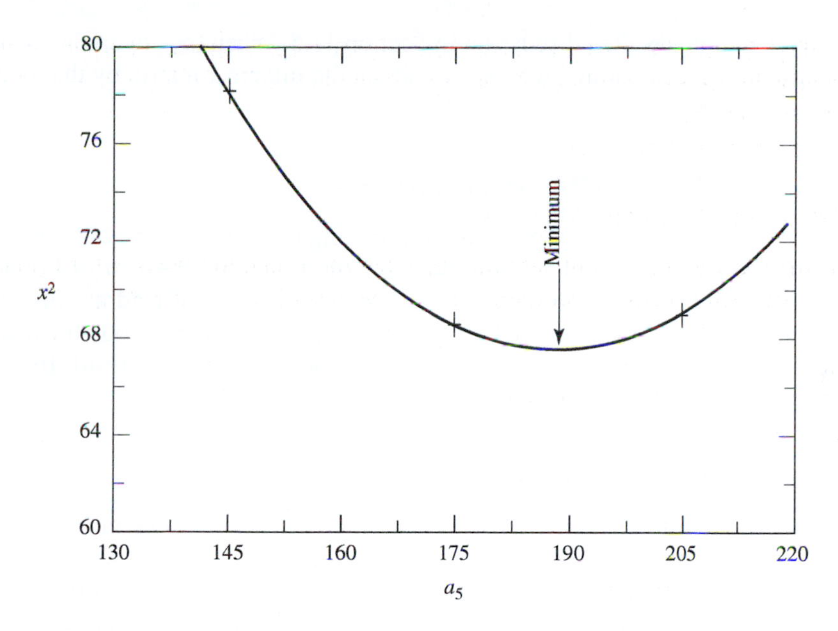

FIGURE 8.3
Plot of χ^2 versus a single parameter a in the region of a local minimum. The location of the minimum is calculated by fitting a parabola through the three indicated data points.

We note that for uncorrelated parameters, Equation (8.11) is equivalent to Equation (7.22) with Equation (7.25) for obtaining the uncertainties from the curvature matrix.

We can also use the quadratic relation to find the approximate location of a χ^2 minimum by considering the equation of a parabola that passes through three points that straddle the minimum, and solving for the value of the parameter at the minimum, as illustrated in Figure 8.3. If we have calculated three values of χ^2, $\chi_1^2 = \chi^2(a_{j1})$, $\chi_2^2 = \chi^2(a_{j2})$, and $\chi_3^2 = \chi^2(a_{j3})$, where $a_{j2} = a_{j1} + \Delta a_j$ and $a_{j3} = a_{j2} + \Delta a_j$, then the value a_j' of the parameter at the minimum of the parabola is given by

$$a_j' = a_{j3} - \Delta a_j \left[\frac{\chi_3^2 - \chi_2^2}{\chi_1^2 - 2\chi_2^2 + \chi_3^2} + \frac{1}{2} \right] \tag{8.12}$$

In addition, we can estimate the errors in the fitting parameters a_j by varying each parameter about its minimum to increase χ^2 by 1 from the minimum value. The variation σ_j in the parameter a_j, which will increase χ^2 by 1 from its value at the minimum of the parabola, is given by

$$\sigma_j = \Delta a_j \sqrt{2(\chi_1^2 - 2\chi_2^2 + \chi_3^2)^{-1}} \tag{8.13}$$

Alternatively, we can attempt to calculate the second derivative of χ^2 at the minimum and find the standard deviation from Equation (8.11).

If the parameters are correlated, the method summarized in Equation (8.13) for determining uncertainties in the parameters is valid only under the condition that, with $a_j = a_j' \pm \sigma_j$, χ^2 be minimized with respect to all other parameters. This condition severely limits the usefulness of this procedure for determining the uncertainties.

We provide a more detailed discussion in Section 11.5. When the covariant terms in the error matrix are important, it is best to obtain the full error matrix by the method described in Section 7.2.

8.2 SEARCHING PARAMETER SPACE

The method of least squares consists of determining the values of the parameters a_j of the function $y(x)$ that yield a minimum for the function χ^2 given in Equation (8.4). For nonlinear fitting problems, there are several ways of finding this minimum value. In Sections 8.3 and 8.4 we discuss approximation methods for finding solutions to the m coupled nonlinear equations in m unknowns that result from the minimization procedure of Equation (8.5).

Starting Values and Local Minima

Fitting nonlinear functions to data samples sometimes seems to be more of an art than a science. In part, this is in the nature of the approximation process, where the speed of convergence toward a solution may depend upon the choice of the method for finding solutions, the choice of starting values for the parameters, and possibly the choice of the step size. To use any of these methods, we must first determine starting values, estimates to be used by the fitting routine for initial calculations of the function and of chi square. For the pure search methods we must also define step sizes, the initial variations of the parameters. Neither starting values nor step sizes, of course, are needed in linear fitting.

Another problem in nonlinear fitting is the existence of multiple solutions or local minima. For an arbitrary function there may be more than one minimum of the χ^2 function within a reasonable range of values for the parameters, and thus, more than one set of solutions of the m coupled equations. An unfortunate choice of starting point may "drive" the solution toward a local minimum rather than to the absolute minimum that we seek. Before attempting a nonlinear least-squares fit, therefore, it is useful to search the parameter space to locate the main minima and identify the desired range of parameters over which to refine the search.

The first step is to find starting values for the parameters. A convenient approach, for which a computer graphics program is very useful, is to make plots of the data with curves calculated from trial values of the parameters. By visual inspection, one can often determine acceptable starting values with little or no further calculations. A basic requirement is that the area under the plotted curve be approximately the same as that under the data.

Another approach is to map the parameter space and search for values of the parameters that approximately minimize χ^2. In the simplest brute-force mapping procedure, the permissible range of each parameter a_j is divided into p equal increments Δa_j so that the m-parameter space is divided into $\Pi_{j=1}^{m} P_j$ hypercubes. The value of χ^2 is then evaluated at the vertices of each hypercube. This procedure yields a coarse map of the behavior of χ^2 as a function of all the parameters a_j. At the vertex for which χ^2 has its lowest value, the size of the grid can be reduced to obtain more precise values of the parameters. For a simple two- or three-parameter fit, the parameters obtained by this procedure may be sufficiently precise that no further searching is required. For more than three parameters, the mapping is rather tedious and displaying the grid map is difficult.

A variation on the regular lattice method is a Monte Carlo search of the m-dimensional space. Trial values of the parameters are generated randomly from uniform distributions of the parameters, selected within predefined ranges, and a value of χ^2 determined for each trial. After several trials, the set of trial values that gives the lowest value of χ^2 can be used as starting values. The general Monte Carlo method was discussed in Chapter 5.

A more sophisticated method of locating the various minima of the χ^2 hypersurface involves traversing the surface from minimum to minimum by the path of lowest value in χ^2, as a river follows a ravine in travelling from lake to ocean. Starting at a point in the m-dimensional space, the search traverses the length of the local minimum, then continues in the same general direction but in a direction that minimizes the new values of χ^2. When a new local minimum is discovered, the search repeats the process until all local minimum have been located in the specified region of the space.

For relatively straightforward fitting problems, it should be sufficient to plot the data, make a reasonable estimate of the parameters to be used as starting values in the search procedure, and perform the fit by one or more of the methods described in the following sections. As a precaution, one should vary the starting values of the parameters to test whether or not the various fits converge to the same values of the parameters, within the expected uncertainties. If the dimensionality of the space is low enough, a grid of starting points may be used. For higher dimensionality, a Monte Carlo method may be used to select random starting points.

Bounds on the Parameters

From a particular set of starting values for the parameters, the search may converge toward solutions that are physically unreasonable. In Example 8.1 negative values for the parameters are not acceptable, and the current trial value of one of the parameters a_2, or a_3, may limit the possibility of determining values of the others. For example, if a_2 becomes very small or 0, a_4 cannot be determined at all. If it is not possible to find starting values for the parameters that prevent the search from wandering into these illegal regions, it may be necessary to place limits on them in the search procedure to keep them within physically allowable ranges. Simple *if then* statements in the routines may be sufficient. Care should be taken that the final value of any parameter is not at one of these artificially imposed limits.

Selection and Adjustment of Step Sizes

There are no hard and fast rules for selecting step sizes for the search methods. Clearly the steps will be different for different parameters and should be related to the slope of the χ^2 function. Very small step sizes result in slow convergence, whereas step sizes that are too large will overshoot the local minima and require constant readjustment to bracket the valleys. In the sample routines in Section 8.7, we choose initial step sizes to be proportional to the starting values of the parameters and readjust them if necessary after each local minimum is found. In the simple grid-search calculation, we adjust the step sizes to be those values that increase χ^2 by approximately 2 from its value at the local minimum.

Condition for Convergence

A change in χ^2 per degree of freedom (χ^2/dof) of less than about 1% from one trial set of parameters to the next is probably not significant. However, because of the problems of local minima and very flat valleys in the parameter space, it may not be sufficient to set an arbitrary condition for convergence, start a search, and let it run to completion. If the starting parameters are not chosen very carefully, the search may stop in a flat valley with an inappropriately large value of χ^2. If this happens, there are several possible ways to proceed. We can choose different starting values and retry the fit, as suggested in the previous sections, or we can set tighter convergence requirements (e.g., $\Delta\chi^2/\text{dof} < 0.1\%$) and rerun the search in the hope that the program will escape from the valley and reach the appropriate minimum. A convenient approach for small problems is to observe the process of the search and to cut it off manually when it appears that a stable minimum has been found. If a suitable minimum cannot be found, then different starting values should be tried. When fitting curves to several similar samples of data, we may find it satisfactory to establish suitable starting parameters, step sizes, and a cutoff criterion for the first set, and employ an automatic method for the remaining sets.

Computer Illustration of Nonlinear Fitting Methods

In the following sections we discuss and illustrate with computer routines four methods of fitting Equation (8.2) to the data of Example 8.1.

Program 8.0. NONLINFT (Appendix E) Common calling routine to test the four different fitting methods. Repeats the calculations until a χ^2-minimum is found. Variables are defined in the program until FITVARS and data input and output are handled in the program unit FITUTIL as in the fitting programs of Chapters 6 and 7. FITFUNC8 calculates the fitting function.

Step sizes for the fit are set initially in the routine FETCHPARAMETERS to be a fraction of the starting values of the parameters. (The step sizes must not be scaled to the parameters throughout the calculation, however, lest they become 0 when a parameter is 0, which would halt the search in that parameter.)

Tables 8.2, 8.3, 8.4, and 8.5 show values of χ^2 and the parameters a_1 through a_5 for several stages of the calculation at the beginning, middle, and end of each of the four types of search. The tables include the time to find the solution relative to the time for the fastest procedure.

Program 8.1. GRIDSEAR (Appendix E) Routine GRIDLS illustrates the grid-search method.

Program 8.2. GRADSEAR (Appendix E) Routine GRADLS illustrates the gradient-search method.

Program 8.3. EXPNDFIT (Appendix E) Routine CHIFIT illustrates fitting by expansion of the fitting function.

Program 8.4. MARQFIT (Appendix E) Routine MARQUARDT illustrates fitting by the gradient-expansion algorithm.

Program 8.5. FITFUN8 (Appendix E) Fitting function and χ^2-calculation for all fits called from Program 8.0.

Program 8.6. MAKEAB8 (Appendix E) Matrix set-up for non-linear fits.

Program 8.7. NUMDERIV (Website) Numerical derivatives.

Program B.1. MATRIX (Appendix E) Matrix products and inversion.

8.3 GRID-SEARCH METHOD

If the variation of χ^2 with each parameter a_j is not very sensitive to the values of the other parameters, then the optimum parameter values can be obtained most simply by minimizing χ^2 with respect to each of the parameters separately. This is the *grid-search* method. The procedure is simply to select starting values of the parameters, find the value of one of the parameters that minimizes χ^2 with respect to that parameter, set the parameter to that value, and repeat the procedure for each parameter in turn. The entire process is then repeated until a stable χ^2 minimum is obtained.

Grid search. The procedure for a grid search may be summarized as follows:

1. Select starting values a_j and step or increment sizes Δa_j for each parameter and calculate χ^2 with the starting parameters.
2. Increment one parameter a_j by $\pm\Delta a_j$ and calculate χ^2, where the sign is chosen so that χ^2 decreases.
3. Repeat step 2 until χ^2 stops decreasing and begins to increase. The increase in χ^2 indicates that the search has crossed a ravine and started up the other side.
4. Use the last three values of a_j (which bracket the minimum) and the associated values of χ^2 to determine the minimum of the parabola, which passes through the three points as illustrated in Figure 8.3. [See Equation (8.12).]
5. Repeat to minimize χ^2 with respect to each parameter in turn.
6. Continue to repeat the procedure until the last iteration yields a predefined negligibly small decrease in χ^2.

The main advantage of the grid-search method is its simplicity. With successive iterations of the search, the absolute minimum of the χ^2 function in parameter space can be located to any desired precision.

The main disadvantage is that, if the variations of χ^2 with the parameters are strongly correlated, then the approach to the minimum may be very slow. Consider, for example, the contour plot of χ^2 as a function of two parameters in Figure 8.4. The χ^2 contours are generally approximately elliptical near the minimum. The degree of correlation of the parameters is indicated by the tilt of the ellipse. If two

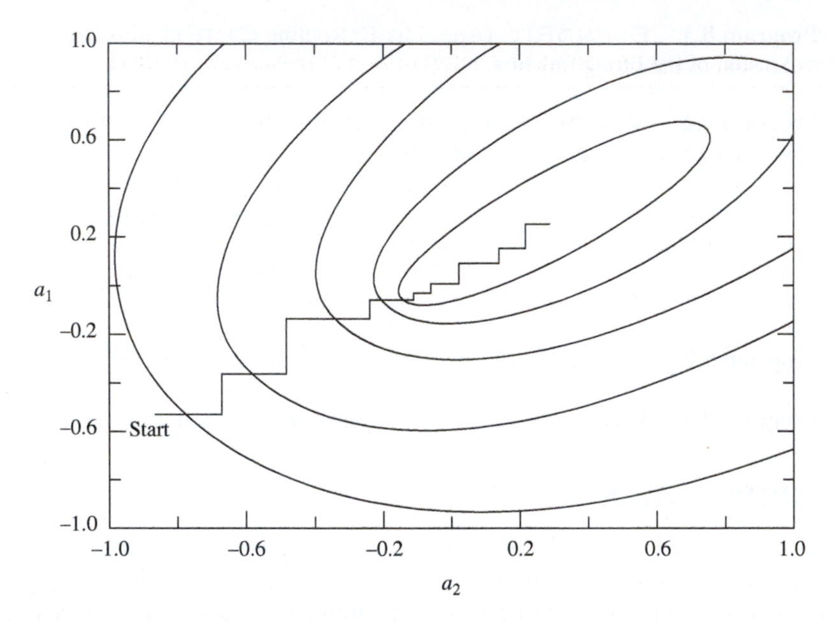

FIGURE 8.4
Contour plot of χ^2 as a function of two highly correlated variables. The zigzag line represents the search path approach to a local minimum by the grid-search method.

parameters are not correlated, so that the variation of χ^2 with each parameter is independent of the variation with the other, then the axes of the ellipse will be parallel to the coordinate axes. Thus, if a grid search is initiated near one end of a tilted ellipse, the search may follow a zigzag path as indicated by the solid line in Figure 8.4 and the search will be very inefficient. Nevertheless, the simplicity of the calculations involved in a grid search often compensates for this inefficiency.

Program 8.1. GRIDSEAR (Appendix E) Routine GRIDLS illustrates the grid-search method.

The main search routine, GRIDLS, is entered with the value of χ^2 (CHISQR) as argument. In a loop over each of the m parameters in turn, the value of the parameter is varied until χ^2 has passed through a local minimum in the parameter. The three most recent values of χ^2 that bracket the minimum are stored in the variables CHISQ1, CHISQ2, and CHISQ3. The best estimate of the parameter at this stage of the calculation is determined from the minimum of the parabola that passes through the three points. The step size (DELTAA(J)) is then adjusted to be that value that increases χ^2 by 2 from its value at the local minimum.

One pass through GRIDLS corresponds to a single zigzag along the path of Figure 8.4. The search is repeated until χ^2 does not change by more than the preset level, CHICUT.

A call to the function SIGPARAB in the program unit FITUTIL at the end of the search returns an estimate of the uncertainty in each parameters in turn from a calculation of the independent variation needed to increase χ^2 by 1 from its minimum value.

TABLE 8.2
Two exponentials plus constant background: grid-search method

Trial	χ^2	a_1	a_2	a_3	a_4	a_5
0	406.6	10.0	900.0	80.0	27.0	225.0
1	143.0	14.5	1332.3	106.8	27.7	207.2
2	96.9	12.6	1233.9	127.9	28.2	198.4
3	79.4	11.6	1155.1	140.2	28.8	192.2
4	72.9	11.2	1100.3	147.0	29.3	189.2
⋮						
16	66.7	11.3	963.5	148.8	32.3	185.3
17	66.7	11.3	962.5	148.2	32.4	185.8
⋮						
39	66.3	10.9	959.3	139.1	33.3	195.4
40	66.2	10.8	959.2	138.9	33.3	195.7
Uncertainties		0.6	28.3	4.5	0.8	5.0

χ^2/dof = 1.23; probability = 12.1%; relative time = 9.1

Note: Stages in the fit to counts from the decay of excited states of silver. The values of χ^2 and the parameters are listed at the beginning, middle, and end of the search. The uncertainties in the parameters correspond to a change of 1 in χ^2 from its value at the end of the search.

Table 8.2 shows values of χ^2 and the parameters a_1 through a_5 for several stages of the calculation at the beginning, middle, and end of the search. The search is relatively slow, but eventually a satisfactory solution is found. Note that the calculated uncertainties correspond to the diagonal terms in the error matrix for uncorrelated parameters. If correlations are considered to be important, the matrix inversion methods discussed in the following sections could be used to find better approximations to the uncertainties.

8.4 GRADIENT-SEARCH METHOD

The search could be improved if the zigzagging direction of travel in Figure 8.4 were replaced by a more direct vector toward the appropriate minimum. In the *gradient-search* method of least squares, all the parameters a_j are incremented simultaneously, with relative magnitudes adjusted so that the resultant direction of travel in parameter space is along the gradient (or direction of maximum variation) of χ^2.

The gradient $\nabla\chi^2$ is a vector that points in the direction in which χ^2 increases most rapidly and has components in parameter space equal to the rate of change of χ^2 along each axis:

$$\nabla\chi^2 = \sum_{j=1}^{n}\left[\frac{\partial\chi^2}{\partial a_j}\hat{a}_j\right] \tag{8.14}$$

where $\hat{a}_j$ indicates a unit vector in the direction of the a_j coordinate axis. In order to determine the gradient, we estimate the partial derivatives numerically as discussed in Appendix A:

$$(\nabla\chi^2)_j = \frac{\partial\chi^2}{\partial a_j} \simeq \frac{\chi^2(a_j + f\Delta a_j) - \chi^2(a_j)}{f\Delta a_j} \tag{8.15}$$

where f is a fraction of the step size Δa_j by which a_j is changed in order to determine the derivative.

The gradient has both magnitude and dimensions and, if the dimensions of the various parameters a_j are not all the same (which is usually the case), the components of the gradient do not even have the same dimensions. Let us define dimensionless parameters b_j by rescaling each of the parameters a_j to a size that characterizes the variation of χ^2 with a_j rather roughly. We shall use the step sizes Δa_j as the scaling constants, so that

$$b_j = \frac{a_j}{\Delta a_j} \tag{8.16}$$

The derivative with respect to b_j then becomes

$$\frac{\partial\chi^2}{\partial b_j} = \frac{\partial\chi^2}{\partial a_j}\Delta a_j \tag{8.17}$$

which may be calculated numerically as

$$\frac{\partial\chi^2}{\partial b_j} \simeq \frac{\chi^2(a_j + f\Delta a_j) - \chi^2(a_j)}{f\Delta a_j}\Delta a_j = \frac{\chi^2(a_j + f\Delta a_j) - \chi^2(a_j)}{f} \tag{8.18}$$

We can then define a dimensionless gradient γ, with unit magnitude and components

$$\gamma_j = \frac{\partial\chi^2/\partial b_j}{\sqrt{\sum_{j=1}^{m}(\partial\chi^2/\partial b_j)^2}} \tag{8.19}$$

In the numerical calculation of Equation (8.18), the quantities Δa_j and f occur only in the argument of χ^2 and not as scale factors.

The direction that the gradient-search method follows is the *direction of steepest descent,* which is opposite of the gradient γ. The search begins by incrementing all parameters simultaneously by an amount Δa_j, with relative value given by the corresponding component γ_j of the dimensionless gradient and absolute magnitude given by the size constant Δa_j:

$$\delta a_j = -\gamma_j \Delta a_j \tag{8.20}$$

The minus sign ensures that the value of χ^2 decreases. The size constant Δa_j of Equation (8.20) is the same as that of Equation (8.16).

There are several possible methods of continuing the gradient search after a first step. The most straightforward is to recompute the gradient after each change in the parameters. One disadvantage of this method is that it is difficult to approach the bottom of the minimum asymptotically because the gradient tends to 0 at the minimum. Another disadvantage is that recomputation of the gradient at each step for small step sizes results in an inefficient search, but the use of larger step sizes makes location of the minimum less precise.

A reasonable variation on the method is to search along one direction of the original gradient in small steps, calculating only the value of χ^2 until χ^2 begins to rise again. At this point, the gradient is recomputed and the search continues in the new direction. Whenever the search straddles a minimum, a parabolic interpretation of χ^2 is used to improve the determination of the minimum.

A more sophisticated approach would be to use second partial derivatives of χ^2 to determine changes in the gradient along the search path:

$$\left.\frac{\partial \chi^2}{\partial a_j}\right|_{a_j+\delta a_j} \simeq \left.\frac{\partial \chi^2}{\partial a_j}\right|_{a_j} + \sum_{k=1}^{m}\left(\frac{\partial^2 \chi^2}{\partial a_j \partial a_k}\delta a_k\right) \tag{8.21}$$

If the search is already fairly near the minimum, this method does decrease the number of steps needed, but at the expense of more elaborate computation. If the search is not near enough to the minimum, this method can actually increase the number of steps required when first-order perturbations on the gradient are not valid.

The efficiency of the gradient search decreases markedly as the search approaches a minimum because the evaluation of the derivative according to the method of Equation (8.18) involves taking differences between nearly equal numbers. In fact, at the minimum of χ^2, these differences should vanish. For this reason, one of the methods discussed in the following sections may be used to locate the actual minimum once the gradient search has approached it fairly closely.

Program 8.2. GRADSEAR (Appendix E) Routine GRADLS illustrates the gradient-search method.

On each entry to the main search routine, GRADLS, the components of the gradient GRADLS(J) are calculated numerically from Equation (8.18) in the procedure CALCGRAD. The argument FRACT of this routine, corresponding to the variable *f* of Equation (8.18), determines the fraction of the step size (DELTAA) used in the numerical calculation of the partial derivative. Each parameter A(J) is then changed by the amount STEPDOWN*DELTAA(J)*GRAD(J), where STEPDOWN is a scaling factor that is set initially in the main program and readjusted after each stage to the size needed to locate the minimum.

The initial values of DELTAA(J) determines to some extent the execution speed of each pass through the routine GRADLS, and the value of CHICUT determines when the search will stop. Because of the small gradient near the χ^2 minimum, it may take many steps to reach a reasonable value of χ^2, and the cutoff, CHICUT, may have to be set to a very low value. For such cases, user intervention can be provided as an alternate method of stopping the search.

At the conclusion of the search, the uncertainties in the parameters are estimated in the function SIGPARAB as in the routine GRADLS.

Table 8.3 shows values of χ^2 and the parameters a_1 through a_5 for several stages of the calculation at the beginning, middle, and end of the search. For Example 8.1, the gradient search is considerably faster than the grid-search approach because all the parameters are varied together at each step. However, the gradient-search method has one disadvantage that is not illustrated. If the starting values of the parameters are too far from the final values, the grid search has a good chance of plodding along until it reaches the correct solution. The gradient search, on the other hand, may tend to get bogged down in local minima that correspond to a long, flat valley in the parameter space.

TABLE 8.3
Two exponentials plus constant background: gradient-search method

Trial	χ^2	a_1	a_2	a_3	a_4	a_5
0	406.6	10.0	900.0	80.0	27.0	225.0
1	82.3	10.6	1061.0	94.0	34.4	254.2
2	72.6	9.8	984.0	98.8	36.8	237.4
3	69.8	9.9	966.9	100.9	36.8	244.6
4	69.3	9.8	953.7	101.6	36.7	242.1
⋮						
19	66.6	8.9	952.2	114.7	35.5	233.6
20	66.5	8.9	954.8	114.9	35.6	233.9
Uncertainties		0.6	26.5	3.8	0.8	7.0

χ^2/dof = 1.23; probability = 11.8%; relative time = 4.0

Note: Stages in the fit to counts from the decay of excited states of silver. The values of χ^2 and the parameters are listed at the beginning, middle, and end of the search. The uncertainties in the parameters corresponding to a change of 1 in χ^2 from its value at the end of the search.

8.5 EXPANSION METHODS

Instead of searching the χ^2 hypersurface to map the variation of χ^2 with parameters, we should be able to find an approximate analytical function that describes the χ^2 hypersurface and use this function to locate the minimum, with methods developed for linear least-squares fitting. The approximations will introduce errors into the calculated values of the parameters, but successive iterations of the analytical method should approach the χ^2 minimum with increasing accuracy. The main advantage of such an approach is that the number of points on the χ^2 hypersurface at which computations must be made will be fewer than for a grid or gradient search. This advantage is somewhat offset by the fact that the computations at each point are considerably more complicated. However, the analytical solution essentially chooses its own step size and, thus, the user is spared the problem of trying to optimize the step size for speed and precision.

Parabolic Expansion of χ^2

In Equation (8.9) we expanded χ^2 to second order in the parameters about a local minimum χ_0^2 where $a_j = a_j'$:

$$\chi^2 \simeq \chi_0^2 + \sum_{j=1}^{m}\left\{\frac{\partial \chi_0^2}{\partial a_j}\delta a_j\right\} + \frac{1}{2}\sum_{k=1}^{m}\sum_{j=1}^{m}\left\{\frac{\partial^2 \chi_0^2}{\partial a_j \partial a_k}\delta a_j \delta a_k\right\} \tag{8.22}$$

which is equivalent to approximating the χ^2 hypersurface by a parabolic surface. Here we define $\delta a_j \equiv a_j - a_j'$, and χ_0^2 is given by

$$\chi_0^2 = \sum\left\{\frac{1}{\sigma_i^2}[y_i - y'(x_i)]^2\right\} \tag{8.23}$$

where $y'(x_i)$ is the value of the function when $\delta a_j = 0$.

Applying the method of least squares, we minimize χ^2 as expressed in Equation (8.22) with respect to the *increments* (δa_j) in the parameters, and solve for the optimum values of these increments to obtain

$$\frac{\partial \chi^2}{\partial(\delta a_k)} = \frac{\partial \chi_0^2}{\partial a_k} + \sum_{j=1}^{m} \left\{ \frac{\partial^2 \chi_0^2}{\partial a_k \partial a_j} \delta a_j \right\} = 0 \quad k = 1, m \tag{8.24}$$

The result is a set of m linear equations in δa_j that we can write as

$$\beta_k - \sum_{j=1}^{m} (\delta a_j \alpha_{jk}) = 0 \qquad k = 1, m \tag{8.25}$$

with

$$\beta_k \equiv -\frac{1}{2} \frac{\partial \chi_0^2}{\partial a_k} \qquad \text{and} \qquad \alpha_{jk} \equiv \frac{1}{2} \frac{\partial^2 \chi_0^2}{\partial a_j \partial a_k} \tag{8.26}$$

The factors $\pm½$ are included for agreement with the conventional definitions of these quantities.

As in Chapter 7, we can treat Equation (8.25) as a matrix equation:

$$\boldsymbol{\beta} = \boldsymbol{\delta a}\, \boldsymbol{\alpha} \tag{8.27}$$

where $\boldsymbol{\beta}$ and $\boldsymbol{\delta a}$ are row matrices and $\boldsymbol{\alpha}$ is a symmetric matrix of order m. We shall find that $\boldsymbol{\alpha}$ is the *curvature matrix* discussed in Section 7.2, so named because it measures the curvature of the χ^2 hypersurface.

Method of Computation

The solution of Equation (8.27) can be obtained by matrix inversion as in Section 7.2:

$$\boldsymbol{\delta a} = \boldsymbol{\beta \epsilon} \qquad \delta a_k = \sum_{j=1}^{m} (\epsilon_{kj} \beta_j) \tag{8.28}$$

where the error matrix $\boldsymbol{\epsilon} = \boldsymbol{\alpha}^{-1}$ is the inverse of the curvature matrix.

If the parameters are independent of one another, that is, if the variation of χ^2 with respect to each parameter is independent of the values of the other parameters, then the cross-partial derivatives a_{jk} $(j \neq k)$ will be 0 in the limit of a very large data sample and the matrix $\boldsymbol{\alpha}$ will be diagonal. The inverse matrix $\boldsymbol{\epsilon}$ will also be diagonal and Equation (8.27) will degenerate into m separate equations:

$$\delta \alpha_j \simeq \frac{\beta_j}{\alpha_{jj}} = \frac{\partial x_0^2}{\partial a_j} \div \frac{\partial^2 x_0^2}{\partial a_j^2} \tag{8.29}$$

Computation of the matrix elements by Equation (8.26) requires knowledge of the first and second derivatives of χ^2 evaluated at the current values of the parameters. Analytic forms of the derivatives are generally quickest to compute, but may be difficult or cumbersome to derive. If it is not convenient or possible to provide analytic forms of the derivatives, then they can be computed by the method of finite differences (see Appendix A). In the following expressions, we use forward differences for efficient calculations. The intervals Δa_j should be chosen to be large enough to avoid roundoff errors but small enough to furnish reasonably accurate values of the derivatives near the minimum:

$$\frac{\partial \chi_0^2}{\partial a_j} \simeq \frac{\chi_0^2(a_j + \Delta a_j, a_k) - \chi_0^2(a_j, a_k)}{\Delta a_j}$$

$$\frac{\partial^2 \chi_0^2}{\partial^2 a_j} \simeq 4\left[\frac{\chi_0^2(a_j, a_k) - 2\chi_0^2(a_j + \delta a_j/2, a_k) + \chi_0^2(a_j + \Delta a_j, a_k)}{(\Delta a_j)^2}\right]$$

$$\frac{\partial^2 \chi_0^2}{\partial a_j \partial a_k} \simeq [\chi_0^2(a_j, a_k) - \chi_0^2(a_j + \Delta a_j, a_k) - \chi_0^2(a_j, a_k + \Delta a_k) + \chi_0^2(a_j + \Delta a_j, a_k + \Delta a_k)]/[\Delta a_j \Delta a_k] \tag{8.30}$$

In actual practice, calculations are faster and, in general, more accurate if the elements of the matrix $\boldsymbol{\alpha}$ are determined from the first-order expansion (to be discussed in the following text), which involves only first derivatives of $y(x)$ with respect to the parameters, rather than the second derivatives of χ^2 as expressed in Equation (8.30).

Fitting Procedure

Within the limits of the approximation of the χ^2 hypersurface by a parabolic extrapolation, we can solve Equation (8.27) directly to yield parameter increments δa_j such that χ^2 should be minimized for $a_j' + \delta a_j$. If the starting point is close enough to the minimum so that higher-order terms in the expansion can be neglected, this becomes an accurate and precise method. But if the starting point is not near enough, the parabolic approximation of the χ^2 hypersurface is not valid and the results will be in error. In fact, if the starting point is so far from the minimum that the curvature of χ^2 is negative, the solution will tend toward a maximum rather than a minimum. During computation, therefore, the diagonal elements α_{jj} of the matrix $\boldsymbol{\alpha}$ must be set positive whether they are or not. The resulting magnitude for δa_j will be incorrect, but the sign will be correct.

Expansion of the Fitting Function

An alternative to expanding the χ^2 function to develop an analytic description for the hypersurface is to expand the fitting function $y(x)$ in the parameters a_j and to use the method of linear least squares to determine the optimum value for the parameter increments δa_j. If we carry out the derivation rigorously and drop higher-order terms, we should achieve the same result as for the expansion of χ^2 to first and second order.

First-Order Expansion

Let us expand the fitting function $y(x)$ in a Taylor series about the point a_j', to first order in the *parameter increments* $\delta a_j = a_j - a_j'$:

$$y(x) \simeq y'(x) + \sum_{j=1}^{m} \left[\frac{\partial y'(x)}{\partial a_j} \delta a_j \right] \tag{8.31}$$

where $y'(x)$ is the value of the fitting function when the parameters have starting point values a'_j and the derivatives are evaluated at the starting point. The result is a linear function in the parameter increments δa_j to which we can apply the method of linear least squares developed in Chapter 7.

In this approximation, χ^2 can be expressed explicitly as a function of the parameter increments δa_j:

$$\chi^2 = \sum \left(\frac{1}{\sigma_i^2} \left\{ y_i - y'(x_i) - \sum_{j=1}^{m} \left[\frac{\partial y'(x_i)}{\partial a_j} \delta a_j \right] \right\}^2 \right) \tag{8.32}$$

Following the method of least squares, we minimize χ^2 with respect to each of the parameter increments δa_j by setting the derivatives equal to 0:

$$\frac{\partial \chi^2}{\partial \delta a_k} = -2 \sum \left(\frac{1}{\sigma_i^2} \left\{ y_i - y'(x_i) - \sum_{j=1}^{m} \left[\frac{\partial y'(x_i)}{\partial a_j} \delta a_j \right] \right\} \frac{\partial y'(x_i)}{\partial a_k} \right) = 0 \tag{8.33}$$

As before, this yields the set of m simultaneous Equations (8.25), which can be expressed as the matrix Equation (8.27):

$$\boldsymbol{\beta} = \boldsymbol{\delta a}\, \boldsymbol{\alpha} \tag{8.34}$$

where β_k is defined as in Equation (8.26) and α_{jk} is given by

$$\alpha_{jk} \simeq \sum \left[\frac{1}{\sigma_i^2} \frac{\partial y'(x_i)}{\partial a_j} \frac{\partial y'(x_i)}{\partial a_k} \right] \tag{8.35}$$

Second-Order Expansion

Suppose we make a Taylor expansion of the fitting function $y(x)$ to second order in the parameter increments δa_j:

$$y(x) \simeq y'(x) + \sum_{j=1}^{m} \left[\frac{\partial y'(x)}{\partial a_j} \delta a_j \right] + \frac{1}{2} \sum_{j=1}^{m} \sum_{k=1}^{m} \left[\frac{\partial^2 y'(x)}{\partial a_j \partial a_k} da_j da_k \right] \tag{8.36}$$

If we include the last term of Equation (8.36) in the expression for χ^2 of Equation (8.32) and again minimize χ^2 by setting to 0 the derivatives with respect to the increments δa_j, we again obtain Equation (8.25), this time with

$$\begin{aligned} \beta_k &\equiv \sum \left\{ \frac{1}{\sigma_i^2} [y_i - y'(x_i)] \frac{\partial y'(x_i)}{\partial a_k} \right\} = -\frac{1}{2} \frac{\partial \chi_0^2}{\partial a_k} \\ \alpha_{jk} &\equiv \sum \frac{1}{\sigma_i^2} \left\{ \frac{\partial y'(x_i)}{\partial a_j} \frac{\partial y'(x_i)}{\partial a_k} - [y_i - y'(x_i)] \frac{\partial^2 y'(x_i)}{\partial a_j \partial a_k} \right\} \\ &= \frac{1}{2} \frac{\partial^2 \chi_0^2}{\partial a_j \partial a_k} \end{aligned} \tag{8.37}$$

The resulting definitions for β_k and a_{jk} are identical to those of Equation (8.26) obtained by expanding the χ^2 function, and the χ^2-expansion method is therefore equivalent to a second-order expansion of the fitting function.

Let us compare Equations (8.37) with the analogous Equations (7.14) and (7.15) for linear least-squares fitting. The definitions of α_{jk} in Equations (8.37) and (7.15) are equivalent in the linear approximation [See Equation (7.22)] and thus $\boldsymbol{\alpha}$ corresponds to the curvature matrix. The definition of β_k in Equation (8.37) is equivalent, in the linear approximation, to the definition of β_k in Equations (7.14) except for the substitution of $y_i - y'(x_i)$ for y_i. We can justify this substitution by noting that the solutions of Equation (8.34) are the parameter increments δa_j, whereas those of Equation (7.14) are the parameters themselves. In essence, we are applying linear least-squares methods to fit the parameter increments to difference data Δy_i between the actual data and the starting values of the fitting $y'(x_i)$:

$$\Delta y_i = y_i - y'(x_i) \tag{8.38}$$

Thus, the expression given in Equation (8.35) for α_{jk} is a first-order approximation to the curvature matrix that is given to second order in Equation (8.37). For linear functions, the second-order term vanishes. It is convenient to use the first-order approximation for fitting nonlinear functions and thus avoid the necessity of calculating the second derivatives in Equation (8.37). We note that this procedure can be somewhat justified on the grounds that, in the vicinity of the χ^2 minimum, we should expect the factor of $y_i - y'(x_i)$ in the expression for α of Equation (8.37) to be close to 0 so that the first term in the expression will dominate.[1]

Program 8.3. EXPNDFIT (Appendix E) Routine CHIFIT illustrates non-linear fitting by expansion of the fitting function. The program is called repeatedly from the main program NONLINFIT, until χ^2 passes through a minimum. EXPNDFIT calls the following routines to set up and manipulate the matrices.

Program 8.6. MAKEAB8 (Appendix E) Sets up the $\boldsymbol{\alpha}$ and $\boldsymbol{\beta}$ matrices.
The routine uses the first-order approximation of Equation (8.35) to calculate the components α_{jk} of the curvature matrix. This is equivalent to neglecting terms in the second derivatives of the fitting function $y(x)$ in the expression for α_{jk} in Equation (8.37). The routines in this program unit use numerical derivatives and therefore differ from those with the same names in Chapter 7, which use analytic derivatives.

Program 8.7. NUMDERIV (website) Numerical derivatives.
Derivatives of χ^2 (XISQ) are calculated numerically by the functions DXISQ_DA, D2XISQ_DA2, and D2XISQ_DAJK in this program unit. To avoid repetitive calculations, the values of the derivatives at each value of x and for the variation of each of the m parameters are calculated once for each trial and stored in arrays. If available, analytic expressions for the derivatives could be substituted directly for the functions to increase the speed and accuracy of the calculation.

Program B.1. MATRIX (Appendix E) Matrix multiplication and inversion.

[1]See Press et al. (1986), page 523.

TABLE 8.4
Two exponentials plus constant background: χ^2 expansion method

Trial	χ^2	a_1	a_2	a_3	a_4	a_5
0	406.6	10.0	900.0	80.0	27.0	225.0
1	86.2	11.1	933.8	140.4	33.8	170.5
2	66.6	10.8	861.2	128.9	33.9	201.7
3	66.1	10.4	958.2	131.2	34.0	205.4
Uncertainties		1.8	49.9	21.7	2.5	30.5

χ^2/dof = 1.22; probability = 12.4%; relative time = 1.0

Note: All stages in the fit to counts from the decay of excited states of silver. The uncertainties in the parameters correspond to the square roots of the diagonal terms in the error matrix.

At the conclusion of the search, the inverse $\boldsymbol{\epsilon}$ of the final value of the curvature matrix $\boldsymbol{\alpha}$ is treated as the error matrix, and the errors in the parameters are obtained from the square roots of the diagonal terms by calls to the function SIGMATRX in the unit FitFunc8. Table 8.4 shows values of χ^2 and the parameters a_1 through a_5 for all stages of the calculation.

8.6 THE MARQUARDT METHOD

Convergence

One disadvantage inherent in the analytical methods of expanding either the fitting function $y(x)$ or χ^2 is that although they converge quite rapidly to the point of minimum χ^2 from points nearby, they cannot be relied on to approach the minimum with any accuracy from a point outside the region where the χ^2 hypersurface is approximately parabolic. In particular, if the curvature of the χ^2 hypersurface is used, as in Equation (8.37) or (8.26), the analytical solution is clearly unreliable whenever the curvature becomes negative. Symptomatic of this problem is the need to set positive the diagonal elements α_{jj} of the matrix α so that all curvatures are treated as if they were positive.

In contrast, the gradient search of Section 8.4 is ideally suited for approaching the minimum from far away, but does not converge rapidly near the minimum. Therefore, we need an algorithm that behaves like a gradient search for the first portion of a search and behaves more like an analytical solution as the search converges. In fact, it can be shown (see Marquardt 1963) that the path directions for gradient and analytical searches are nearly perpendicular to each other, and that the optimum direction is somewhere between these two vectors.

One advantage of combining these two methods into one algorithm is that the simpler first-order expansion of the analytical method will certainly suffice because the expansion need only be valid in the immediate neighborhood of the minimum. Thus, to calculate the curvature matrix α, we can use the approximation of Equation (8.35) and ignore the second derivatives of Equation (8.37).

Gradient-Expansion Algorithm

A convenient algorithm (see Marquardt 1963), which combines the best features of the gradient search with the method of linearizing the fitting function, can be obtained by increasing the diagonal terms of the curvature matrix $\boldsymbol{\alpha}$ by a factor $1 + \lambda$ that controls the interpolation of the algorithm between the two extremes. Equation (8.34) becomes

$$\boldsymbol{\beta} = \boldsymbol{\delta a}\,\boldsymbol{\alpha}' \quad \text{with} \quad \alpha'_{jk} = \begin{cases} \alpha_{jk}(1+\lambda) & \text{for } j = k \\ \alpha_{jk} & \text{for } j \neq k \end{cases} \tag{8.39}$$

If λ is very small, Equations (8.39) are similar to the solution of Equation (8.34) developed from the Taylor expansion. If λ is very large, the diagonal terms of the curvature matrix dominate and the matrix equation degenerates into m separate equations

$$\beta_j \simeq \lambda \delta a_j \alpha_{jj} \tag{8.40}$$

which yield the vector increment $\boldsymbol{\delta a}$ in the same direction as the vector $\boldsymbol{\beta}$ of Equation (8.37) (or opposite to the gradient of χ^2).

The solution for the parameter increments δa_j follows from Equations (8.39) after matrix inversion

$$\delta a_j = \sum_{k=1}^{m} (\beta_k \epsilon'_{jk}) \tag{8.41}$$

where the β_k are given by Equation (8.37) and the matrix $\boldsymbol{\epsilon}'$ is the inverse of the matrix $\boldsymbol{\alpha}'$ with elements given by Equations (8.39).

The initial value of the constant factor λ should be chosen small enough to take advantage of the analytical solution, but large enough that χ^2 decreases. Because this algorithm approaches the gradient-search method with small steps for large λ, there should exist a value of λ such that $\chi^2(a + \delta a) < \chi^2(a)$. The recipe given by Marquardt is:

1. Compute $\chi^2(a)$.
2. Start initially with $\lambda = 0.001$.
3. Compute δa and $\chi^2(a + \delta a)$ with this choice of λ.
4. If $\chi^2(a + \delta a) > \chi^2(a)$, increase λ by a factor of 10 and repeat step 3.
5. If $\chi^2(a + \delta a) < \chi^2(a)$, decrease λ by a factor of 10, consider $a' = a + \delta a$ to be the new starting point, and return to step 3, substituting a' for a.

For each iteration it may be necessary to recompute the parameter increments δa_j from Equation (8.41), and the elements α_{jk} and β_j of the matrices, several times to optimize λ. As the solution approaches the minimum, the value of λ will decrease and the program should locate the minimum with a few iterations. A lower limit may be set for the value λ, but in practice this limit will seldom be reached.

TABLE 8.5
Two exponentials plus constant background: Marquardt method

Trial	χ^2	a_1	a_2	a_3	a_4	a_5
0	406.6	10.0	900.0	80.0	27.0	225.0
1	82.9	11.0	933.5	139.3	33.9	173.9
2	66.4	10.8	960.1	130.6	33.8	201.2
3	66.1	10.4	958.3	131.4	33.9	205.0
Uncertainties		1.8	49.9	21.7	2.5	30.5

χ^2/dof = 1.22; probability = 12.4%; relative time = 1.0

Note: All stages in the fit to counts from the decay of excited states of silver. The uncertainties in the parameters correspond to the square roots of the diagonal terms in the error matrix.

TABLE 8.6
Elements of the error matrix (Marquardt method)

1/*k*	1	2	3	4	5
1	3.38	−3.69	27.98	−2.34	−49.24
2	−3.69	2492.26	81.89	−69.21	−3.90
3	27.98	81.89	468.99	−44.22	−615.44
4	−2.34	−69.21	−44.22	6.39	53.80
5	−49.24	−3.90	−615.44	53.80	929.45

Note: Error matrix from a fit to the radioactive silver data. The diagonal terms are the variances σ_k^2 and the off-diagonal terms are the covariances σ_{kl}^2 of the parameters a_k.

Program 8.4. MARQFIT (Appendix E) Routine MARQUARDT illustrates fitting by the gradient-expansion algorithm.
The procedure uses the same program units as those in Program 8.3, and is identical to that program except for the adjustment of the diagonal elements α_{jj} of the matrix $\boldsymbol{\alpha}$ by the variable LAMBDA according to Equation (8.39).

At the conclusion of the search, the inverse $\boldsymbol{\epsilon}$ of the final value of the curvature matrix $\boldsymbol{\alpha}$ is treated as the error matrix, and the errors in the parameters are obtained from the square roots of the diagonal terms by calls to the function SIGMATRX in the unit FitFunc8. Table 8.5 shows values of χ^2 and the parameters a_1 through a_5 for all stages of the calculation. Table 8.6 shows the error matrix from the fit.

8.7 COMMENTS

Although the Marquardt method is the most complex of the four fitting routines, it is also the clear winner for finding fits most directly and efficiently. It has the strong advantage of being reasonably insensitive to the starting values of the parameters, although in a peak-over-background example (Chapter 9), it does have difficulty when the starting parameters of the function for the peak are outside reasonable ranges. The Marquardt method also has the advantage over the grid- and gradient-search methods of providing an estimate of the full error matrix and better calculation of the diagonal errors.

The routines of Programs 8.3 and 8.4 were tested with both numerical and analytical derivatives. Typical search paths with numerical derivatives are shown in Tables 8.4 and 8.5. For the sample problem with the assumed starting conditions, the minimum χ^2 was found in only a few steps by either method with essentially no time difference. Both methods are reasonably insensitive to starting values of parameters in which the fit is linear, but can be sensitive to starting values of the nonlinear parameters. Program 8.4 had remarkable success over a broad range of starting values, whereas Program 8.3 required better definition of the starting values of the parameters and generally required many more iterations.

The uncertainties in the parameters for these fits were calculated from the diagonal terms in the error matrices and are, in general, considerably larger than the uncertainties obtained in the grid- and gradient-search methods. Because the latter errors were obtained by finding the change in each parameter to produce as change of χ^2 of 1 from the minimum values, without reoptimizing the fit, there is a strong suggestion that correlations among the parameters play an important role in fitting Figure 8.1. This point of view is supported by examination of the error matrix from the method 4 fit (Table 8.6), which shows large off-diagonal elements.

With poorly selected starting values, the searches may terminate in local minima with unacceptably high values of χ^2 and, therefore, with unacceptable final values for the parameters. Termination in the sample programs is controlled simply by considering the reduction in χ^2 from one iteration to the next and stopping at a preselected difference. With this method, it is essential to check the results carefully to be sure that the absolute minimum has indeed been found.

SUMMARY

Nonlinear function: One that cannot be expressed as a sum of terms with the coefficients of the terms.

Minimum of χ^2 (parabolic approximation):

$$a_j' = a_{j3} - \Delta a_j \left[\frac{\chi_3^2 - \chi_2^2}{\chi_1^2 - 2\chi_2^2 + \chi_3^2} + \frac{1}{2} \right]$$

Estimate of standard deviation from $\Delta\chi^2 = 1$:

$$\sigma_j = \Delta a_j \sqrt{2(\chi_1^2 - 2\chi_2^2 + \chi_3^2)^{-1}}$$

Grid search: Vary each parameter in turn, minimizing χ^2 with respect to each parameter independently. Many successive iterations are required to locate the minimum of χ^2 unless the parameters are independent; that is, unless the variation of χ^2 with respect to one parameter is independent of the values of the other parameters.

Gradient search: Vary all the parameters simultaneously, adjusting relative magnitudes of the variations so that the direction of propagation in parameter space is along the direction of steepest descent of χ^2.

Direction of steepest descent: Opposite the gradient $\nabla\chi^2$:

$$(\nabla\chi^2)_j = \frac{\partial\chi^2}{\partial a_j} \simeq \frac{\chi^2(a_j + f\Delta a_j) - x(a_j)}{f\Delta a_j}$$

$$\delta a_j = \frac{-((\partial\chi^2/\partial a_j)\Delta a_j^2)}{\sqrt{\sum_{j=1}^{m}((\partial\chi^2/\partial a_j)\Delta a_j)^2}}$$

Parabolic expansion of χ^2:

$$\boldsymbol{\delta\alpha} = \boldsymbol{\beta\epsilon} \qquad \delta a_k = \sum_{j=1}^{m}(\epsilon_{kj}\beta_j)$$

with

$$\beta_k \equiv -\frac{1}{2}\frac{\partial\chi^2}{\partial a_k} \quad \text{and} \quad \alpha_{jk} \equiv \frac{1}{2}\frac{\partial^2\chi^2}{\partial a_j \partial a_k}$$

Linearization of the fitting function:

$$\beta_k \equiv \sum\left\{\frac{1}{\sigma_i^2}[y_i - y(x_i)]\frac{\partial y(x_i)}{\partial a_k}\right\} = -\frac{1}{2}\frac{\partial\chi_0^2}{\partial a_k}$$

$$\alpha_{jk} \equiv \sum\frac{1}{\sigma_i^2}\left\{\frac{\partial y(x_i)}{\partial a_j}\frac{\partial y(x_i)}{\partial a_k} - [y_i - y'(x_i)\frac{\partial^2 y(x_i)}{\partial a_j \partial a_k}\right\}$$

$$= \frac{1}{2}\frac{\partial^2\chi^2}{\partial a_j \partial a_k}$$

Gradient-expansion algorithm—the Marquardt method: Make λ just large enough to insure that χ^2 decreases:

$$\alpha'_{jk} = \begin{cases} \alpha_{jk}(1+\lambda) & \text{for } j = k \\ \alpha_{jk} & \text{for } j \neq k \end{cases}$$

$$\alpha_{jk} \simeq \sum\left[\frac{1}{\sigma_i^2}\frac{\partial y(x_i)}{\partial a_j}\frac{\partial y(x_i)}{\partial a_k}\right] \qquad \beta_k = -\frac{1}{2}\frac{\partial\chi^2}{\partial a_k}$$

$$\delta a_j = \sum_{k=1}^{m}(\beta_k \epsilon'_k)$$

Uncertainty in parameter a_j: $\alpha_{aj} = e_{jj}$ corresponds to $\Delta\chi^2 = 1$.

EXERCISES

8.1. Use an interpolation method (see Appendix A) to find the equation of the parabola that passes through the three points (x_1, y_1), (x_2, y_2), and (x_3, y_3). Find the value of x at the minimum of the parabola and thus verify Equation (8.12).

8.2. From the results of Exercise 8.1, verify Equation (8.13).

8.3. The following data represent histogram bin counts across a Lorentzian peak:

x_i	1.824	1.828	1.832	1.836	1.840	1.844	1.848	1.852	1.856	1.860
y_i	558	679	696	736	834	812	899	817	767	657

(*a*) Use the grid-search method to fit the equation $y(x) = AP_L(x; \mu, \Gamma)$ to the data and find the maximum-likelihood value of μ, where $P_L(x; \mu, \Gamma)$ is the Lorentzian function of Equation (2.32) and the known parameters are $A = 75$ and $\Gamma = 0.055$. Assume that x is given at the lower edge of each histogram bin and that the errors in y are statistical. Find the uncertainty in μ.

Suggested procedure: (i) Calculate χ^2 at the peak of the distribution and at a value on each side. (ii) Find the minimum of a parabola that passes through the three points. (iii) Repeat the procedure with three points centered on the minimum χ^2 until the value of μ has been determined to ± 0.001.

(*b*) Repeat the procedure for a two-parameter fit, with Γ as the second unknown.

8.4. Consider the histogram of measured time intervals displayed in Figure 1.2. The numbers of events in the bins bounded by $t = 0.59$ to 0.70s.

2, 2, 11, 6, 12, 8, 4, 3, 1, 1, 0

Fit a Gaussian curve [Equation (2.23)] to these data by the least-squares method to find μ, σ, and the amplitude of the curve A. Bins with fewer than seven events should be merged to improve the reliance on Gaussian statistics. Compare the parameters obtained from the fit with those determined by taking the mean and standard deviation of the data.

8.5. The following data correspond to counts recorded in Example 6.2 with the addition of an unknown randomly fluctuating background term a_1. Use the Marquardt method to fit the equation $C = a_1 + a_2/d^2$ to these data to find the parameters a_1 and a_2 and the full error matrix. Assume statistical uncertainties.

i	1	2	3	4	5	6	7	8	9	10
d_i(m)	0.20	0.25	0.30	0.35	0.40	0.45	0.50	0.60	0.75	1.00
C_i	944	688	467	366	316	317	264	251	214	184

8.6. Use the method of least squares to fit the five-parameter equation $y(x) = a_1 + a_2x + a_3G(x; a_4, a_5)$ to the following data where $a_4 = \mu$, $a_5 = \sigma$, and $G(x; \mu, \sigma)$ is the Gaussian curve of Equation (2.23).

i	1	2	3	4	5	6	7	8	9	10
x_i	1.0	1.1	1.2	1.3	1.4	1.5	1.6	1.7	1.8	1.9
y_i	31	25	24	30	34	37	31	30	64	54
i	11	12	13	14	15	16	17	18	19	20
x_i	2.0	2.1	2.2	2.3	2.4	2.5	2.6	2.7	2.8	2.9
y_i	95	94	78	79	43	54	58	52	46	41

Use the Marquardt method and find an estimate of the error matrix. The value of x is given at the lower edge of each bin. Assume statistical uncertainties.

8.7. To check the inverse-square relationship expressed in Coulomb's law,

$$F = kQ_1Q_2/r^2$$

Students in an undergraduate laboratory measured the force of electrostatic repulsion between two charged conducting spheres as a function of the distance between the centers of the spheres.

They applied the same potential to each sphere so that each carried the same charge. Because of the mutual repulsion of the charges on the conducting spheres, the effective separation of the two charge distributions is not simply the separation of the centers of the spheres. The resulting reduction in the repulsive force is a function of the separation r of the spheres and their radii a, given approximately by the correction factor

$$f = 1 - 4(a/r)^3$$

where $a = 1.9$ cm in this experiment. Thus, the relation between the mutual force on the spheres and their separation, including the correction factor, can be expressed as

$$F_{coulomb} = \left[1 - 4\left(\frac{a}{r}\right)^3\right]\frac{kQ_0Q_1}{r^2}$$

The students used a torsion balance to study the variation of the repulsive force, so that the force was proportional to the measured torsion angle. The relation between the torsion angle θ and the separation r of the centers of the spheres, including the correction factor, can be rewritten as a "fitting equation"

$$\theta = A[1 - 4(a/r)^e]$$

with unknown parameters, the scale factor A and the exponent e.

The students obtained the following measurements of the torsion angle (θ in degrees) as a function of the separation between the centers of the spheres (r in cm).

r_i	5.0	6.0	7.0	8.0	9.0	10.0	12.0	14.0	16.0	18.0	20.0
θ_i	264	233	179	136	111	84	63	53	33	30	27

Assume that the uncertainty in the angle is $\pm 1°$.

(*a*) Use one of the nonlinear fitting methods to determine the two parameters e and A of the fitting equation, and their uncertainties.

(*b*) Make a better estimate of the uncertainty in θ by considering the uncertainty required to give χ^2 = number of degrees of freedom.

(*c*) What effect does this change have on the uncertainties in the fitted parameters?

CHAPTER 9

FITTING COMPOSITE CURVES

9.1 LORENTZIAN PEAK ON QUADRATIC BACKGROUND

Many fitting problems involve determining the parameters of a resonant peak or peaks, superimposed upon a background signal. Examples may be found in various types of spectroscopic studies where the objective is to determine the properties of one or more resonant states.

EXAMPLE 9.1 We consider a problem from nuclear or particle physics illustrated by the 4000-event histogram of Figure 9.1, which shows a large peak on a smoothly varying background. We shall assume that the data have been drawn from a distribution that includes a resonant state described by the Lorentzian distribution, and that the background can be described by a second-degree polynomial in the energy E.[1] We shall attempt to fit Equation (9.1) to the data to determine the amplitude A_0, the resonant energy E_0, and the full width at half maximum Γ.

$$y(E) = a_1 + a_2 E + a_3 E^2 + A_0 \frac{\Gamma/(2\pi)}{(E - E_0)^2 + (\Gamma/2)^2} \tag{9.1}$$

We note that Equation (9.1) is linear in the parameters a_1, a_2, and a_3, but not in the parameters E_0 and Γ.

---

[1]These "data" were actually generated by the Monte Carlo method described in Chapter 5. The parameters used in the generation are listed in the second column of Table 9.1.

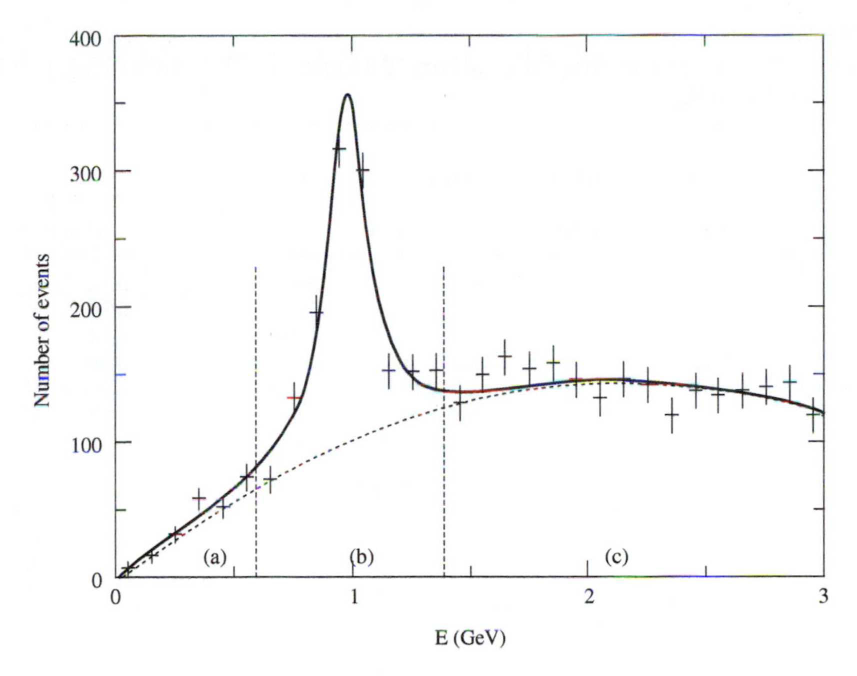

FIGURE 9.1
Histogram data in bins of 0.10 GeV of the 4000 simulated events generated from Equation (9.13) with parameters listed in column 2 of Table 9.1. The solid curve illustrates a fit of Equation (9.1) to the data. The dashed curve indicates the polynomial background.

We used the Marquardt method with numerical derivatives to fit Equation (9.1) to the histogram of Figure 9.1, because this is clearly the most flexible and convenient of the four methods considered in Chapter 8. The amplitudes of the polynomial function (a_1 through a_3), the amplitude of the Lorentzian peak ($a_4 = A_0$), and the mean E_0 and half-width Γ of the Lorentzian function (a_5 and a_6) were treated as free parameters of the fit. Starting values for a_5 and a_6 were obtained by inspecting the histogram of Figure 9.1; starting values for the other parameters, the coefficients of the various terms, were obtained by trial and error. Because the Marquardt method is exact for a function that is linear in the parameters, convergence of the fit is relatively insensitive to starting values of a_1 through a_4. The method is more sensitive to starting values for the Lorentzian parameters (E_0 and Γ). If starting values were too far from the obvious parameters of the peak, the program would coast to a halt in a shallow local minimum with obviously incorrect values for the parameters, and with a higher than expected value of χ^2. Starting values for all fits are listed in column 3 of Table 9.1.

Results of this six-parameter fit to the distribution in Figure 9.1 are summarized in column 4 of Table 9.1 and the curve calculated from Equation (9.1) with the parameters found in the fit is plotted on the histogram of Figure 9.1. The dashed curve shows the contribution of background under the peak. The χ^2 probability of the fit (7.9%) is low, but acceptable.

Because one of the objectives of the analysis of Example 9.1 is to determine E_0, the mean of the peak function of Equation (9.1), we must be careful in the

TABLE 9.1
Results of least-squares fits of Equations (9.1) and (9.13) to data displayed in Figures 9.1 and 9.2

	Values used to generate data	Starting values for fit	Six-parameter fit (Figure 9.1)	Six-parameter fit (Figure 9.2-inset)	Nine-parameter fit (Figure 9.2)
dof			24	54	51
χ^2			34.3	72.9	56.0
$P_{\chi 2}$			7.9%	4.4%	29.4
Num	4000		3944	3927	3994
a_1	1.0	1	2.2 ± 2.6	-2.2 ± 1.3	-2.1 ± 1.1
a_2	45.0	1	136.0 ± 8.1	73.9 ± 3.7	73.7 ± 3.6
a_3	$-10.$	1	-31.6 ± 3.1	-18.0 ± 1.4	-18.0 ± 1.4
			Peak 1		
a_4 (A_0)	20.0	1	79.8 ± 7.0	33.9 ± 2.7	28.8 ± 3.0
a_5 (E_0)	1.0	1	0.9838 ± 0.0068	0.9912 ± 0.0050	0.9968 ± 0.0044
a_6 (Γ_0)	0.1	0.1	0.197 ± 0.024	0.139 ± 0.015	0.108 ± 0.017
			Peak 2		
a_7 (A_1)	3.5	1	—	—	5.3 ± 2.2
a_8 (E_1)	0.8	0.825	—	—	0.824 ± 0.017
a_9 (Γ_1)	0.12	0.05	—	—	0.083 ± 0.034

choice of the value of the independent variable that we use in the fit. On the histogram of Figure 9.1, the value of E_i at the left-hand edge of selected bins is indicated, but for the fit we used the value of E at the center of each bin. If we had used values of E_i from the left-hand edge of the histogram bins, the value for E_0 from the fit would have been too low by half a bin width. For wide bins and a rapidly varying fitting function, it might be advisable to select the value of E_i for each bin by weighting according to the steepness of the function.

Note that the problem of selecting the absolute value of the abscissa corresponding to the ordinate value was not important in the determination of the mean lifetimes in Example 8.1 because lifetimes are determined effectively from differences, rather than absolute values, of the independent variable. We must, however, always take care when we plot results of a fit that the curve is not displaced half a bin width from the data.

Program 9.1 LORINFIT (Appendix E) illustrates use of the Marquardt method to fit a Lorentzian peak on a quadratic background.

9.2 AREA DETERMINATION

When dealing with problems of peaks and backgrounds, we may wish to determine not only the position and width of a peak in a spectrum, but also the number of events or area of the peak, which may measure the intensity of a transition or the strength of a reaction. When peaks are not well separated, or when the contribution from background is substantial, least-squares fitting can provide a consistent method of extracting such information from the data.

The importance of consistency should not be underestimated. Whether or not the method chosen is the best possible method, as long as it involves a well-understood and clearly specified procedure, other experimenters will be able to check and compare the results safe in the knowledge that their comparisons are justified and meaningful. The method of least squares is considered to be an unbiased estimator of the fitting parameters and all parameters are presumed to be estimated as well as possible. This assumption is based on the validity of both the fitting function in describing the data and the least-squares method. If we try to fit the data with an incorrect fitting function, or try to fit data with uncertainties that do not follow the Gaussian distribution, then the fitting procedure may not yield optimum results.

Although we refer to the number of events as the area of a peak or plot, the true area is, of course, the number of events multiplied by the data interval or histogram bin width. Thus, to find the area A_p of the peak from the results of the fit in Example 9.1, we calculate

$$A_P = \int_{-\infty}^{\infty} A_0 \frac{\Gamma/(2\pi)}{(E - E_0)^2 + (\Gamma/2)^2} \tag{9.2}$$

Because we used the normalized form of the Lorentzian function, the integral is just the coefficient a_4 obtained in the search $A_P = A_0 = a_4$. The area of the peak on the histogram is the product of the number of events N_P in the peak and the width ΔE of the histogram bin

$$A_P = N_P \times \Delta E,$$

so the number of events in the peak is given by

$$N_P = A_P/\Delta E \tag{9.3}$$

The result from Example 9.1 is $N_P = (79.8 \pm 7.0)/0.1 = (798 \pm 70)$ events.

Alternatively, we might plot the background curve on the graph

$$y_b(E) = a_1 + a_2 E + a_3 E^2 \tag{9.4}$$

and count the number of events in the peak above the background in a selected range encompassing the peak. We have indicated such a range by vertical dotted lines at $E_0 - 2\Gamma$ and $E_0 + 2\Gamma$ in Figure 9.1. With this method we should be obliged to estimate and correct for events outside the selected region.

Uncertainties in Areas under Peaks

If we calculate the area of the peak from Equations (9.2) and (9.3), then the uncertainty should be estimated from the uncertainties in the parameters by the error propagation equation. We have used this method to obtain the uncertainty in the number of events of the peak of Figure 9.1 in the calculation that follows Equation (9.3).

The uncertainty σ_A in the area under a peak can also be estimated by considering the uncertainty in the parent distribution. If the data are distributed according to the Poisson distribution, the uncertainty in the area A_p is given by $\sigma_a^2 \simeq A_p$. If we obtain the area by counting the number of events above background, then the variance of the difference will be the sum *(not the difference)* of the variance of the total area under the peak and the variance of the subtracted background A_b:

$$\sigma_P^2 = \sigma_t^2 + \sigma_b^2 = A_t + A_b$$

where the subscripts *p, b,* and *t* correspond to peak, background, and total (= peak + background). In order to keep $s_t = A_t$ as small as possible, we should count events only in that region where the peak-to-background ratio is large and make corrections for the tails of the distribution.

Area under a Curve with Poisson Statistics

Curiously enough, if the data are distributed around each data point according to the Poisson distribution, as in a counting experiment, the method of least squares consistently *underestimates* the area under a fitted curve by an amount approximately equal to the value of χ^2. To show this, let us consider fitting such data with an arbitrary peak, represented by $bf_p(x; \mu, \sigma)$ plus a polynomial background similar to Example 9.1:

$$y(x) = a + bf(x; \mu, \sigma) \tag{9.5}$$

where we have simplified the background to a single term a for clarity.

Using the method of least squares, we define χ^2 to be the weighted sum of the squares of deviations of the data from the fitted curve

$$\chi^2 = \sum\left[\frac{1}{\sigma_i^2}(y_i - a - bf(x; \mu, \sigma))^2\right] \tag{9.6}$$

and obtain the solution by minimizing χ^2 simultaneously with respect to each of the parameters. The required derivatives with respect to the two parameters a and b, in which the function is linear, are

$$\begin{aligned} \frac{\partial\chi^2}{\partial b} &= -2\sum\left[\frac{1}{\sigma_i^2}(y_i - a - bf(x; \mu, \sigma))f(x; \mu, \sigma)\right] = 0 \\ \frac{\partial\chi^2}{\partial a} &= -2\sum\left[\frac{1}{\sigma_i^2}(y_i - a - bf(x; \mu, \sigma))\right] = 0 \end{aligned} \tag{9.7}$$

We can write χ^2 in terms of the derivatives of Equation (9.7) as

$$\chi^2 = \sum\left[\frac{y_i}{\sigma_i^2}(y_i - a - bf(x; \mu, \sigma))\right] + \frac{1}{2}\left(a\frac{\partial\chi^2}{\partial a} + b\frac{\partial\chi^2}{\partial b}\right) \tag{9.8}$$

and setting the derivatives to 0 gives

$$\chi^2 = \sum\left[\frac{y_i}{\sigma_i^2}(y_i - a - bf(x; \mu, \Gamma))\right] \tag{9.9}$$

If the data represent the number of counts per unit time in a detector, then they are distributed according to the Poisson distribution and we can approximate $\sigma_i^2 \simeq y_i$. Equation (9.9) becomes

$$\begin{aligned}\chi^2_{\min} &\simeq \Sigma[y_i(a + bf(x;\mu,\Gamma))] \\ &= \text{area(data)} - \text{area(fit)}\end{aligned} \tag{9.10}$$

Thus, we observe that the area under the total fit is underestimated by an amount equal to $\chi^2_{\min}$.

For this derivation we require only that the fitting function consist of a sum of terms, each one of which is multiplied by a coefficient

$$y(x) = \sum_{j=1}^{m} a_j f_j(x) \tag{9.11}$$

The function $f_j(x)$ can contain any number of other parameters in nonlinear form, but may not contain any of the coefficients a_j. Even reparameterizing the function of Equation (9.5) [or Equation (9.1)] and minimizing χ^2 with respect to the area explicitly would not affect the discrepancy between the actual and estimated areas.

Note that for data that are distributed with a constant uncertainty $\sigma_i = \sigma$, the second equation of Equations (9.7) is sufficient to ensure that $\Sigma y(x_i) = \Sigma y_i$. It is the assumption of a Poisson distribution for the data $\sigma_i^2 = y_i$ that yields the discrepancy between the actual and estimated areas.

If the agreement between the fit and the data should be exact, $\chi^2 = 0$, then the estimated and actual areas would be equal. For a fitting function that is a good representation of the data, the value of χ^2 will approximately equal the number of degrees of freedom, so that if there are many bins and a few parameters to be determined, the average discrepancy will be about 1 per bin. Thus, the correction may be negligible for distributions with large numbers of events.

We would like to find ways to reduce the discrepancy. The fact that we know the approximate value of the discrepancy in the total histogram is, in itself, not very helpful because we do not know how to allocate the discrepancy between peak and background. We might find the ratio of the integral A_p of the peak [Equation (9.2)] to the integral A of the complete function Equation (9.1) and scale to the total number of events in the plot to estimate the number of events in the peak. This method assumes that the correction is proportional to the area. Another possibility is to make separate fits to the peak and background regions of the plot, so that we can try to assign the estimated correction separately to the two regions of the plot.

One obvious way of reducing the discrepancy between the area of the measured and fitted data is to reduce the value of χ^2 at the minimum so that the correction is small. A method of accomplishing this reduction, which is not universally accepted but which can be justified by practical considerations, is the technique of smoothing the data, averaging in some mathematically acceptable way over adjacent bins. (See Appendix A.5). Under any smoothing process there can be no overall gain in information, and a net improvement of the fit to the area must be offset by an increased uncertainty in the estimation of other parameters, such as the width and position of the peak. But smoothing will decrease the value of χ^2 at the minimum and thereby reduce the bias in the estimation of the area.

Referring to Table 9.1, we observe that the areas under the three fitted curves differ from the area under the data sample (4000 events), although the differences do not agree with the predicted values ($\chi^2_{\min}$), perhaps because of the complexity of the nonlinear fitting process. Linear least squares polynomial fits to appropriate data, such as the background distributions in Example 9.1, yield the expected differences between the area of the data and the fitted curves. See Exercise 9.1.

9.3 COMPOSITE PLOTS

Single Peak and Background

For a fitting function $y(x)$ that is separable into a peak $y_p(x)$ plus a background $y_b(x)$, such as Equation (9.1), it may be convenient to consider at least some facets of the fitting procedure separately. The least-squares procedure for minimizing χ^2 with respect to each of the parameters a_j,

$$\frac{\partial}{\partial a_j}\sum\left\{\frac{1}{\sigma_i^2}[y_i - y_b(x_i) - y_P(x_i)]^2\right\}^{=0} \tag{9.12}$$

can be considered equally well in terms of fitting the sum of the curves $y(x)$ to the total yield y_i or of fitting one function $y_p(x)$ to the difference spectrum $y_i' = y_i - y_b(x_i)$. The only provision is that the uncertainties in the data points of $\sigma_i' = \sigma_i$ must be the same in both calculations.

If the background curve can be assumed to be a slowly varying function under the peak, as in Figure 9.1, and may reasonably be interpolated under the peak from fitting on both sides, it may be preferable to fit the background curve $y_b(x)$ outside the region of the peak and to fit the peak function $y_p(x)$ only in the region of the peak.

Such a procedure might help isolate special problems that result from fitting with an incorrect peak or background. The χ^2 function measures not only the deviations of the parameters from an ideal fit, but also the discrepancy between the form chosen for the shape of the fitting function $y(x)$ and the parent distribution of the data. If the shape of the fitting function does not represent that of the parent distribution exactly, the value of χ^2 may have large contributions from local data regions. By fitting separate regions of a plot, it may be possible to discover whether the disagreement is in the background or the peak region. In the histogram of Figure 9.1, our interest is in the properties of the peak function, and not in the background, which we parameterize with a simple power series in E. However, the value of χ^2 for the fit is calculated for the entire plot and includes contributions from discrepancies between the background and the fitted curve, as well as between the peak and curve. We may be able to isolate problems to one or the other region by separating the fit into two parts.

Another reason for making separate fits to regions of a plot is to search for starting values for an overall fit. For example, when fitting a function that consists of peak functions plus background function, it may be useful first to fit the regions outside the peaks to get starting values for the background parameters and then to fit separately the region close to each peak, to find starting values for the peak parameters.

As an example, assume that we wish to find starting values for the fit of Equation (9.1) to the data of Figure 9.1. The following procedure could be used:

1. Separate the curve into three regions (*a*), (*b*), and (*c*) as indicated by the two vertical lines on Figure 9.1.
2. Fit the background polynomial $y_1(x) = a_1 + a_2E + a_3E^2$ simultaneously to regions below and above the peak to obtain provisional values for the parameters a_1 through a_3.
3. Fit the entire function of Equation (9.1) to the central region, with the fixed values of a_1 through a_3 obtained in step 2 to obtain values for the parameters a_4, a_5, and a_6.
4. Fit the entire function of Equation (9.1) simultaneously to regions (*a*) and (*c*), with the starting values of the parameters a_4 through a_6 set to the values obtained in steps 2 and 3 to obtain new values of the parameters a_1 through a_3.

If the parameters continue to change significantly on each iteration, the process can be repeated from step 2 as required. Alternatively, it may be sufficient to skip step 3 and to fit for all parameters after step 2.

In fitting the peak and background functions over different parts of the spectrum, it is important to note that the complete function $y(E)$ of Equation (9.1) must be fitted to both regions; that is, in the region outside the peak where the background is being fitted, the calculation of the tail of the peak must be included, and underneath the peak, the background terms must be included.

Multiple Peaks

Separation of closely spaced peaks is an important problem in many research fields. Although we should not attempt to extract information from our data by sorting in bins that are smaller than the uncertainties in our measurements, and should not use bin widths that are so narrow that the numbers of events in the bins are too small to satisfy Gaussian statistics, we also should not err in the other direction and risk suppressing important details. Selecting optimum bin sizes is critical. For some data samples, different bin widths for different regions of the data sample may be appropriate.

EXAMPLE 9.2 We have noted that, although the 4.4% probability for the fit to the data of Example 9.1 is rather low, it could be acceptable. However, because the data were plotted in rather coarse bins ($\Delta E = 0.1$ GeV), some information may have been suppressed. To check this possibility, we plotted the data in smaller bins ($\Delta E = 0.05$ GeV) as illustrated in Figure 9.2. (Note that in plotting Figure 9.2 we have eliminated some bins from the lower and upper edges of the histogram in order to enhance the display; all 60 bins are included in the fits.)

Plotted in smaller bins, the large peak near $E = 1.00$ GeV appears to be considerably narrower than indicated in Figure 9.1. There is also a suggestion of a possible excess of events in the bin centered at $E = 0.825$ GeV on the low-energy side of the main peak. As illustrated by the curve on Figure 9.2, a fit of the two-peak Equation (9.13) to the narrow-bin data, seems to confirm the existence of a second peak. To

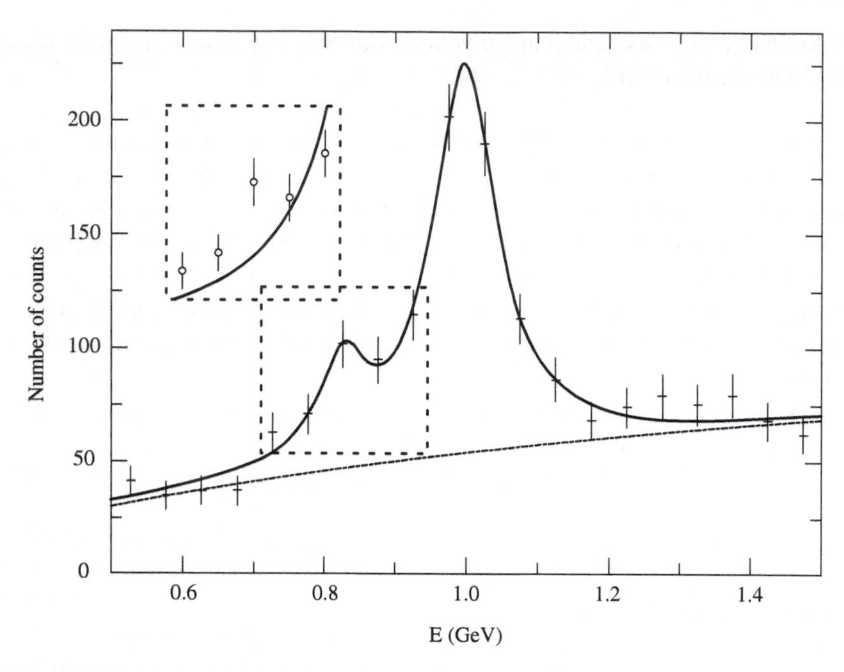

FIGURE 9.2
Histogram data in bins of 0.05 GeV of the 4000 simulated events shown in Figure 9.1. The solid curve illustrates a fit of Equation (9.13) to the data. The inset illustrates, in the region of the smaller peak, a fit of the single-peak Equation (9.1) to the entire data sample.

obtain this fit, we chose as starting values for the mass and width of the second peak, 0.825 and 0.05 GeV, respectively,

$$
\begin{aligned}
y(E) = a_1 + a_2E + a_3E^2 + A_0 \frac{\Gamma_0/(2\pi)}{(E - E_0)^2 + (\Gamma_0/2)^2} \\
+ A_1 \frac{\Gamma/(2\pi)}{(E - E_1)^2 + (\Gamma/2)^2}
\end{aligned}
\tag{9.13}
$$

suggested by examination of Figure 9.2.

Results of the fit are listed in column 6 of Table 9.1. The 29.4% chi-squared probability for this fit is a marked improvement from 4.4% for the single-peak fit. The inset on Figure 9.2 shows the region of the smaller peak with a curve calculated by fitting the single-peak Equation (9.1) to the entire data sample of Figure 9.2. Parameters determined in this fit are listed in column 5 of Table 9.1.

We can estimate the statistical significance of the smaller peak in Example 9.2 by counting the total number of events above the single-peak background (shown in the inset) and considering whether or not the excess is consistent with a statistical fluctuation. There are 102 events in the peak bin over a background of 69.5 events, corresponding to a fluctuation of $(102 - 69.5)\backslash\sqrt{69.5} = 3.9$ standard deviations in the background signal. Referring to Table C.2, we infer that there is a $(1 - 0.99990) = 0.00010$, or 0.01% probability that we should obtain a result this large, or larger, from a statistical fluctuation. Thus, the smaller peak appears to be very well established.

But we should wait before rushing into publication; our analysis is not finished. We calculated the probability of finding a 3.9 standard deviation fluctuation in a particular bin. However, there are 60 bins in this data sample, and the fluctuation could have appeared in any of them. The probability that a 3.9 standard deviation would *not* appear in any of the 60 pairs is 0.9999^{60}, so the probability of observing the fluctuation in any of the bin pairs is $1 - 0.9999^{60} \sim 0.6\%$. This probability is low enough to give us considerable confidence that the smaller peak is not a fluctuation. If we had some a priori reason, such as a theoretical prediction or evidence from another experiment, to believe that the smaller peak should be located in the particular energy region where it appears, then the argument against a statistical fluctuation would be even more compelling.

While there appears to be firm statistical support for a second peak in the data of Example 9.2, that support depends strongly on our understanding of the contributions in the region of the second peak from the smooth background distribution and the tail of the large peak. If, for example, background counts were 10% higher, decreasing the excess by 10%, the fluctuation would decrease from 3.9 to 2.9 standard deviations and the probability of a fluctuation of this magnitude in any bin would increase from about 1% to 20%, a considerably less compelling number.

Are there further tests we can make on our data sample to help us understand the significance of our result? For problems such as this, where the statistical significance of a result may be in question, the Monte Carlo method (Chapter 5) provides a powerful tool for more detailed examination. We shall use this technique in Chapter 11 to make a simple statistical test of these data. A full Monte Carlo program, which incorporates all the known or estimated details involved in the creation of the data sample, is invaluable in the planning and analysis of a serious experiment.

SUMMARY

Background subtraction:

$$y_P(x) = y(x) - y_b(x) \qquad (y_P \longrightarrow \text{peak};\ y_b \longrightarrow \text{background})$$

Uncertainty in area of peak:

$$\sigma_{A_P}^2 = \sigma_A^2 + \sigma_{A_b}^2 \qquad (\simeq A + A_B \text{ for Poisson statistics})$$

Area under fitted peak curve:

$$A_P = \int_{-\infty}^{+\infty} y_P(x)\, dx$$

Discrepancy in area under a curve with Poisson statistics:

$$\chi^2_{\min} = \sum \left[\frac{y_i}{\sigma_i^2} (y_i - y(x_i)) \right] \simeq \text{area(data)} - \text{area(fit)}$$

EXERCISES

9.1 The following data are drawn from the background distribution illustrated by the dashed curve in Figure 9.1 The data points correspond to the numbers of counts in 15 histogram bins, which are 0.2 GeV wide, each *centered on* the indicated value of E.

E	0.1	0.3	0.5	0.7	0.9	1.1	1.3	1.5	1.7	1.9	2.1	2.3	2.5	2.7
N	4	30	49	71	87	91	120	136	147	133	130	118	142	122

Plot a of the data.

Use a linear-fitting technique, such as those described in Chapter 7, to fit a second-order polynomial to these data. Assume statistical uncertainties in the counts. Compare the number of events in the histogram to the number determined by the fit. Is the difference consistent with the prediction of Equation (9.10)?

9.2 Find the area of the peak in Figure 9.1 by counting the area between the vertical dotted lines and subtracting the estimated background. Refer to the data in Exercise 9.4. Estimate the correction for the tails. Estimate the uncertainty in your determination of the area.

9.3 Refer to the data of Exercise 8.6. Fit the histogram by the method outlined in Section 9.3 with separate fits of the background second-order polynomial to the regions outside the peak and of the Gaussian function to the region of the peak.

9.4 The accompanying table lists the numbers of events in the histogram bins of Example 9.1 from $E = 0.0$ to 3.0 GeV in steps of 0.05 GeV.

(*a*) Fit Equation (9.1) to the data to obtain the parameters for this distribution. Compare to the values of the parameters listed in column 5 of Table 9.1.

(*b*) Repeat the fit with adjacent bins merged (i.e., combine bins 1 and 2, bins 3 and 4, etc.) and observe the effect on the value of χ^2, the determination of the area of the peak, and the determination of the mean and half-width of the peak. Assume statistical uncertainties.

7	2	6	12	15	18	31	29	27	27	41	35	37	37	63	71	102	95	115	202
190	113	86	68	74	79	75	79	68	62	69	81	79	85	87	68	70	89	77	70
71	62	85	62	73	70	59	61	77	61	62	73	67	71	75	66	73	71	71	49

CHAPTER 10

DIRECT APPLICATION OF THE MAXIMUM-LIKELIHOOD METHOD

The least-squares method is a powerful tool for extracting parameters from experimental data. However, before a least-squares fit can be made to a data set that consists of individual measurements or events, the events must be sorted into a histogram, which may obscure some detailed structure in the data. Because the least-squares method was derived from the principle of maximum likelihood, it might be better in some instances to use the maximum-likelihood method directly to compare experimental data to theoretical predictions, without the necessity of *binning* data into histograms with the corresponding loss of information.

We have already used the method in Chapter 4 to find estimates for the mean and standard deviation of data obtained in repeated measurements of a single variable, where we have assumed that the measurements were distributed according to Gaussian probability. Now, we extend the method to other distribution functions and to multiparameter fits. Maximum-likelihood methods can be applied directly to many "curve fitting" problems, and such fitting is almost as easy to use as the least-squares method, and considerably more flexible. However, the direct maximum-likelihood method requires computations for each *measured event,* rather than for each *histogram bin* as in least-squares fitting, and therefore the technique may be too slow for very large data samples.

Direct maximum-likelihood calculations have an advantage over the least-squares method for two particular types of problems: (1) low-statistics experiments with insufficient data to satisfy the requirement of Gaussian statistics for individual histogram bins and (2) experiments in which the fitting function corresponds to a different probability density function for each measured event so that binning the data leads to a reduction in information and a loss of sensitivity in determining the parameters. If the data set is sufficiently large, then the least-squares method can be applied to problems of either type, and that method is generally preferred in view of its smaller computing requirement. At any rate, it is not possible to extract more than minimal information from a very small data set, so we should expect the direct maximum-likelihood method to be most useful for intermediate problems with modest data samples.

10.1 INTRODUCTION TO MAXIMUM LIKELIHOOD

The basic maximum-likelihood procedure is relatively simple. Assume that we have a collection of N events corresponding to the measurement of an independent variable x_i and a dependent variable y_i, where i runs from 1 to N. We wish to obtain the parameters, $a_1, a_2, \ldots, a_m$, of a fitting function $y(x_i) \equiv y(x_i; a_1, a_2, \ldots, a_m)$ from these data. For each event, we convert $y(x_i)$ to a normalized probability density function

$$P_i \equiv P(x_i; a_1, a_2, \ldots, a_m) \tag{10.1}$$

evaluated at the observed value x_i. The likelihood function $\mathcal{L}(a_1, a_2, \ldots, a_m)$ is the product of the individual probability densities

$$\mathcal{L}(a_1, a_2, \ldots, a_m) = \prod_{i=1}^{N} P_i \tag{10.2}$$

and the maximum-likelihood values of the parameters are obtained by maximizing $\mathcal{L}(a_1, a_2, \ldots, a_m)$ with respect to the parameters.

In many experiments, the probability density function P_i will be made up of two components: a theoretical factor corresponding to the underlying principle being tested and an experimental factor corresponding to the biases introduced by experimental conditions.

EXAMPLE 10.1 In Example 5.7 we presented a Monte Carlo program for studying biases that could arise in an experiment to measure the mean life of the short-lived K_s^0 meson (or kaon). The example includes details of the experiment and Figure 5.4 illustrates schematically the experimental apparatus.

In brief, the experiment involves measuring the distance between the point of production and point of decay of the kaon, determining the meson's velocity, and calculating the meson's time of flight from production to decay. After correction for bias introduced by the finite size of the experimental apparatus, the mean life of the kaon could be determined from measurements of many such events.

The dashed rectangle on Figure 5.4 indicates the region in which events are collected, the fiducial region for the experiment. We select decay vertices only within this region to assure precise measurements of both the separation of the two vertices and the trajectories of secondary particles from decay of the kaon. These latter measurements determine the momentum, and thus the velocity, of the kaon. Loss of events that do not

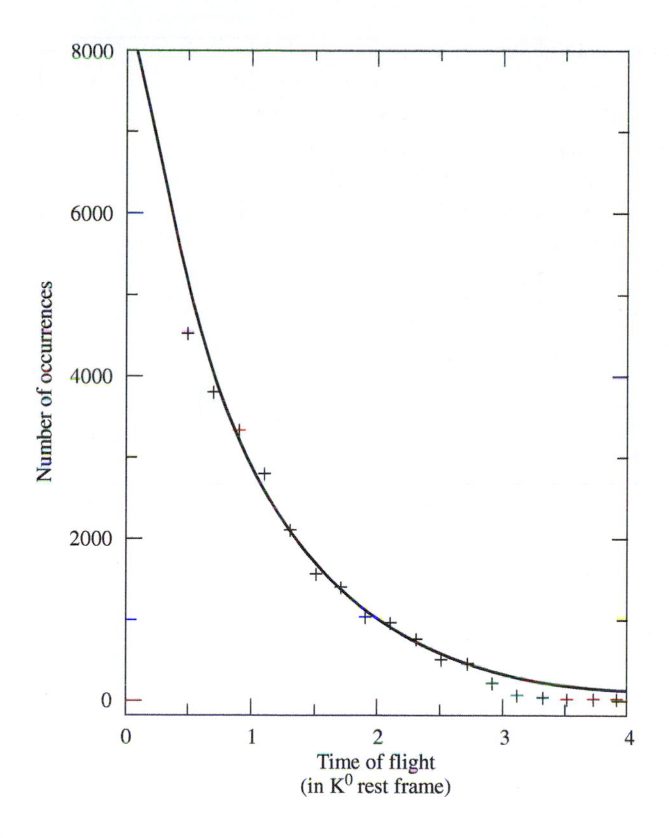

FIGURE 10.1
Frequency distribution of times of flight for 23,565 events that survived fiducial cuts in a 40,000-event Monte Carlo generation, as a function of the proper time (in units of 10^{-10} s). The exponential curve was calculated from the nominal value $\tau_K = 0.894 \times 10x^{-10}$ s to represent the expected distribution of the 40,000 generated events.

fall within the fiducial region bias the final calculation of the mean life and therefore we must understand the biases and make corrections.

In the following examples, we assume that the coordinates of the two vertices and the magnitude of the momentum of the decaying kaon have been determined.

We used the Monte Carlo program of Example 5.7, with the mean life of the kaon set to its nominal value of $\tau_K = 0.894 \times 10^{-10}$s, to generate 40,000 events in order to study the efficiency of the detector with reasonably high precision. It is important that the statistical uncertainties introduced in the determination of the efficiency function be negligible compared to the statistical and other uncertainties in the actual experiment. The distribution of the 23,565 generated events that survived fiducial cuts is shown as crosses in Figure 10.1 with the expected exponential distribution of the total 40,000-event sample shown as a smooth curve.

In Figure 10.2 we have plotted the resulting efficiency as a function of the times of flight of the kaons (the *proper time*) in their individual rest frames, with the efficiency function defined as the ratio of observed to expected events [or the point-by-point ratio $\epsilon(T) = N'(T)/N(T)$] from Equation (5.31). The dotted line in Figure 10.2 illustrates the region over which the efficiency reasonably may be assumed to be 100%.

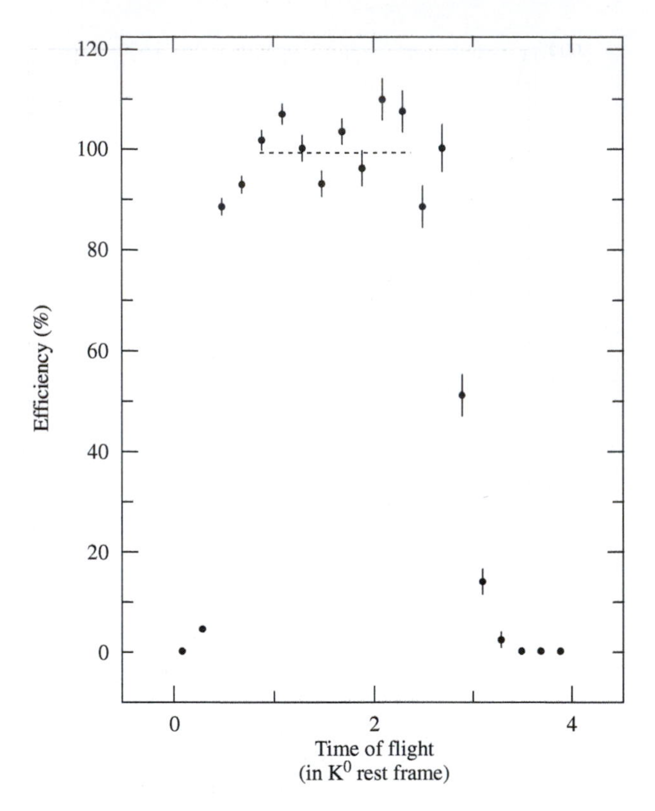

FIGURE 10.2
Efficiency function $\epsilon(T) = N(T)/N(T)$, calculated from the ratio of observed events (crosses) to expected events (smooth curve in Figure 10.1). The dotted line illustrates the region over which the efficiency reasonably may be assumed to be 100%.

We also used the Monte Carlo program, with different random-number seeds and the same nominal value of τ_K, to generate a small "data set" of 1000 events, of which 598 survived the fiducial cut, to use in testing our analysis procedures.

We shall discuss several aspects of the analysis of such data in the following examples.

EXAMPLE 10.1a: Least-squares Method Figure 10.3 shows on a semilogarithmic plot the distribution, as crosses (x), of the 598 events that survived the fiducial cuts from the total sample of 1000 events generated in Example 10.1. The straight line shows the expected distribution if there had been no efficiency losses. In order to extract the mean life of the kaon from these data, we apply the efficiency function illustrated in Figure 10.2 to correct for losses. The corrected data points are plotted in Figure 10.3 as data points with vertical error bars corresponding to the statistical uncertainties in the data, scaled by the efficiency factor. (Uncertainties in the correction factor were negligible.) The efficiency was assumed to be 100% in the region indicated by the horizontal dotted line in Figure 10.2. The very large error bars on "corrected" points at the two ends of the plot result from scaling low-statistics data points and illustrate the problem of using data in regions of low efficiency. Generally, it is

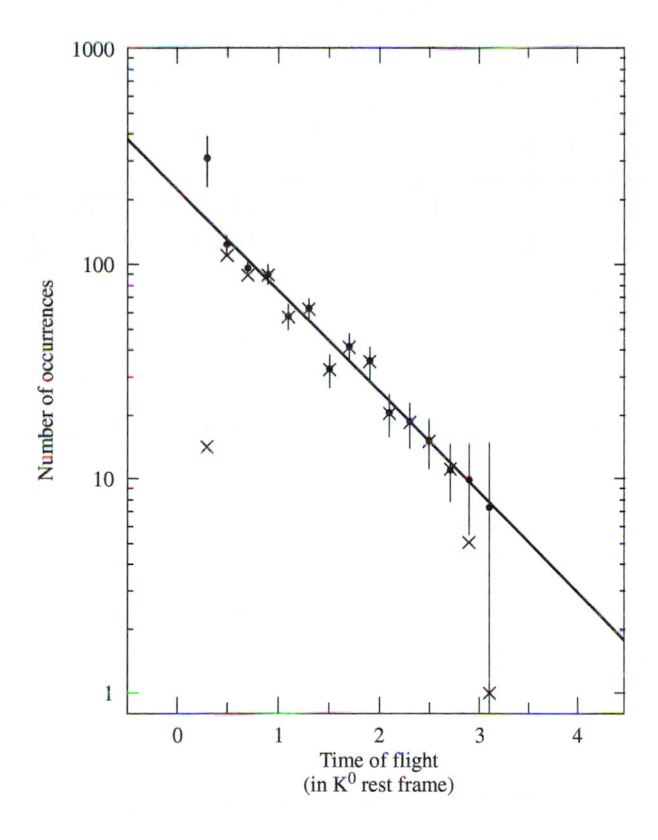

FIGURE 10.3
Semilogarithmic plot of the frequency distribution of 598 events that survived fiducial cuts from a 1000-event (Monte Carlo) data sample. The uncorrected data are shown as crosses; the data corrected for efficiency losses are shown as data points with error bars. The straight line shows the result of a linear least-squares fit to the corrected semilogarithmic data.

wise to eliminate points that require such large corrections from the sample, because they contribute little to the overall result and depend heavily on the corrections.

From the linear slope of the logarithmic plot, illustrated by the straight line through the data points, we obtain an "experimental" mean life $\tau = (0.925 \pm 0.058)$. Alternatively, we could have used a nonlinear least-squares fitting technique to determine τ directly from a linear plot of the data.

Direct Maximum Likelihood

Most actual experiments are more complex and have efficiency functions that are considerably more complicated than the one illustrated by our example. For such problems, application of direct maximum likelihood may be the preferable method for finding the best estimate of the parameters. To apply this method, we must define a probability function for each recorded event.

The probability of observing a single event that survives for a time t_i is

$$P_i = A_i p(t_i; \tau) \tag{10.3}$$

The first factor A_i represents the *detection efficiency,* or probability that the particle will decay within a predefined *fiducial volume* within our apparatus, so that a satisfactory measurement can be made of its flight time. This factor depends upon the coordinates of the production and decay vertices of the decaying particle, its momentum vector, and the geometry of the fiducial volume. The second factor $p(t_i; \tau)$ is proportional to the probability that a particle of mean lifetime τ will decay between time t_i and $t_i + dt$ and is therefore proportional to $e^{-t_i/\tau}$. Equation (10.3) becomes

$$P_i = A_i e^{-t_i/\tau} \tag{10.4}$$

It might appear that the two factors in Equation (10.3) are independent, so that the detection efficiency factor is independent of the decay probability, but, as we have observed in the previous example, this is not generally true. Because of the finite size of our measuring apparatus, we may preferentially lose events that survive for very short times so that we can't make precise measurements of their flight paths, as well as those that survive for very long times and therefore decay outside the acceptable limits of our detectors. Losses of both types depend upon the mean life that we are attempting to determine, the "τ" in the second factor of Equation (10.3). For each particle that is observed to decay within the apparatus, we can define a *potential path length* as the distance it would travel if it had not decayed. Because each decaying particle has a different potential path length, we must calculate geometric factors to correct for those particles that decay outside the detector. The correction factors will depend on the parameters and will be a function of the production and decay coordinates and the momentum vectors of each decaying particle. Clearly, one element of good experiment design should be to minimize the dependence of these geometric correction factors on the parameters sought in the experiment.

Normalization for Maximum Likelihood

The factor A_i in Equation (10.4) corresponds to a normalization for each measurement to assure unit probability for observing in this experiment *any* event that has the mean life, coordinates, and kinematics of the observed decaying particle. To determine the normalizing factor A_i we refer to Figure 5.4 and consider the fiducial volume of our apparatus, indicated by the dashed rectangle. From each particle's production coordinates and momentum vector, we can determine the minimum distance d_1 that the particle must travel to enter the region and the maximum distance d_2 it can travel before leaving the region. (We can, of course, observe some events outside the fiducial volume, but we reject them because they cannot be measured precisely.) These minimum and maximum distances d_1 and d_2 must be converted to times of flight t_1 and t_2 in the rest frame of the decaying particles, and the normalizing factors A_i can then be determined from the condition

$$\int_{t_1}^{t_2} P_i dt_i = A_i \int_{t_1}^{t_2} e^{-t_i/\tau} dt_i = 1 \tag{10.5}$$

With this normalization, the individual event probability P_i of Equation (10.4) becomes the probability density for observing a single event. The normalized joint probability or the likelihood function for observing N such events in our experiment is just the product of the individual probability functions:

$$\mathscr{L}(\tau) = \prod_{i=1}^{N} P_i = \prod_{i=1}^{N} A_i e^{-t_i/\tau} \tag{10.6}$$

Parameter Search

Our object is to find the value of the parameter τ that maximizes this likelihood function. Because the probability of observing any particular event is less than 1, the product of a large number of such probabilities (one for each measured event) may be a very small number, and may, in fact, be too small for the computer to handle. To avoid problems, it is usually preferable to maximize the logarithm of the likelihood function

$$M = \ln \mathscr{L} \tag{10.7}$$

rather than the likelihood function itself, so that the product of Equation (10.6) becomes a sum. The logarithms should be reasonable, negative numbers. For our particular example, the logarithm of the likelihood function of Equation (10.6) is given by

$$M(\tau) = \ln[\mathscr{L}(\tau)] = \sum \left[\ln A_i - \frac{t_i}{\tau} \right] \tag{10.8}$$

with A_i defined by Equation (10.5). Note that A_i is a function of the unknown parameter τ, as well as of the production coordinates, momentum vector, and fiducial volume, and must be calculated separately for each event, *and for every trial value of* τ.

In general, this problem, like the corresponding nonlinear least-squares fitting problem, cannot be solved in closed form. However, either the grid- or gradient-search method of minimizing the χ^2 function discussed in Chapter 8 can be adopted directly. It is only necessary to search for a maximum of M (or a minimum value of $-M$) with the same routines we used in Chapter 8 to find a minimum of χ^2.

We may note a correspondence between the quantity $M(\tau)$, determined in Equation (10.7) from the likelihood function for *individual events,* and the goodness-of-fit parameter χ^2, determined by Equation (8.7) from the likelihood function $P(a)$ for *binned data:*

$$\chi^2 = -2\ln[\mathscr{L}(\tau)] + \text{constant} \tag{10.9}$$

In the limit of a large number of events, the two methods must yield the same value τ' for the maximum-likelihood estimate of the parameter τ. In both cases the likelihood function will be a Gaussian function of the parameter near the optimum value

$$\mathscr{L}(\tau) \propto \exp\left(-\frac{(\tau - \tau')^2}{2\sigma^2}\right) \tag{10.10}$$

so we can expect $M(\tau)$, like $\chi^2(\tau)$, to vary quadratically with the parameter τ in the vicinity of τ'.

EXAMPLE 10.1b Let us consider the simplest form of this problem. Assume that the unknown mean lifetime is sufficiently short so that our apparatus is large enough to include many lifetimes and, therefore, the loss of particles that decay at very long times is negligible. Let us also assume that our equipment can detect particles at very short as well as very long times. Then the limits on the normalization integral of Equation (10.5) become $t_1 = 0$ and $t_2 = \infty$ and A_i is the same for every event and is given by $A_i = 1/t$. The likelihood function becomes

$$\mathcal{L}(\tau) = \prod A_i e^{-t_i/\tau} = \prod \frac{e^{-t_i/\tau}}{\tau} \tag{10.11}$$

with logarithm

$$M(\tau) = \ln[\mathcal{L}(\tau)] = -\frac{1}{\tau}\sum t_i - N \ln \tau \tag{10.12}$$

We can obtain the maximum of Equation (10.12) by taking the derivative of $M(\tau)$ with respect to τ and setting it to 0:

$$\begin{aligned} \frac{dM(t)}{d\tau} &= \frac{d}{d\tau}\left\{-\frac{1}{\tau}\sum t_i - N \ln \tau\right\} \\ &= \frac{1}{\tau^2}\sum t_i - \frac{N}{\tau} = 0 \end{aligned} \tag{10.13}$$

The solution is $\tau = \Sigma t_i/N$; that is, the maximum-likelihood estimate of the mean life is just the mean of the individual lifetime measurements. We should have reached the same result if we had found the maximum of $\mathcal{L}(t)$ from Equation (10.11).

EXAMPLE 10.1c Suppose that we repeat the experiment, but with poorer experimental resolution so that we cannot distinguish the decay vertex (x_2, y_2, z_2) from the creation vertex (x_1, y_1, z_1) unless they are separated by a distance d_1. For simplicity, we assume that the decaying particles are all produced with the same velocity, so that the lower cutoff distance d_1 translates into the same lower cutoff in time t_1 for all events. (In an actual experiment, of course, the decaying particles would be produced with various velocities, so that the calculated lower cutoff time t_1 would vary from event to event.)

For this example, the normalization integral of Equation (10.5) becomes

$$A_i \int_{t_1}^{\infty} e^{-t_i/\tau} dt_i = 1 \tag{10.14}$$

which gives

$$A_i = \frac{e^{t_1/\tau}}{\tau} \tag{10.15}$$

The likelihood function becomes

$$\mathcal{L}(\tau) = \prod_{i=1}^{N} A_i e^{-t_i/\tau} = \prod_{i=1}^{N} \frac{e^{t_1/\tau}}{\tau} e^{-t_i/\tau} = \prod_{i=1}^{N} \frac{e^{(t_1 - t_i)/\tau}}{\tau} \tag{10.16}$$

so that

$$M = \ln \mathscr{L} = \sum \frac{[t_1 - t_i]}{\tau} - \sum \ln \tau \tag{10.17}$$

Setting

$$\frac{dM(\tau)}{d\tau} = 0 \tag{10.18}$$

gives

$$\frac{d}{d\tau} \sum \left\{ \frac{[t_1 - t_i]}{\tau} - \ln \tau \right\} = - \sum \left\{ \frac{t_1 - t_i}{\tau^2} \right\} - \frac{N}{\tau} = 0 \tag{10.19}$$

or

$$\tau = \frac{\Sigma[t_i - t_1]}{N} = \frac{\Sigma t_i}{N} - t_1 \tag{10.20}$$

As we should expect, the lifetime τ would have been overestimated if we had neglected to take account of the cutoff at short times.

EXAMPLE 10.1d Let us consider a more realistic problem in which we have both short and long cutoffs on the observable path. We also assume that the unstable particles are produced at various locations within the target and with various momentum vectors **p**.

For this example, we must calculate the normalization integral, Equation (10.5), separately for each event with individual values for t_1 and t_2 determined from the minimum and maximum distance cutoffs, d_1 and d_2, respectively. The resulting expression for the likelihood function is

$$\mathscr{L}(\tau) = \prod_{i=1}^{N} A_i e^{-t_i/\tau} = \prod_{i=1}^{N} \left[\frac{e^{-t_i/\tau}}{\tau[e^{-t_1/\tau} - e^{-t_2/\tau}]} \right] \tag{10.21}$$

with

$$M(\tau) = \ln[\mathscr{L}(\tau)]$$

Setting to zero the derivative of $M(\tau)$ with respect to τ gives us the equation for the maximum-likelihood value of τ. However, the resulting equation cannot be solved analytically for τ although it could be solved by interpolation (see Appendix A). We choose, rather, to maximize $M(\tau)$ by a one-dimensional grid-search method because search methods are more generally applicable to maximum-likelihood problems and can readily be extended to multiple parameter problems.

10.2 COMPUTER EXAMPLE

Sample Maximum Likelihood Fit

We use the program MAXLIKE to select and analyze the 598 events that survived the fiducial area cuts, from the 1000-event uncorrected data sample generated in

Example 10.1a. The events were generated with $\tau_K = 0.894 \times 10^{-10}$ s and the distribution of the selected events is illustrated by the crosses in Figure 10.3.

Program 10.1 MAXLIKE (Appendix E) A grid-search method to maximize the logarithm of the likelihood function of Equation (10.21). The routines have been written specifically for Example 10.1d.
STARTUP sets the range of the parameter TAU for the search.
FETCHDATA assigns the input data file, reads the limits of the fiducial region (d_1 and d_2), reads data for individual events.
SEARCH sets and increments TAU and calls LOGLIKE, which returns the logarithm of the likelihood function M. Compares each calculated value of M to the preceding value. Terminates the search when M stops increasing and starts to decrease, indicating that M has passed through a local maximum. At termination, fits a parabola to the last three points to find a better estimate of TAU at the maximum.
LOGLIKE calls LOGPROB to find the logarithm of the probability density for each event; sums to calculate the logarithm of the likelihood function.
LOGPROB calculates the logarithm of the probability density for an event.
ERROR calculates the uncertainty SIGTAU in TAUATMIN, the maximum likelihood value of the parameter TAU, by finding the change in TAU needed to decrease M by $\Delta M = 1/2$.
PLOTLIKECURVE (Not listed) calculates and plots the shape of the likelihood function in the region of the maximum. Plots a Gaussian curve with mean and standard deviation equal to TAUMIN and DTAU.

Grid-Search Solution

At each step the program increments τ by a preset amount $\Delta\tau$ and repeats the calculation until $M(\tau)$ has passed through a maximum and has started to decrease. The program fits a parabola to the three points that bracket the maximum to find the value τ' at the maximum of $M(\tau)$. For a more detailed problem, the program could be written to repeat the calculation with smaller values of $\Delta\tau$ to find a better estimate of τ', as in the fitting examples in Chapter 8. Either the grid- or gradient-search method of Chapter 8 could be adapted to solve multiparameter problems.

Results of the Fit

We analyzed the data set twice: first with data selected in the nominal fiducial region (10 cm to 40 cm), which gave $\tau' = (0.943 \pm 0.059) \times 10^{-10}$s for the 598 events that survived the cut, and then, to test the sensitivity of the calculation to our choice of fiducial region, with data selected in the less-appropriate fiducial region with $d_1 = 10$ cm and $d_2 = 20$ cm, which gave $\tau' = (0.78 \pm 0.14) \times 10^{-10}$s for the 373 events that survived this cut. Plots of the relative values of the likelihood function versus trial values of the parameter τ are shown as crosses in Figure 10.4a for the data selected in the nominal fiducial region and in Figure 10.4b for data selected in the less-appropriate fiducial region. As expected, the incorrect fiducial region clearly selects

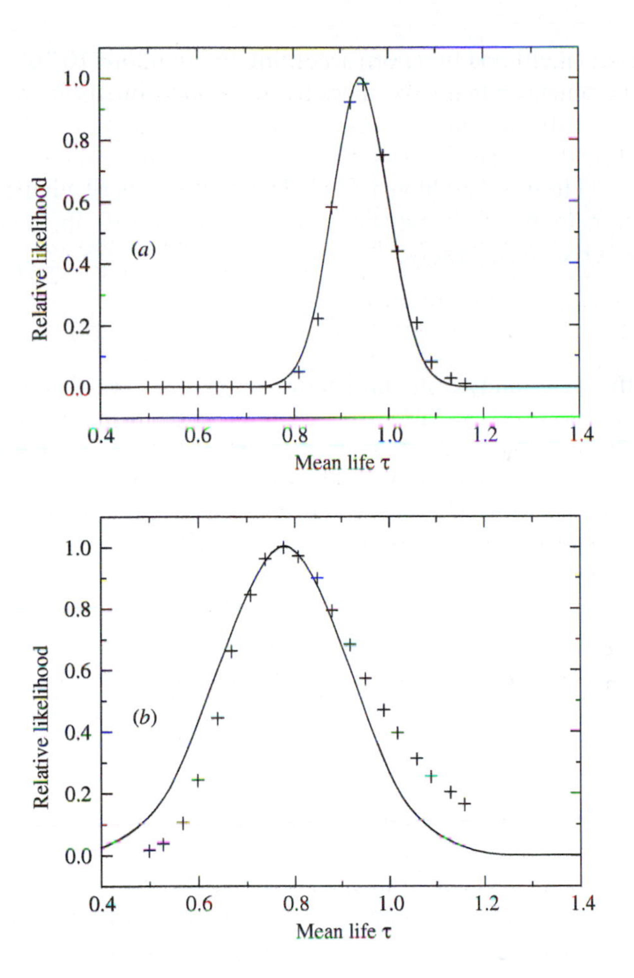

FIGURE 10.4
Relative values of the likelihood function versus trial values of the parameter for events that passed the fiducial cuts for the decay vertex. The data points are indicated by crosses; the smooth Gaussian curves were calculated from Equation (10.10) with the values of the means and standard deviations obtained in the two fits. (*a*) Nominal fiducial cuts: 10 − 40 cm; 598 events survived; $\tau' = 0.943 \times 10^{-10}$ s, $\sigma = 0.059 \times 10^{-10}$ s. (*b*) Incorrect fiducial cuts: 10 − 20 cm; 373 events survived; $\tau' = 0.78 \times 10^{-10}$ s, $\sigma = 0.14 \times 10^{-10}$ s.

fewer events and, therefore, gives a less-precise result. In an actual experiment, we should have to consider a trade-off between the number of surviving events in the sample, and the precision with which those surviving events could be measured, and choose our fiducial region to maximize the overall quality of the result.

We observed that, for a sufficiently large event sample, the likelihood function should become Gaussian in the parameters in the vicinity of a χ^2 minimum (or a maximum of the likelihood function) according to Equation (10.10), where τ' is the value of the parameter τ that maximizes the likelihood function. We show on Figures 10.4a and 10.4b Gaussian curves calculated from Equation (10.10), with τ' and σ determined by the respective fits. Both the data points and the Gaussian curves have been scaled to unit height at $\tau = \tau'$. The data points of Figure 10.4a closely follow the curve; in the lower statistics example in Figure 10.4b, the data points depart from the curve considerably.

Uncertainties

To estimate the uncertainty σ in our determination of τ', we found the change in τ necessary to decrease M by $\Delta M = 1/2$ from its value at the maximum τ' (corresponding to an increase of χ^2 by 1 or a change of $e^{-1/2}$ in the likelihood function $\mathscr{L}$). Because the likelihood function for the larger sample (Figure 10.4a) closely followed the Gaussian form, our estimate of the uncertainty should be satisfactory. However, the smaller sample (Figure 10.4b) was skewed from the Gaussian, so that our estimate of the standard deviation might be somewhat low. For multiparameter fits it is often useful to plot contours of χ^2 (or of M) as a function of pairs of the parameters to study the uncertainties. (See Chapter 11.)

There are several other ways to estimate the uncertainty in a parameter after performing a maximum-likelihood fit. If the distribution of the likelihood function is sufficiently close to a Gaussian, we can find σ_τ from Equation (8.11):

$$\sigma_\tau^2 = \left(\frac{\partial^2 M(\tau)}{\partial \tau^2}\right)^{-1} \tag{10.22}$$

If it is not possible to calculate Equation (10.22) exactly (although it is possible for our example), we can find the second derivative by taking finite differences as discussed in Appendix A.

If the likelihood function does not follow the Gaussian distribution, we can try a numerical integration of the likelihood function to find limiting values that include ~68.3% of the total area, corresponding to the 1 standard deviation limit. Alternatively, we may use a method suggested by Orear (1958) who points out that, for small event samples, where the likelihood function may not be very Gaussianlike, it may be preferable to calculate an average value of the second derivative through the equation

$$\overline{\frac{\partial^2 M}{\partial a^2}} = \frac{\int [\partial^2 M/\partial a^2]\, \mathscr{L}(a)\, da}{\int \mathscr{L}(a)\, da} \tag{10.23}$$

where a is the unknown parameter and the integrals are over the allowable range of the parameter. This procedure has the advantage over the method of Equation (10.22) of giving more weight to the tails of the distribution in cases where they drop off more slowly than those of a Gaussian curve.

Another method of determining the uncertainties in the parameters is to use a Monte Carlo calculation to produce simulated data sets, comparable to our measured data, and to use the method outlined in Chapter 11 for determining confidence levels for our results. This method has the advantage that it depends only on the assumptions made in the Monte Carlo generation, and not on any statistical expectations about the shape of the likelihood function. In many experiments, especially those with low statistics, it provides the most reliable estimate of parameter uncertainties.

Goodness of Fit

One disadvantage of the direct maximum-likelihood method is that it does not provide a convenient test of the quality of the fit. The value at the peak of the likelihood function itself is not useful because it represents only the maximized probability for obtaining our particular experimental result and we have no way of predicting the expected probability.

An estimate of the goodness of fit can be obtained by making a histogram of the data and comparing it to a prediction based on our best estimate of the parameters. A Monte Carlo simulation of the experiment may be required to calculate the predicted distribution, with a χ^2 test to compare the data to the prediction.

It is not always clear just which data variable should be histogrammed for this purpose. We would like to find that variable on which the parameters depend most strongly. For our sample problem, the lifetime τ in the rest frames of the particles is an obvious choice, because that is the variable we would choose if we were to solve the problem by the least-squares method. However, it might be wise to try plots of several variables to be sure that the fit is satisfactory. To test, we could generate with our Monte Carlo program a large sample of events based on the parameters discovered in each search, apply the fiducial cuts, and calculate χ^2 from the agreement between the Monte Carlo results and our data sample. We should be aware that, because we did not actually minimize χ^2 for the experimental distribution with respect to the parameters, a satisfactory value of χ^2 may be at best an indication that nothing is drastically wrong with the solution.

SUMMARY

Normalized probability density function:

$$P_i \equiv P(x_i, a_1, a_2, \ldots, a_m)$$

Likelihood function:

$$\mathscr{L}(a_1, a_2, \ldots, a_m) = \prod_{i=1}^{N} P_i$$

Single-event probability density: $P_i = A_i \cdot p(x_i; a)$ where A_i is the detection efficiency and $p(x_i; a)$ is proportional to the interaction probability

Logarithm of likelihood function: $M = \ln \mathscr{L} = \Sigma \ln P_i$

Maximization of $\mathscr{L}$ *or of* M: $\partial \mathscr{L}/\partial a_j = 0$ or $\partial M/\partial a_j = 0$ for all a_j

Gaussian form of likelihood function for large data sample:

$$\mathcal{L}(a_j) \propto \exp\left(-\frac{(a_j - a_j')^2}{2\sigma^2}\right)$$

Uncertainties in parameters:

$$\sigma_j^2 = \left(\frac{\partial^2 M(a_j)}{\partial a_j^2}\right)^{-1}$$

Method for low statistics:

$$\overline{\frac{\partial^2 M}{\partial a^2}} = \frac{\int [\partial^2 M/\partial a^2]\, \mathcal{L}(a)\, da}{\int \mathcal{L}(a)\, da}$$

EXERCISES

10.1. In a scattering experiment, the angles of the scattered particles are measured and the cosines of the angles in the center-of-mass rest frame of the incident and target particles are calculated and recorded. Fifty such measurements, drawn from the distribution $y(x) = a_1 + a_2 \cos^2\theta$, are listed in the table. Use the direct maximum-likelihood method to determine the values of the parameters a_1 and a_2. Note that it is necessary to convert the distribution function $y(x_i)$ to a normalized probability function and that the normalization constant will be different for each pair of trial values of a_1 and a_2.

−0.999	−0.983	−0.956	−0.946	−0.933	−0.925	−0.916	−0.910
−0.881	−0.739	−0.734	−0.717	−0.715	−0.675	−0.665	−0.649
−0.621	−0.537	−0.522	−0.508	−0.499	−0.471	−0.460	−0.419
−0.403	−0.311	−0.305	−0.281	−0.170	−0.162	−0.063	0.214
0.438	0.444	0.508	0.586	0.638	0.677	0.721	0.730
0.768	0.785	0.790	0.793	0.877	0.896	0.931	0.938
0.948	0.993						

Because of the small amount of data, the uncertainties in the parameters a_1 and a_2 are so large that the values of the parameters are not very meaningful. Therefore, to complete the problem, you should use the Monte Carlo program written for Exercise 5.8 to generate 500 events and use your calculation to find the parameters from those data.

10.2. Students in an undergraduate physics laboratory determined the mass of the Λ hyperon by measuring graphically the energies and the momentum vectors of the proton and π meson into which the Λ hyperons decayed. Because of the large uncertainties in the measurements, the calculated square of the masses of the decaying particles forms a truncated Gaussian distribution that is limited on the low-mass side by $(M_p + M_\pi)^2 = 1.1617\ (\text{GeV}/c^2)^2$, but is not limited on the high-mass side. The following 50 numbers represent squares of the calculated masses in units of $(\text{GeV}/c^2)^2$.

1.2981	1.2618	1.2145	1.2539	1.4230	1.3963	1.3701	1.2303	1.3655	1.2042
1.3190	1.2086	1.2118	1.2078	1.2726	1.2438	1.1838	1.1666	1.1908	1.1922
1.2525	1.3615	1.1855	1.2697	1.2044	1.3397	1.4317	1.2713	1.2203	1.2817
1.2046	1.2856	1.1980	1.2595	1.1721	1.2608	1.1689	1.4838	1.1743	1.2954
1.2586	1.2655	1.2316	1.2372	1.2969	1.2015	1.2000	1.1677	1.2080	1.1893

Use the direct maximum-likelihood method to fit a truncated Gaussian to these data to determine the maximum-likelihood value of the mass of the squared particle. A search in two-parameter space will be required since neither the mean nor the width of the distribution is known.

Note that it is necessary to calculate numerically the normalization of the truncated Gaussian for each pair of trial values of the mean and standard deviation of the Gaussian function. It is advisable to set up a table of the integral of the standard Gaussian and to use interpolation to find the desired normalizations. A simple automatic or manual grid search will suffice for maximizing the likelihood function.

10.3. Use Program 5.4 (available on the website) to generate 1000 sample kaon decay events with nominal mean life $\tau = 0.894 \times 10^{-10}$ s.

(*a*) Plot a histogram of the times of flight of all the generated kaons in their own rest frames (proper times).

(*b*) Use Program 10.1 (available on the website), with nominal fiducial cuts on your data ($d_1 = 10.0$ cm and $d_2 = 40$ cm) to repeat the analysis of Example 10.1d to find the maximum likelihood solution τ' for the kaon mean life. Plot a histogram of the events that survive the cuts.

(*c*) With the value of τ', which you determined in part (*b*), and random number seeds that are different from those used in part (*a*), generate 20,000 events to serve as your estimate of the parent distribution. Apply the nominal fiducial cuts to these data and plot a histogram of the data in the same bins as you used in part (*b*).

(*d*) Calculate χ^2 for the agreement between your "experimental" histogram and the surviving events from the "parent" distribution. If the numbers of events in your bins of the parent distribution are large enough, their uncertainties can be ignored in this calculation. If they are not, you must use the combined statistical errors of the two distributions when calculating χ^2.

CHAPTER

11

TESTING THE FIT

11.1 χ^2 TEST FOR GOODNESS OF FIT

The method of least squares is based on the hypothesis that the optimum description of a set of data is one that minimizes the weighted sum of the squares of the deviation of the data y_i from the fitting function $y(x_i)$. The sum is characterized by the variance of the fit s^2, which is an *estimate* of the variance of the data σ^2. For a function $y(x_i)$, which is linear in m parameters and is fitted to N data points, we have

$$s^2 = \frac{1}{N-m} \frac{\Sigma\{(1/\sigma_i^2)[y_i - y(x_i)]^2\}}{(1/N)\Sigma(1/\sigma_i^2)} = \frac{1}{N-m} \sum w_i [y_i - y(x_i)]^2 \tag{11.1}$$

where the factor $\nu = N - m$ is the number of degrees of freedom for fitting N data points (implied in the unlabeled sums) with m parameters and the weighting factor for each measurement is given by

$$w_i = \frac{1/\sigma_i^2}{(1/N)\Sigma(1/\sigma_i^2)}, \tag{11.2}$$

the inverse of the variance σ_i^2 that describes the uncertainties in each point, normalized to the average of all the weighting factors.

The variance of the fit s^2 is also characterized by the statistic χ^2 defined in Equation (7.5) for polynomials:

$$\chi^2 \equiv \sum \left\{ \frac{1}{\sigma_i^2} [y_i - y(x_i)]^2 \right\} \tag{11.3}$$

with

$$y(x_i) = \sum_{k=1}^{m} a_k f_k(x_i)$$

The relationship between s^2 and χ^2 can be seen most easily by comparing s^2 with the reduced chi-square χ_ν^2,

$$\chi_\nu^2 = \frac{\chi^2}{\nu} = \frac{s^2}{\langle \sigma_i^2 \rangle} \tag{11.4}$$

where $\langle \sigma_i^2 \rangle$ is the weighted average of the individual variances

$$\langle \sigma_i^2 \rangle = \frac{(1/N)\Sigma((1/\sigma_i^2)\sigma_i^2)}{(1/N)\Sigma(1/\sigma_i^2)} = \left[\frac{1}{N}\sum \frac{1}{\sigma_i^2}\right]^{-1} \tag{11.5}$$

and is equivalent to σ^2 if the uncertainties are all equal, $\sigma_i = \sigma$.

The parent variance of the data σ^2 is a characteristic of the dispersion of the data about the parent distribution and is not descriptive of the fit. The estimated variance of the fit s^2, however, is characteristic of both the spread of the data and the accuracy of the fit. The definition of χ^2, as the ratio of the estimated variance s^2 to the parent variance σ^2 times the number of degrees of freedom ν, makes it a convenient measure of the goodness of fit.

If the fitting function is a good approximation to the parent function, then the estimated variance s^2 should agree well with the parent variance σ^2, and the value of the reduced chi-square should be approximately unity, $\chi_\nu^2 = 1$. If the fitting function is not appropriate for describing the data, the deviations will be larger and the estimated variance will be too large, yielding a value of χ_ν^2 greater than 1. A value of χ_ν^2 less than 1 does not necessarily indicate a better fit, however; it is simply a consequence of the fact that there exists an uncertainty in the determination of s^2, and the observed values of χ_ν^2 will fluctuate from experiment to experiment. A value of χ_ν^2 that is very small may indicate an error in the assignment of the uncertainties in the measured variables.

Distribution of χ^2

The probability distribution function for χ^2 with ν degrees of freedom is given by

$$p_\chi(x^2; \nu) = \frac{(x^2)^{1/2(\nu-2)} e^{-x^2/2}}{2^{\nu/2}\Gamma(\nu/2)} \tag{11.6}$$

The chi-square distribution of Equation (11.6) is derived in many texts on statistics[1] but we shall simply quote the results here.

The gamma function $\Gamma(n)$ is equivalent to the factorial function $n!$ extended to nonintegral arguments. It is defined for integral and half-integral arguments by the values at arguments of 1 and ½ and a recursion relation:

[1]See Pugh and Winslow (1966), Section 12-5.

$$\Gamma(1) = 1 \qquad \Gamma(\tfrac{1}{2}) = \sqrt{\pi} \qquad \Gamma(n-1) = n\Gamma(n)$$

For integral values of n

$$\Gamma(n+1) = n! \qquad n = 0, 1, \ldots$$

For half-integral values of n

$$\Gamma(n+1) = n(n-1)(n-2)\cdots(\tfrac{3}{2})(\tfrac{1}{2}\sqrt{\pi}) \qquad n = \tfrac{1}{2}, \tfrac{3}{2}, \tfrac{5}{2}, \ldots \qquad ((11.7)$$

Calculating factorial functions can lead to computer overflow problems. For computational purposes it is convenient to replace the factorial form of the gamma function by a form of Stirling's approximation[2]:

$$\Gamma[n] = \sqrt{2\pi} e^{-n} n^{(n-1/2)} (1 + 0.0833/n) \qquad (11.8)$$

This approximation, which is accurate to ~0.1% for all $n \geq \tfrac{1}{2}$, avoids both the problems of overflow in calculating factorials and the necessity of testing and choosing the appropriate form for integral or half-integral argument. The trade-off is computer speed. Calculating exponentials may be slower than calculating factorials, but high speed usually is not required for nonrepetitive calculations.

If the function of the parent population is denoted by $y_0(x)$, the value of χ_0^2 determined from the parameters of the parent function

$$\chi_0^2 = \sum\left\{\frac{1}{\sigma_i^2}[y_i - y_0(x_i)]^2\right\} \qquad (11.9)$$

is distributed according to Equation (11.6) with $\nu = N$ degrees of freedom. If the function $y(x)$ used in the determination of χ^2 contains m parameters, the value of χ^2 calculated from Equation (11.3) is distributed according to Equation (11.6) with $\nu = N - m$ degrees of freedom.

More useful for our purposes than the probability density distribution $p_\chi(x^2; \nu)$ of Equation (11.6) is the integral probability $P_\chi(\chi^2; \nu)$ between $x^2 = \chi^2$ and $\chi^2 = \infty$:

$$P_\chi(\chi^2; \nu) = \int_{\chi^2}^{\infty} P_\chi(x^2; \nu)\, dx^2 \qquad (11.10)$$

Equation (11.10) describes the probability that a random set of n data points drawn from the parent distribution would yield a value of χ^2 equal to or greater than the tabulated value.

Program 11.1. CHI2PROB (Appendix E) χ^2-probability.
CHIPROBDENS computation of the function $p_\chi(\chi^2; \nu)$ [Equation (11.6)] using function GAMMA to approximate the gamma function.
CHIPROB Numerical calculation of the integral, Equation (11.10), by Simpson's rule. If variable overflow is a problem, double-precision variables could be employed.

---

[2]"Review of Particle Properties" (1986), p. 53.

The calculation returns the integral to an accuracy of about ±0.1%. The trade-off on accuracy versus speed of computation is controlled by the value of the constant DX, the integration step.

For the special case of 1 degree of freedom, $\nu = 1$, the χ^2-probability density function of Equation (11.6) takes the form

$$p_\chi(x^2; \nu) = e^{-x^2/2}/(2\pi x^2)^{1/2}$$

which is difficult to integrate numerically near $x = 0$. However, the integral is finite, and the function can be expanded in a Taylor series about $x = 0$ and integrated analytically. We use that technique for $\nu = 1$ and $\chi^2 < 2$.

Similarly, for $\nu = 2$, where the function takes the form

$$p_\chi(x^2; \nu) = e^{-x^2/2}/2$$

the analytic form of the integral is used.

For a fitting function that is a good approximation to the parent function, the experimental value of χ^2_ν should be close to one and the probability from Equation (11.10) should be approximately 0.5. For poorer fits, the values of χ^2_ν will be larger and the associated probability will be smaller. There is an ambiguity in interpreting the probability because χ^2_ν is a function of the quality of the data as well as the choice of parent function, so that even correct fitting functions occasionally yield large values of χ^2_ν. However, the probability of Equation (11.10) is generally either reasonably close to 0.5, indicating a reasonable fit, or unreasonably small, indicating a bad fit. In fact, for most purposes, the reduced chi-square χ^2_ν is an adequate measure of the probability directly. The probability will be reasonably close to 0.5 so long as χ^2_ν is reasonably close to 1; that is, less than about 1.5.

Example 11.1. Consider the solution of the problem of fitting two exponential curves plus a linear background to the data from the radioactive silver decay of Example 8.1. The fit (see Table 8.5) gave $\chi^2 = 66.1$ for 54 degrees of freedom, or $\chi^2_\nu = 1.22$, with $P_\chi(\chi^2; \nu) = 12.4\%$. We can interpret this result in the following way. Assume that the parameters we found are, indeed, the parameters of the parent distribution. Then, suppose that we were to repeat our experiment many times, drawing many different data samples from that parent distribution. Our result indicates that in 12.4% of those experiments we should expect to obtain fits that are no better than that listed in Table 8.5.

11.2 LINEAR-CORRELATION COEFFICIENT

Let us assume that we have made measurements of pairs of quantities x_i and y_i. We know from the previous chapters how to fit a function to these data by the least-squares method, but we should stop and ask whether the fitting procedure is justified and whether, indeed, there *exists* a physical relationship between the variables x and y. What we are asking here is whether or not the variations in the observed values of one quantity y are *correlated* with the variations in the measured values of the other quantity x.

For example, if, as in Example 6.1, we were to measure the potential difference across segments of a current-carrying wire as a function of the segment length, we should find a definite and reproducible correlation between the two quantities. But if we were to measure the potential of the wire as a function of time, even though there might be fluctuations in the observations, we should not find any significant reproducible long-term relationship between the pairs of measurements.

On the basis of our discussion in Chapter 6, we can develop a quantitative measure of the degree of correlation or the probability that a linear relationship exists between two observed quantities. We can construct a linear-correlation coefficient r that will indicate quantitatively whether or not we are justified in determining even the simplest linear correspondence between the two quantities.

Reciprocity in Fitting x Versus y

Our data consist of pairs of measurements (x_i, y_i). If we consider the quantity y to be the dependent variable, then we want to know if the data correspond to a straight line of the form

$$y = a + bx \tag{11.11}$$

We have already developed the analytical solution for the coefficient b, which represents the slope of the fitted line given in Equation (6.12):

$$b = \frac{N\Sigma x_i y_i - \Sigma x_i \Sigma y_i}{N\Sigma x_i^2 - (\Sigma x_i)^2} \tag{11.12}$$

where the weighting factors in σ_i have been omitted for clarity. If there is no correlation between the quantities x and y, then there will be no tendency for the values of y to increase or decrease with increasing x, and, therefore, the least-squares fit must yield a horizontal straight line with a slope $b = 0$. But the value of b by itself cannot be a good measure of the degree of correlation because a relationship might exist that included a very small slope.

Because we are discussing the interrelationship between the variables x and y, we can equally well consider x as a function of y and ask if the data correspond to a straight-line form

$$x = a' + b'y \tag{11.13}$$

The values of the coefficients a' and b' will be different from the values of the coefficients a and b in Equation (11.11), but they are related if the variables x and y are correlated.

The analytical solution for the inverse slope b' is similar to that for b in Equation (11.12):

$$b' = \frac{N\Sigma x_i y_i - \Sigma x_i \Sigma y_i}{N\Sigma y_i^2 - (\Sigma y_i)^2} \tag{11.14}$$

If there is no correlation between the quantities x and y, then the least-squares fit must yield a horizontal straight line with a slope $b' = 0$.

If there is a complete correlation between x and y, then there exists a relationship between the coefficients a and b of Equation (11.11) and between a' and b' of Equation (11.13). To see what this relationship is, we rewrite Equation (11.13):

$$y = -\frac{a'}{b'} + \frac{1}{b'}x = a + bx \tag{11.15}$$

and equate coefficients

$$a = -\frac{a'}{b'} \qquad b = \frac{1}{b'} \tag{11.16}$$

We see from Equation (11.16) that $bb' = 1$ for complete correlation. If there is no correlation, both b and b' are 0 and Equations (11.16) do not apply. We. therefore define, as a measure of the degree of linear correlation, the experimental linear-correlation coefficient $r \equiv \sqrt{bb'}$:

$$r \equiv \frac{N\Sigma x_i y_i - \Sigma x_i \Sigma y_i}{[N\Sigma x_i^2 - (\Sigma x_i)^2]^{1/2}[N\Sigma y_i^2 - (\Sigma y_i)^2]^{1/2}} \tag{11.17}$$

The value of r ranges from 0, when there is no correlation, to ± 1, when there is complete correlation. The sign of r is the same as that of b (and b'), but only the absolute magnitude is important.

The correlation coefficient r cannot be used directly to indicate the degree of correlation. A probability distribution for r can be derived from the two-dimensional Gaussian distribution, but its evaluation requires a knowledge of the correlation coefficient ρ of the parent population. A more common test of r is to compare its value with the probability distribution for the parent population that is completely uncorrelated; that is, for which $\rho = 0$. Such a comparison will indicate whether or not it is probable that the data points could represent a sample derived from an uncorrelated parent population. If this probability is small, then it is more probable that the data points represent a sample from a parent population where the variables are correlated.

For a parent population with $\rho = 0$, the probability that any random sample of uncorrelated experimental data points would yield an experimental linear-correlation coefficient equal to r is given by[3]

$$p_r(r; \nu) = \frac{1}{\sqrt{\pi}} \frac{\Gamma[(\nu + 1)/2]}{\Gamma(\nu/2)} (1 - r^2)^{(\nu-2)/2} \tag{11.18}$$

where $\nu = N - 2$ is the number of degrees of freedom for an experimental sample of N data points. The gamma function for integral and half-integral values was defined in Equation (11.7).

Integral Probability

A more useful distribution than that of Equation (11.18) is the probability $P_c(r; N)$ that a random sample of N uncorrelated experimental data points would yield an

[3]For a derivation see Pugh and Winslow (1966), Section 12-8.

experimental linear-correlation coefficient as large as or larger than the observed value of $|r|$. This probability is the integral of $p_r(r; \nu)$ for $\nu = N - 2$:

$$P_c(r; N) = 2\int_{|r|}^{1} p_x(r; \nu)\, dx \qquad \nu = N - 2 \tag{11.19}$$

With this definition, $P_c(r; N)$ indicates the probability that the observed data could have come from an uncorrelated ($\rho = 0$) parent population. A small value of $P_c(r; N)$ implies that the observed variables are probably correlated.

Because Equation (11.19) cannot be integrated analytically, the function must be integrated either by making a series expansion of the argument and integrating term by term or by performing a numerical integration. With fast computers, the latter method is more convenient and generally applicable to such problems.

Program 11.2 LCORLATE (Appendix E) Correlation probability computations. LCORPROB computes the probability of Equation (11.19) by numerical integration. Input variables RCORR and NOBSERV correspond to the value of the experimental linear-correlation coefficient and the number of observations, respectively. (The number of degrees of freedom is the number of observations minus 2.) The program uses the following routines: LINCORREL computes the function $p_r(r; \nu)$ of Equation (11.18) using the approximation of Equation (11.8) for the gamma function (calculated by the function GAMMA in the program unit GENUTIL). Because LINCORREL is intended to be used as an argument to the integration routine SIMPSON, it can have only one argument. The parameter ν is passed in the global variable PSIMPS by the calling routine.
LINCORPROB computes $P_c(r; \nu)$ of Equation (11.19) by numerically integrating LINCORREL by Simpson's rule. The calculation returns the integral to an accuracy of about ± 0.01. The trade-off on accuracy versus speed of computation is controlled by the value of the constant DX, the integration step.

Example 11.2. For the data of Example 6.1, the linear-correlation coefficient r can be calculated from Equation (11.17) with the data of Table 6.1:

$$r = \frac{9 \times 779.3 - 450.0 \times 12.44}{\sqrt{(9 \times 28{,}500 - 450.0^2) \times (9 \times 21.32 - 12.44^2)}} = 0.9998$$

The probability for determining, from an uncorrelated population with $9 - 2 = 7$ degrees of freedom, a value of r equal to or larger than the observed value, can be calculated from Equation (11.19) (see Table C.3). The result $P_c(r; N) < 0.001\%$ indicates that it is extremely improbable that the variables x and V are linearly uncorrelated. Thus, the probability is high that the variables are correlated and the linear fit is justified.

Similarly, in the experiment of Example 6.2, the linear-correlation coefficient can be calculated from Equation (11.17) by including the weighting factors $\sigma_i^2 = y_i$ as in Table 6.2, so that, for example, N is replaced by Σw_i and Σx_i is replaced by $\Sigma w_i x_i$, and so forth:

$$r = \frac{0.03570 \times 81.02 - 0.1868 \times 10}{\sqrt{(0.03570 \times 1.912 - 0.1868^2) \times (0.03570 \times 3693 - 10^2)}} = 0.9939$$

Again, the probability $P_c(r; N)$ for $r = +0.9938$ with $\nu = 10 - 2 = 8$ degrees of freedom is very small ($< 0.001\%$), indicating that the change in counting rate C is linearly correlated to a high degree of probability with $x = 1/r^2$, the inverse square of the distance between the source and counter.

11.3 MULTIVARIABLE CORRELATIONS

If the dependent variable y_i is a function of more than one variable,

$$y_i = a + b_1 x_{i1} + b_2 x_{i2} + b_3 x_{i3} + \cdots \tag{11.20}$$

we might investigate the correlation between y_i and each of the independent variables x_{ij} or we might also enquire into the possibility of correlation between different variables x_{ij}. Here, we use the first subscript i to represent the observation, as in the previous discussions, and the second subscript j to represent the particular variable under investigation. The variables x_{ij} could be different variables, or they could be functions of x_i, $f(x_i)$, as in Chapter 7. We shall rewrite Equation (11.17) for the linear-correlation coefficient r in terms of another quantity s_{jk}^2.

We define the *sample covariance* s_{jk}^2:

$$s_{jk}^2 \equiv \frac{1}{N-1} \sum [(x_{ij} - \bar{x}_j)(x_{ik} - \bar{x}_k) \tag{11.21}$$

where the means $\bar{x}_j$ and $\bar{x}_k$ are given by

$$\bar{x}_j \equiv \frac{1}{N} \sum x_{ij} \quad \text{and} \quad \bar{x}_k = \frac{1}{N} \sum x_{ik} \tag{11.22}$$

and the sums are taken over the range of the subscript i from 1 to N. The weights have been omitted for clarity. With this definition, the sample variance for one variable s_j^2,

$$s_j^2 \equiv s_{jj}^2 = \frac{1}{N-1} \sum (x_{ij} - \bar{x}_j)^2 \tag{11.23}$$

is analogous to the sample variance s^2 defined in Equation (1.9):

$$s^2 = \frac{1}{N-1} \sum (x_i - \bar{x})^2 \tag{11.24}$$

It is important to note that the sample variances s_j^2 defined by Equation (11.23) are measures of the ranges of variation of the variables and not of the uncertainties in the variables.

Equation (11.21) can be rewritten for comparison with Equation (11.17) by substituting the definitions of Equation (11.22):

$$\begin{aligned} s_{jk}^2 &\equiv \frac{1}{N-1} \sum [(x_{ij} - \bar{x}_j)(x_{ik} - \bar{x}_k)] \\ &= \frac{1}{N-1} \sum (x_{ij} x_{ik} - \bar{x}_j \bar{x}_k) \\ &= \frac{1}{N-1} \sum \left(x_{ij} x_{ik} - \frac{1}{N} \sum x_{ij} \sum x_{ik} \right) \end{aligned} \tag{11.25}$$

If we substitute x_{ij} for x_i and x_{ik} for y_i in Equation (11.17), we can define the *sample linear-correlation coefficient* between any two variables x_i and x_k as

$$r_{jk} = \frac{s_{jk}^2}{s_j s_k} \tag{11.26}$$

with the covariances and variances s_{jk}^2, s_j^2, and s_k^2 given by Equations (11.23) and (11.25). Thus, the linear-correlation coefficient between the jth variable x_j and the dependent variable y is given by

$$r_{jy} = \frac{s_{jy}^2}{s_j s_y} \tag{11.27}$$

Similarly, the linear-correlation coefficient of the parent population of which the data are a sample is defined as

$$\rho_{jk} = \frac{\sigma_{jk}^2}{\sigma_j \sigma_k} \tag{11.28}$$

where σ_j^2, σ_k^2, and σ_{jk}^2 are the true variances and covariances of the parent population. These linear-correlation coefficients are also known as product-moment correlation coefficients.

With these definitions we can consider either the correlation between the dependent variable and any other variable r_{jy} or the correlation between any two variables r_{jk}.

Polynomials

In Chapter 7 we investigated functional relationships between y and x of the form

$$y = a_0 + a_1 x + a_2 x^2 + a_3 x^3 + \cdots \tag{11.29}$$

In a sense, this is a variation on the linear relationship of Equation (11.20) where the powers of the single independent variable x are considered to be various variables $x_j = x^j$. The correlation between the independent variable y and the mth term in the power series of Equation (11.29), therefore, can be expressed in terms of Equations (11.23) through (11.27):

$$\begin{aligned} r_{my} &= \frac{s_{my}^2}{s_m s_y} \\ s_m^2 &= \frac{1}{N-1}\left[\Sigma x_i^{2m} - \frac{1}{N}(\Sigma x_i^m)^2\right] \\ s_y^2 &= \frac{1}{N-1}\left[\Sigma y_i^2 - \frac{1}{N}(\Sigma y_i)^2\right] \\ s_{my}^2 &= \frac{1}{N-1}\left[\Sigma x_i^m y_i - \frac{1}{N}\Sigma x_i^m \Sigma y_i\right] \end{aligned} \tag{11.30}$$

Weighted Fit

If the uncertainties in the data points are not all equal ($\sigma_i \neq \sigma$), we must include the individual standard deviations σ_i as weighting factors in the definition of variances, covariances, and correlation coefficients. From Chapter 6 the prescription for introducing weighting is to multiply each term in the sum by $1/\sigma_i^2$.

The formula for the correlation remains the same as Equations (11.26) and (11.27), but the formulas of Equations (11.21) and (11.23) for calculating the variances and covariances must be modified:

$$s_{jk}^2 \equiv \frac{1/(N-1)\Sigma[(1/\sigma_i^2)(x_{ij}-\bar{x}_j)(x_{ik}-\bar{x}_k)]}{(1/N)\Sigma(1/\sigma_i^2)}$$
$$s_j^2 \equiv s_{jj}^2 = \frac{1/(N-1)\Sigma[(1/\sigma_i^2)(x_{ij}-\bar{x}_j)^2]}{(1/N)\Sigma(1/\sigma_i^2)} \tag{11.31}$$

where the means $\bar{x}_j$ and $\bar{x}_k$ are also weighted means

$$\bar{x}_j = \frac{\Sigma x_{ij} w_i}{N} = \frac{\Sigma(x_{ij}/\sigma_i^2)}{\Sigma(1/\sigma_i^2)}$$

The weighting factors

$$w_i = \frac{1/\sigma_i^2}{(1/N)\Sigma(1/\sigma_i^2)} \tag{11.32}$$

for each data point are the inverse of the variances σ_i^2 that describe the uncertainties in each point, normalized to the average of all the weighting factors.

Multiple-Correlation Coefficient

We can extrapolate the concept of the linear-correlation coefficient, which characterizes the correlation between two variables at a time, to include multiple correlations between groups of variables taken simultaneously. The linear-correlation coefficient r of Equation (11.17) between y and x can be expressed in terms of the variances and covariances of Equation (11.31) and the slope b of a straight-line fit given in Equation (11.12):

$$r^2 = \frac{s_{xy}^4}{s_x^2 s_y^2} = b\frac{s_{xy}^2}{s_y^2} \tag{11.33}$$

In analogy with this definition of the linear-correlation coefficient, we define the *multiple-correlation coefficient* R to be the sum over similar terms for the variables of Equation (11.20):

$$R^2 \equiv \sum_{j=1}^{n}\left(b_j\frac{s_{jy}^2}{s_y^2}\right) = \sum_{j=1}^{n}\left(b_j\frac{s_j}{s_y}r_{jy}\right) \tag{11.34}$$

The linear-correlation coefficient r is useful for testing whether one particular variable should be included in the theoretical function that is fitted to the data. The multiple-correlation coefficient R characterizes the fit of the data to the entire function. A comparison of the multiple-correlation coefficient for different functions is therefore useful in optimizing the theoretical functional form.

We shall discuss in the following sections how to use these correlation coefficients to determine the validity of including each term in the polynomial of Equation (11.29) or the series of arbitrary functions of Equation (11.20).

11.4 *F* TEST

As noted in Section 11.1, the χ^2 test is somewhat ambiguous unless the form of the parent function is known, because the statistic χ^2 measures not only the discrepancy between the estimated function and the parent function, but also the deviations between the data and the parent function simultaneously. We would prefer a test that separates these two types of information so that we can concentrate on the former type. One such test is the F test, which combines two different methods of determining a χ^2 statistic and compares the results to see if their relation is reasonable.

F Distribution

If two statistic χ_1^2 and χ_2^2, which follow the χ^2 distribution, have been determined, the ratio of the reduced chi-squareds, $\chi_{\nu 1}^2$ and $\chi_{\nu 2}^2$, is distributed according to the F distribution[4]

$$f = \frac{\chi_1^2/\nu_1}{\chi_2^2/\nu_2} \tag{11.35}$$

with probability density function

$$P_f(f; \nu_1, \nu_2) = \frac{\Gamma[(\nu_1 + \nu_2)/2]}{\Gamma(\nu_1/2)\Gamma(\nu_2/2)} \left(\frac{\nu_1}{\nu_2}\right)^{\nu_1/2} \frac{f^{1/2(\nu_1 - 2)}}{(1 + f\nu_1/\nu_2)^{1/2(\nu_1+\nu_2)}} \tag{11.36}$$

where ν_1 and ν_2 are the numbers of degrees of freedom corresponding to χ_1^2 and χ_2^2. By the definition of χ_ν^2 [see Equation (11.4)], a ratio of ratios of variances

$$\frac{\chi_{\nu_1}^2}{\chi_{\nu_2}^2} = \frac{s_1^2/\sigma_1^2}{s_2^2/\sigma_2^2} \tag{11.37}$$

is also distributed as F, where s_1 and s_2 are experimental estimates of standard deviations σ_1 and σ_2 pertaining to some characteristic of the same or different distributions.

As with our tests of χ^2 and the linear-correlation coefficient r, we shall be more interested in the integral probability

$$P_F(F; \nu_1, \nu_2) = \int_F^\infty p_f(f; \nu_1, \nu_2)\, df \tag{11.38}$$

which describes the probability of observing such a large value of F from a random set of data when compared to the correct fitting function. The integral function $P_F(F; \nu_1, \nu_2)$ is tabulated and graphed in Table C.5 for a wide range of F, ν_1, and ν_2.

---

[4] See Pugh and Winslow (1966), Section 12-7, for a derivation.

A word of caution is in order concerning the use of these tables. Because the statistic F in Equation (11.35) is defined as the ratio of two determinations of χ^2 without specifying which must be in the numerator, we can define two statistics F_{12} and F_{21},

$$F_{12} = \frac{\chi^2_{\nu 1}}{\chi^2_{\nu 2}} \qquad F_{21} = \frac{\chi^2_{\nu 2}}{\chi^2_{\nu 1}} = \frac{1}{F_{12}} \tag{11.39}$$

which must both be distributed according to the F distribution.

If in some experiment our calculations yield a particular value of F_{12}, we can use Table C.5 to determine whether such a large value is less than 5% probable (Table C.6 and Figure C.6) or less than 1% probable (Table C.7 and Figure C.7). If the test value is less than the tabulated values, we must also make sure that it is not too small. To do this, we compare the value

$$F_{21} = 1/F_{12} \tag{11.40}$$

to the same tables and graphs, noting that the values of ν_1 and ν_2 are reversed. The values of ν_1 and ν_2 specified in Table C.5 correspond to the degrees of freedom for the numerator and denominator of Equation (11.39), respectively.

Example 11.3. Suppose that $F_{12} = 0.2$ with $\nu_1 = 2$ and $\nu_2 = 10$. For Table C.6, the observed value of F_{12} may be as high as 4.10 and still be exceeded by about 5% of random observations. Similarly, we compare $F_{21} = 1/F_{12} = 5.0$ with the 5% point for $\nu_1 = 10$ and $\nu_2 = 2$, which has a value of 19.4. Because the values of F_{12} and F_{21} are well within the 5% limits, we can have confidence in the fit.

What we are estimating in this example is the probability $P_F(F_{12}; \nu_1, \nu_2)$ that F_{12} is not too large and the probability $P_F(1/F_{12}; \nu_2, \nu_1)$ that F_{12} is not too small. It is tempting to simplify this procedure by assuming that

$$P_F(1/F_{12}; \nu_2, \nu_1) = P_F(F_{12}; \nu_1, \nu_2) \tag{11.41}$$

so that our test consists of determining F such that

$$P_F(F; \nu_1, \nu_2) = 0.05$$

with the requirement that

$$F > F_{12} > 1/F$$

This approximation is valid for reasonably large values of ν_1 and ν_2 but not for small values of either, as in the preceding example, where we have $4.10 > F_{12} > 1/19.4$.

Multiple-Correlation Coefficient

There are two types of F tests that are normally performed on least-squares fitting procedures. One is designed to test the entire fit and can be related to the multiple-correlation coefficient R. The other, to be discussed later, tests the inclusion of an additional term in the fitting function.

If we consider the sum of squares of deviations S_y^2 associated with the spread of the data points around their mean (omitting factors of $1/\sigma_i^2$ for clarity),

$$S_y^2 = \Sigma(y_i - \bar{y})^2 \tag{11.42}$$

this is a statistic that follows the χ^2 distribution with $N - 1$ degrees of freedom (only one parameter $\bar{y}$ must be determined from the N data points). It is a characteristic of quantities that follow the χ^2 distribution that they may be expressed as the sum of other quantities that also follow the χ^2 distribution such that the number of degrees of freedom of the original statistic is the sum of the numbers of degrees of freedom of the terms in the sum.

By suitable manipulation and rearrangement, it can be shown that S_y^2 can be expressed as the sum of the two terms,

$$\begin{aligned} S_y^2 = \sum(y_i - \bar{y})^2 &= \sum_{j=1}^{m}\left[(y_i - \bar{y})\sum_{j=1}^{m} a_j(f_j - \bar{f}_j)\right] + \sum_{j=1}^{m}(y_i - \sum a_j f_j)^2 \\ &= \sum_{j=1}^{m}[a_j \sum[(y_i - \bar{y})(f_j - \bar{f}_j)]] + \sum[y_i - y(x_i)]^2 \end{aligned} \tag{11.43}$$

where the fitting function is of the form

$$y(x_i) = \sum_{j=1}^{m} a_j f_j(x_i) \tag{11.44}$$

and we have

$$\bar{f}_j = \frac{1}{N}\sum f_j(x_i) \tag{11.45}$$

The left-hand side of Equation (11.43) is distributed as χ^2 with $N - 1$ degrees of freedom. The right-hand term is our definition of χ^2 from the Equation (11.3) and has $N - m$ degrees of freedom. Consequently, the middle term must be distributed according to the χ^2 distribution with $m - 1$ degrees of freedom.

By comparison with our definition of the multiple-correlation coefficient R in Equation (11.34), we can express this middle tern as a fraction R^2 of the statistic S_y^2:

$$\sum_{j=1}^{m} a_j \sum[(y_i - \bar{y})(f_j - \bar{f}_j)] = R^2 \sum(y_i - \bar{y})^2 \tag{11.46}$$

Equation (11.43) becomes

$$\sum(y_i - \bar{y})^2 = R^2 \sum(y_i - \bar{y})^2 + (1 - R^2)\sum(y_i - \bar{y})^2 \tag{11.47}$$

or

$$S_y^2 = R^2 S_y^2 + (1 - R^2) S_y^2 \tag{11.48}$$

where, as before, both terms on the right-hand side are distributed as χ^2, the first with $m - 1$ degrees of freedom and the second with $N - m$ degrees of freedom.

Thus, the physical meaning of the multiple-correlation coefficient becomes evident. It divides the total sum of squares of deviations S_y^2 into two parts. The first fraction $R^2 S_y^2$ is a measure of the spread of the dependent and independent variable data space. The second fraction, $(1 - R^2) S_y^2$, is the sum of squares of the deviations about the regression and represents the agreement between the fit and the data.

From the definition of Equation (11.35), we can define a ratio F_R of the two terms in the right-hand side of Equation (11.47) that follow the F distribution with $\nu_1 = m - 1$ and with $\nu_2 = N - m$ degrees of freedom,

$$F_R = \frac{R^2/(m-1)}{(1-R^2)/(N-m)} = \frac{R^2}{(1-R^2)} \times \frac{(N-m)}{(m-1)} \tag{11.49}$$

From this definition of F_R in terms of the multiple-correlation coefficient R, it is clear that a large value of F_R corresponds to a good fit, where the multiple correlation is good and $R \simeq 1$. The F test for this statistic is actually a test that the coefficients are 0 ($a_j = 0$). So long as F_R exceeds the test value for F, we can be fairly confident that our coefficients are nonzero. If, on the other hand, $F_R < F$, we may conclude that at least one of the terms in the fitting function is not valid, is decreasing the multiple correlation by its inclusion, and should have a coefficient of 0.

Test of Additional Term

Because of the additive nature of functions that obey the χ^2 statistics, we can form a new χ^2 statistic by taking the difference of two other statistics that are distributed as χ^2. In particular, if we fit a set of data with a fitting function with m terms, the resulting value of chi-square associated with the deviations about the regression $\chi^2(m)$ has $N - m$ degrees of freedom. If we add another term to the fitting function, the corresponding value of chi-square $\chi^2(m+1)$ has $N - m - 1$ degrees of freedom. The difference between these two must follow the χ^2 distribution for 1 degree of freedom.

If we form the ratio of the difference $\chi^2(m) - \chi^2(m+1)$ to the new value $\chi^2_\nu(m+1)$, we can form a statistic F_χ that follows the F distribution with $\nu_1 = 1$ and $\nu_2 = N - m - 1$:

$$F_\chi = \frac{\chi^2(m) - \chi^2(m+1)}{\chi^2(m+1)/(N-m-1)} = \frac{\Delta\chi^2}{\chi^2_\nu} \tag{11.50}$$

This ratio is a measure of how much the additional term has improved the value of the reduced chi-square and should be small when the function with $m + 1$ terms does not significantly improve the fit over the function with m terms. Thus, we can be confident in the relative merit of the new terms if the value of F_χ is large. As for F_R, this is really a test of whether the coefficient for the new term is 0 ($a_{m+1} = 0$). If F_χ exceeds the test value for F, we can be fairly confident that the coefficient should not be 0 and the term, therefore, should be included. Table C.5 and Figure C.5 are useful for testing F_χ. They give the value of F corresponding to various values of the probability $P_F(F;\, 1, \nu_2)$ and various values of ν_2 for the case where $\nu_1 = 1$. Thus, rather than evaluating F for critical values of the probability (for example, 5% or 1%), we can evaluate the probability corresponding to the observed value of F_χ.

A calculation of F_χ could be built into a linear regression program and the resulting value compared to a supplied test value F, to indicate whether or not the last term in the series is justified, and therefore, to determine how many terms in the series should be included in the fit. However, it is probably safer, except possibly in a large, well debugged production run involving fitting polynomials to many similar data sets, to examine the individual values of χ^2 along with F_χ and to adjust the number of terms in the calculation manually. One should, however, be aware that the important figure of merit for added terms is the difference of the two values of χ^2 divided by the new value χ^2_ν of the *reduced* chi-square.

11.5 CONFIDENCE INTERVALS

The object of data fitting is to obtain values for the parameters of the fitted function, and the uncertainties in the parameters. The quality of the fit is indicated by χ^2 and its associated probability, and the uncertainties give the probabilities that our values of the fitted parameters are good estimates of the parent parameters. Whether we estimate our parameters by the least-squares method or by direct application of the maximum-likelihood method, as discussed in Chapter 10, we must always estimate the uncertainty in our parameters to indicate numerically our confidence in our results.

Generally, we assume Gaussian statistics and quote the standard deviation σ in a result, where σ appears in the Gaussian probability density function

$$p_G(x; \mu, \sigma) = \frac{1}{\sigma\sqrt{2\pi}} \exp\left[-\frac{1}{2}\left(\frac{x-\mu}{\sigma}\right)^2\right] \tag{11.51}$$

and determines the width of the distribution. As noted in Chapter 2, approximately 68.3% of the events of the Gaussian distribution fall within $\pm\sigma$ of the mean μ and approximately 95.4% fall within $\pm 2\sigma$.

Confidence Level for One-Parameter Fit

One way of looking at the 1 standard deviation limit is to consider that, in a series of repeated experiments, there is approximately a 68% chance of obtaining values within $\pm\sigma$ of the mean μ. Of course, we usually do not know μ, and perhaps not σ either, but have determined experimentally only $\bar{x}$ and s, our estimate of the parameters. However, as long as our experimental estimates $\bar{x}$ and s are reasonably close to the true values μ and σ, we can state that there is approximately a 68% probability that the true value of the measured parameter lies between $\bar{x} - s$ and $\bar{x} + s$, or that at the 68.3% *confidence level*, the true value of the parameter lies between these two limits.

We may wish to quote results in terms of other confidence levels. For example, we refer to the $\pm 2\sigma$ limit as the 95.4% confidence interval, or we may quote a 99% or 99.9% confidence level for a high-precision experiment. The conventional 1σ and 2σ limits are based on the Gaussian distribution, which may or may not apply to the data in question, and even an experimental distribution that nominally follows Gaussian statistics is apt to deviate in the tails.

For any distribution, represented by the normalized probability density function, $p_x(x; \mu)$, we determine the probability that a measurement of the parameter will fall between $\bar{x} - a$ and $\bar{x} + b$ by the integral

$$P_x = \int_{\bar{x}-a}^{\bar{x}+b} p_x(\bar{x}; x)\, dx \tag{11.52}$$

and could quote a confidence level of P_x that the "true" value of the measured parameter is between these two values. Note that we have not specified a region that is

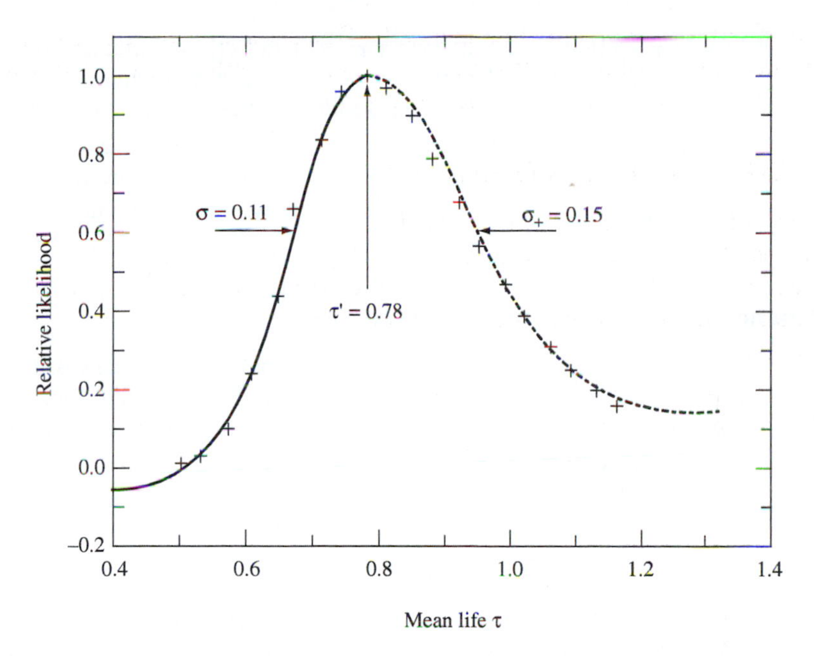

FIGURE 11.1
Relative values of the likelihood function versus trial values of the parameter for the 373-event sample of Example 10.1d. The data points (from Figure 10.4b) are indicated by crosses; the solid and dashed curves represent the results of fitting Gaussian curves separately to the two sides of the distribution. Parameters determined in the two fits are indicated on the graph. All measurements are in units of 10^{-10} s.

symmetrical about the mean. The uncertainties in our measurements may not be symmetrical, although the asymmetry may be hidden if we assume Gaussian statistics in our calculations. For example, the routines for finding uncertainties in parameters found by least-squares fitting (Chapters 7 and 8) generally assume a Gaussian distribution of the parameters and hence produce a single number for the uncertainties.

Example 11.4. As an example of an asymmetrical probability distribution, consider the 373-event data sample of Example 10.1d. In Figure 10.4b we plot as crosses the scaled values of the likelihood function for these data as a function of trial values of the parameter τ. The data points exhibit a marked asymmetry about the mean τ'. The dashed curve was calculated from Equation (10.10) with parameters obtained from the fit.

To make a better determination of σ from this curve, we considered the regions on each side of the mean separately and estimated two separate standard deviations, σ_L and σ_R, with the aid of Equation (1.11). To reduce the effect of the right-hand side tail on the value of σ_r, we imposed a cutoff at $\tau = 1.6$ and used only those data points below the cutoff in this calculation.

A composite curve formed of two Gaussians with the same mean τ but different values of σ is shown as the solid curve in Figure 11.1. It would be reasonable to consider the two values of σ obtained in this way as appropriate estimates of the uncertainty in τ, so that we could report $\tau' = 0.78^{+0.15}_{-0.11}$, as indicated by the arrows on Figure 11.1 rather than $\tau' = 0.78 \pm 0.14$ as we did in Chapter 10. This is equivalent to finding the two positions at which the logarithm of the likelihood function has decreased by $\Delta M = 1/2$ as discussed in Section 10.2. Clearly this result is somewhat subjective if either side of the curve does not follow the Gaussian form. For this example, the value of σ_R depends on how much of the tail is included in the calculation.

Confidence Levels for Multiparameter Fits

The definition of the confidence level in a one-parameter experiment is generally straightforward. We can plot our data and observe if the distribution is Gaussian and estimate directly from the distribution of the probability that the true result lies between two specified values. When two or more variables have been determined and those variables exhibit some correlation, the definition of the confidence level becomes a little more difficult. Consider, for example, the determination of the mean lifetimes τ_1 and τ_2 of two unstable silver isotopes of Example 8.1. The problem was treated in Chapter 8 as a five-parameter problem, with parameters a_4 and a_5 corresponding to the two mean lifetimes, τ_1 and τ_2, respectively, and parameters a_1, a_2, and a_3 corresponding to the amplitudes of a uniform background and the two decaying states. The parameters of most interest in the experiment are a_4 and a_5, and we want to define a joint confidence interval for those two variables.

Figure 11.2 shows two sets of contours for the variation of χ^2 as a function of a_4 and a_5 from the least-squares fit by the Marquardt method discussed in Chapter 8. The small contours, drawn with solid lines, were calculated by holding the parameters a_1, a_2, and a_3 fixed at their optimum values (see Table 8.5) and varying a_4 and a_5 to obtain increases in χ^2 of 1, 2, and 3 from the minimum value. The large contours, shown as dashed lines, were calculated by allowing a_1, a_2, and a_3 to vary to minimize χ^2 for each pair of values of a_4 and a_5. The contour plots cover very different ranges because of the correlations of the displayed parameters, a_4 and a_5, with the remaining parameters a_1 through a_3. The tilt of the closed figures on each plot indicates the degree of correlation of parameters a_4 and a_5 with each other. In an ideal experiment, the contours are ellipses in the region of the χ^2 minimum and if a_4 and a_5 are not correlated, then, with suitable scaling of the axes, the ellipses are circles.

Which plot should we use? Additionally, how do we determine a confidence interval; that is, a region of the a_4-a_5 space in which we estimate there is, for example, a ~68% probability of finding the true values of the two parameters?

First, we should note that, because the fitting function, Equation (8.2), is not linear in the parameters, the methods of testing described in the previous sections strictly do not apply. However, we are much more likely to run into nonlinear fitting problems than the easier linear problems, so we shall continue with this example. At any rate, the function is linear in parameters a_1 through a_3, and we could make a linear expansion of it, over a limited region, in the parameters a_4 and a_5. In fact, this was the basis of a method of fitting nonlinear functions in Chapter 8.

Then, we should use the larger of the two contour diagrams to define our confidence intervals. That implies that if we wish to find the standard deviation of a_4

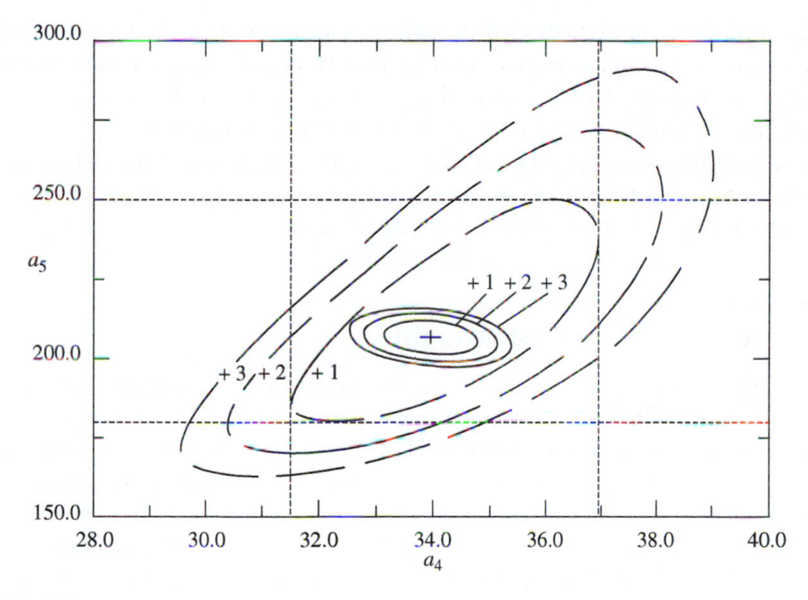

FIGURE 11.2
Two sets of contours for the variation of χ^2 with parameters a_4 and a_5 in the region of the χ^2 minimum. Data are from the least-squares fit by the Marquardt method discussed in Chapter 8. The small contours, drawn with solid lines, were calculated by holding parameters a_1 through a_3 fixed at their optimum while varying a_4 and a_5 to obtain increases in χ^2 of 1, 2, and 3 from the minimum values. The large contours, shown as dashed lines, were calculated by allowing a_1, a_2, and a_3 to vary to minimize χ^2 for each pair of values of a_4 and a_5.

from the contour plot, we should consider the full range of the outer limit of the $\Delta\chi^2 = 1$ contour, and not the intersection of that contour with the a_4 axis. This is equivalent to allowing a_5 to assume its best values for each chosen value of a_4, as we have already assumed for the parameters a_1 through a_3. The two dashed vertical lines indicate the two limits on a_4 that include the 1 standard deviation, or 68.3% of the probability, and the two horizontal lines indicate the 1 standard deviation limits for a_5.

How do we know that the vertical lines enclose 68.3% of the probability? By allowing the four parameters a_1, a_2, a_3, and a_5 to find their optimum values for each chosen value of a_4 and varying a_4, we have separated our χ^2 fitting problem into two parts: a fit of N data points to $m - 1$ parameters with $N - m - 1$ degrees of freedom and a variation of $\Delta\chi^2$ with a_4 about the minimum χ^2, with 1 degree of freedom. As we observed in the previous section, the two variations separately must follow their appropriate χ^2 distributions, so our variation of $\Delta\chi^2$ obeys the χ^2 probability distribution for 1 degree of freedom. If we look at the integrated probability distribution P_χ for 1 degree of freedom [Table C.4, or calculated from Equation (11.10)], we see that $\chi^2 \geq 1$ corresponds to 31.7% of the probability, or $\Delta\chi^2 < 1$ corresponds to 68.3%. Similarly, if we wish to find the limits for 2 standard deviations, we should find the limits of a_4 on the $\Delta\chi^2 = 4$ contour, with all other parameters optimized.

To find the 1 standard deviation region encompassed by the *joint variation* of two parameters, a_4 and a_5, with all other parameters optimized, we must draw the contour corresponding to that value of $\Delta\chi^2$ for 2 degrees of freedom that includes 68.3% of the probability. Referring again to Table C.4 or Equation (11.10), we find that we should draw the contour for $\Delta\chi^2 = 2.30$, and for the 2 standard deviation contour, we should choose $\Delta\chi^2 = 6.14$. Joint confidence intervals with more than two parameters are often of interest, but are difficult to display and are represented best by two-dimensional projections of contours for pairs of variables.

Confidence Level for a Predicted Value

Suppose the predicted value of a physical quantity is $\mu = 1000.0$, and we have made a measurement and obtained the value $\bar{x} = 999.4 \pm 2.0$. At what confidence level is the predicted value consistent with our measurement? The question could be rephrased as, "What is the probability of obtaining from the predicted parent distribution a distribution that is as bad as the one we got, or worse?" Because the shape of the parent distribution was not predicted, but only the value of the mean, we must use our value of the standard deviation, $\sigma \simeq 2.0$, as an estimate of that of the parent distribution. If the distribution is known to follow Gaussian statistics, then the required confidence is twice the integral of the standard Gaussian probability function from $x = \delta$ to ∞, where $\delta = |\mu - \bar{x}|/\sigma = |1000.0 - 999.4|/2.0$.

Now, suppose that the predicted value was necessarily positive—an intensity, for example. Then, we might again assume a Gaussian distribution, but only for positive values of the variable x, and therefore our confidence integral becomes the integral of the standard Gaussian from δ to ∞. However, because the total probability must be normalized to 1, we again multiply the integral by 2 so that the probability or confidence level is the same for both problems.

The method of determining the confidence level thus depends on the type of problem as well as the probability function that is applicable to the problem. For distributions that are symmetrical about their means, such as the Gaussian distribution, we generally consider the probability of obtaining a result that is the specified number of standard deviations from the mean, without regard to sign, unless a particular sign is excluded by the physical problem. For distributions such as the chi-square and Poisson distributions, which are only defined for positive values of their arguments, it is conventional to find a "one-sided" probability as in the case of the χ^2 distribution where we quote the probability of obtaining a value as large as or larger than the value we obtained for a given number of degrees of freedom.

11.6 MONTE CARLO TESTS

A Monte Carlo calculation can help us understand the statistical significance of our results and possibly obtain a better estimate of some of the parameters of the experiment. As a by-product, the Monte Carlo program may also help us identify biases in our analysis procedure.

Suppose, for example, that we have measured a quantity x that is predicted to have a value μ. From our experiment we obtain the value $\bar{x}$ for our estimate of μ.

We want to find the probability of obtaining from a series of similar experiments a value $\bar{x}$ that differs from the predicted value μ by

$$\Delta x \geq |\mu - \bar{x}| \tag{11.53}$$

We can set up a Monte Carlo program to simulate our experiment and to generate events with the parameters predicted by the theoretical principle that we are testing and with the same cuts as those imposed by our experimental apparatus. Such a program can be quite complex, but it may already exist at the time of analysis, if, for example, a Monte Carlo program was written to help plan the experiment. Or it might be possible to use some geometric and kinematic quantities from the actual experiment and only generate those parts of each event that are affected by the parameters in question.

After the Monte Carlo program has been written and debugged, we can simulate repeated experiments with the same parent parameters and the same number of final measurements as in our real experiment. The data from each of these simulated experiments can be processed by our regular analysis program to obtain a group of "experimental" values of $\bar{x}$, and from the distribution of these values we can estimate the required probability.

Example 11.5. Let us use the Monte Carlo method to try to learn more about the significance of the small peak in our data of Example 9.2. Examination of Figure 9.2 leaves no doubt about the existence of a large peak at ~1.0 GeV. Without the fitted curve, the smaller peak near 0.8 GeV would be considerably less striking and further analysis might be helpful. (We note that, if the small peak were indeed spurious, we should have to refit the large peak to obtain a better estimate of its mean energy and width.) In Chapter 9, we estimated the probability to be about 0.01% that the smaller peak is just a fluctuation in a single bin above the single-peak background, with a probability of about 0.6% of such a fluctuation occurring in any one of the 60 bins into which the data were sorted. These are quite compelling numbers. Can we support them with a more detailed calculation by the Monte Carlo method?

We adapted to the study of this problem the Monte Carlo program and the least-squares fitting program, which were used to generate and analyze the data in Chapter 9. With the Monte Carlo program, we simulated the experiment according to Equation (9.1) to generate 4000 single-peak events in each of 1000 trial "experiments." The mean energy (E_0), half-width (Γ), and amplitude of the larger peak, and the amplitudes (a_1 through a_3) of the quadratic background, were set to the values obtained in the six-parameter fit, listed in Table 9.1.

To each set of trial data we fitted Equation (9.13), using identical procedures to those used in Chapter 9, with the exception that, starting values for the parameters of the smaller peak (a_7, a_8, and a_9) were set to the values obtained in the nine-parameter fits of Chapter 9, listed in column 6 of Table 9.1. We selected those fits that yielded parameters of the lower peak consistent with the values determined in Chapter 9 by imposing the following conditions: (1) We required that both the chi-square probability and the amplitude of the smaller peak (a_7) be equal to or greater than the corresponding fitted values listed for the nine-parameter fit in Table 9.1; (2) We required that the central energy of that peak be within plus or minus one histogram bin (0.05 GeV) of the values obtained in that fit.

From the 1000 generated experiments, 5 survived these cuts, or 0.5% of the total trials. This number considerably exceeds the rough estimate of 0.01% made in

TABLE 11.1
Results of generating 4000-event "experiments" from Equations (9.1) and (9.13) with parameters from fits listed in Table 9.1. We used several values of the amplitude A_1 of the smaller peak to test the sensitivities of our analysis to small and possibly spurious peaks.

A_1	Equation	Number of experiments	Number of successes
3.50	9.13	100	61
1.75	9.13	100	18
0.875	9.13	100	5
0.000	9.1	1000	~5

Chapter 9 for a single bin fluctuation. Tests made with other starting values and cuts for the smaller peak yielded similar numbers of survivors.

To check our procedure, we also generated and analyzed 100 two-peak trial "experiments" from Equation (9.13), with the parameters of the smaller peak set to the values from the nine-parameter fit listed in Table 9.1. From these 100 trials, 61, or 61%, survived the cuts. When we repeated the analysis with the amplitude of the smaller peak reduced by a factor of 2 (i.e., $a_4/2$), the success rate dropped to 18%, and a further reduction by another factor of 2 ($a_4/4$) reduced the success rate to 5%. The results of analyses are summarized in Table 11.1.

These results offer strong support for the existence of the smaller peak, and indicate that in a 4000-event experiment we might detect with reasonable probability a peak with only one-fourth the amplitude of the current smaller peak. Clearly, a Monte Carlo simulation should play an important role in planning this type of experiment. A carefully planned Monte Carlo program may be much better (and easier) than a detailed theoretical analysis for finding an answer to the question "How much data will be needed to establish (or disprove) the existence of a specified feature in a distribution."

We offer a final word of caution on using the Monte Carlo technique to study the statistical significance of experimental results. For Examples 9.2 and 11.5, we used a very simple problem to illustrate this technique. Yet, there are many opportunities for errors, which can lead to erroneous conclusions about the significance of our Chapter 9 data. In a larger study, it would be very easy to make a simple mistake that might lie undetected in the program and have a subtle effect on the results. It is important to test the program under a variety of conditions, and to examine results at intermediate stages before drawing conclusions from the result. In particular, if the results of the program lead to conclusions that violate intuition about the experiment, we should check and recheck the calculation. The Monte Carlo method is very powerful, and can enable us to solve very difficult statistical problems in a straightforward manner, but like all powerful tools, it must be used with care.

SUMMARY

Variance of the fit:

$$s^2 = \frac{1}{N-M}\frac{\Sigma\{(1/\sigma_i^2)[y_i - y(x_i)]^2\}}{(1/N)\Sigma(1/\sigma_i^2)} = \frac{1}{N-m}\sum w_i[y_i - y(x_i)]^2$$

Weighting factors:

$$w_i = \frac{1/\sigma_i^2}{(1/N)\Sigma(1/\sigma_i^2)}$$

Relationship between s^2 and χ^2:

$$\chi_\nu^2 = \frac{\chi^2}{\nu} = \frac{s^2}{\langle\sigma_i^2\rangle}$$

where

$$\langle\sigma_i^2\rangle = \left[\frac{1}{N}\sum\frac{1}{\sigma_i^2}\right]^{-1}$$

Probability $P_\chi(\chi^2; \nu)$ that any random set of N data points will yield a value of chi-square as large as or larger than χ^2:

$$P_\chi(\chi^2; \nu) = \int_{\chi^2}^{\infty}\frac{z^{1/2(\nu-2)}e^{-z/2}}{2^{\nu/2}\Gamma(\nu/2)}\,dz$$

Linear-correlation coefficient:

$$r \equiv \frac{N\Sigma x_i y_i - \Sigma x_i \Sigma y_i}{[N\Sigma x_i^2 - (\Sigma x_i)^2]^{1/2}[N\Sigma y_i^2 - (\Sigma y_i)^2]^{1/2}}$$

Probability $P_c(r, N)$ that any random sample of uncorrelated experimental data points would yield an experimental linear-correlation coefficient as large as or larger than $|r|$:

$$P_c(r; \nu + 2) = \int_{|r|}^{1}\frac{1}{\sqrt{\pi}}\frac{\Gamma[(\nu+1)/2]}{\Gamma(\nu/2)}(1 - r^2)^{(\nu-2)/2}$$

Sample covariance:

$$s_{jk}^2 \equiv \frac{1/(N-1)\Sigma[(1/\sigma_i^2)(x_{ij} - \bar{x}_j)(x_{ik} - \bar{x}_k)]}{(1/N)\Sigma(1/\sigma_i^2)} \quad \text{with} \quad \bar{x}_j = \frac{\Sigma(x_{ij}/\sigma_i^2)}{\Sigma(1/\sigma_i^2)}$$

Sample variance: $\sigma_j^2 = \sigma_{jj}^2$

Sample linear-correlation coefficient:

$$r_{jk} = \frac{s_{jk}^2}{s_j s_k}$$

Multiple-correlation coefficient:

$$R^2 \equiv \sum_{j=1}^{n}\left(b_j\frac{s_{jy}^2}{s_y^2}\right) = \sum_{j=1}^{n}\left(b_j\frac{s^j}{s_y}r_{jy}\right)$$

F test:

$$F = \frac{\chi^2_{\nu 1}}{\chi^2_{\nu 2}}$$

$$P_F(F; \nu_1, \nu_2) = \int_F^\infty p_f(f; \nu_1, \nu_2)\, df$$

F test for multiple-correlation coefficient R (for $\nu = N - m$*):*

$$F_R = \frac{R^2/(m-1)}{(1-R^2)/(N-m)} = \frac{R^2}{(1-R^2)} \times \frac{(N-m)}{(m-1)}$$

F test for χ^2 *validity of adding* $(m+1)$*th term:*

$$F_\chi = \frac{\chi^2(m) - \chi^2(m+1)}{\chi^2(m+1)/(N-m-1)} = \frac{\Delta\chi^2}{\chi^2_\nu}$$

Confidence limits: $1\sigma \longrightarrow 68.3\%$; $2\sigma \longrightarrow 95.4\%$; $3\sigma \longrightarrow 99.7\%$

EXERCISES

11.1. Discuss the meaning of χ^2 and justify the relationship between it and the sample variance $s^2 = \chi^2_\nu$.

11.2. Compare the exact calculation of the gamma function $\Gamma(n)$ of Equation (11.7) with the approximate calculation of Equation (11.8) for $n = 1/2, 1, 5/2, 4, 9/2, 10$.

11.3. From Equation (11.6), show that the χ^2-probability density for 1 degree of freedom can be written as

$$p(x^2) = \frac{e^{-x^2/2}}{\sqrt{2\pi x^2}}$$

Calculate to 1% the probability of obtaining a value of χ^2 that is less than 2.00 by expanding the function in a Taylor series and integrating term by term.

11.4. For a typical number of degrees of freedom ($\nu \simeq 10$), find, by numerically integrating Equation (11.6), the range of probability $P_\chi(\chi^2, \nu)$ for finding χ^2 as small as 0.5 or as large as 1.5. Use the approximation for the gamma function of Equation (11.8).

11.5. By numerically integrating Equation (11.6), find the probability of finding a value of $\chi^2_\nu = 1.5$ with $\nu = 100$ degrees of freedom. (Note that double-precision variables must be used.) Would you consider this to be a reasonably good fit?

11.6. Express the linear-correlation probability density of Equation (11.18) in terms of the approximation for the gamma function of Equation (11.8).

11.7. Work out the details of the calculation of the linear-correlation coefficients r for Examples 6.1 and 6.2.

11.8. If a set of data yields a zero slope $b = 0$ when fitted with Equation (11.11), what can you say about the linear-correlation coefficient r? Justify this value in terms of the correlation between x_i and y_i.

11.9. Find the linear-correlation coefficient r_1 between the independent variable T_i and the dependent variable V_i for the data of Example 7.1.

11.10. Find the correlation coefficient r_2 between T_i^2 and V_i for the data of Example 7.1. Does the correlation justify the use of a quadratic term?

11.11. Express the multiple correlation R in terms of x_{ij}, y_i, and their averages.

11.12. Evaluate the multiple-correlation coefficient R for the data of Example 7.1.

11.13. Is a large value of F good or bad? Explain.

11.14. If we wish to set as an arbitrary criterion a probability of 0.01 for the F_χ test, what would be the reasonable average value for F test?

11.15. What different aspects of a fit do the F_R and F_χ tests represent?

11.16. Apply the F_χ test for the quadratic term to the data of Example 7.1 and state your conclusions. (Refer to Table 7.4.)

11.17. Show the intermediate steps in the derivation of Equation (11.43).

11.18. Estimate from Figure 11.2 the 90% confidence limit for each of the two mean lifetimes (a_4 and a_5) of Example 8.1 when all variables are allowed to find their optimum values.

APPENDIX A

NUMERICAL METHODS

There are several reasons why we might want to fit a function to a data sample, and several different techniques that we might use. If we wish to estimate parameters that describe the parent population from which the data are drawn, then the maximum-likelihood or least-squares method is best. If we wish to interpolate between entries in data tables to find values at intermediate points or to find numerically derivatives or integrals of tabulated data, then an interpolation technique will be more useful. Additionally, if we wish to obtain intermediate values between calculated coordinate pairs in order to plot a smooth curve on a graph, then we may wish to use a spline fitting method. In this appendix we shall summarize some standard methods for treating the latter two types of problems, as well as some methods of finding the roots of nonlinear functions, a different sort of interpolation problem.

A.1 POLYNOMIAL INTERPOLATION

With modern fast computers, the need for interpolating within tables to find intermediate values of tabulated functions has reduced markedly. Nevertheless, there are situations in which it may be convenient to represent a complicated function by a simple approximation over a limited range. For example, in a large Monte Carlo calculation, where computing time is a significant consideration, we may approximate a complex function by a simpler polynomial that can be calculated quickly. Alternatively, we may save time by creating a probability integral once at the beginning of the program. and interpolating to find values of x corresponding to the randomly chosen values of y.

For many purposes a linear or quadratic interpolation is satisfactory; that is, we fit a straight line to two coordinate pairs, or a parabola to three, and use the equation of the fitted polynomial to find values of y at nearby values of x. Higher orders may be necessary for functions that have strong variations, but in general, it is better and more convenient to represent a function over a limited region by a series of low-order approximations.

Lagrange's Interpolation Method

Here is a method that is easy to remember and can be used to expand a function to any order. We know it works because of the theorem that states that if you can find any nth-degree polynomial that passes exactly through $n + 1$ points, then you have found the one and only nth-degree polynomial that passes through those points. Think about it. It is obvious for $n = 1$ (2 points).

Let us start with an easy problem. Suppose we have two coordinate pairs (x_0, y_0) and (x_1, y_1), and we want to find the straight line that passes through both of them. We write a function of the form

$$P(x) = y_0 A_0(x) + y_1 A_1(x) \tag{A.1}$$

and search for a function $A_0(x)$ that is 1 when $x = x_0$ and 0 when $x = x_1$, and a function $A_1(x)$ that is 1 when $x = x_1$ and 0 when $x = x_0$. We can guess the form. If we write $A_0(x)$ as a fraction and set its numerator to $(x - x_1)$, then $A_0(x)$ will be 0 for $x = x_1$ and will be $(x_0 - x_1)$ for $x = x_0$. But we want $A_0(x) = 1$ for $x = x_0$, so the denominator of A_0 must be $(x_0 - x_1)$. We can make similar arguments for $A_2(x)$ and thus write as our interpolation equation

$$P(x) = y_0 \frac{(x - x_1)}{(x_0 - x_1)} + y_1 \frac{(x - x_0)}{(x_1 - x_0)} \tag{A.2}$$

Suppose we want a parabola that passes through three points. Then we simply write

$$P(x) = y_0 A_0(x) + y_1 A_1(x) + y_2 A_2(x) \tag{A.3}$$

and, following the previous arguments, write

$$\begin{aligned} P(x) = {} & y_0 \frac{(x - x_1)(x - x_2)}{(x_0 - x_1)(x_0 - x_2)} + y_1 \frac{(x - x_0)}{(x_1 - x_0)} \frac{(x - x_2)}{(x_1 - x_2)} \\ & + y_2 \frac{(x - x_0)}{(x_2 - x_0)} \frac{(x - x_1)}{(x_2 - x_1)} \end{aligned} \tag{A.4}$$

The expansion to higher orders should be obvious. The kth term in an nth order expansion is given by the following product in which the $j = k$ term must be omitted:

$$\prod_{j=0}^{n} \frac{(x - x_j)}{(x_k - x_j)} y_k \quad (\text{excluding } j = k) \tag{A.5}$$

Note that the intervals in x need not be equally spaced. The interpolation for a well-behaved function $y = f(x)$ is completely general.

Newton's Divided Differences

Although the Lagrange interpolation method is especially easy to derive and provides a convenient way of interpolating between points in a function or table, it is not very convenient for repetitive calculations. It is not very convenient as an expansion either, because increasing the order of the expansion requires adding another factor to each term as well as adding another term. What we require is a more familiar form—a discrete analog of the Taylor expansion. For this we turn to Newton's method of divided differences.

There are several forms of the divided differences expansion, roughly characterized by the method we choose to define the differences, forward, backward, or about a central point. We shall restrict ourselves here to forward differences; that is, we calculate the variation of y with respect to x by taking increments in the positive x direction.

Again, consider a set of data points, (x_0, y_0), (x_1, y_1), (x_2, y_2), ... Let us assume that we wish to make a linear interpolation from x_0 to some point x with a first-degree polynomial. We define the zeroth divided difference as the function itself $f(x)$ evaluated at $x = x_0$:

$$f[x_0] \equiv f(x_0) = y_0 \tag{A.6}$$

The first divided difference is defined to be

$$f[x_0, x_1] \equiv \frac{f[x_1] - f[x_0]}{(x_1 - x_0)} \tag{A.7}$$

which is the slope of a linear function. Then, for a linear function,

$$f[x, x_0] = f[x_0, x_1] \tag{A.8}$$

or

$$\frac{f[x_0] - f[x]}{(x_0 - x)} = \frac{f[x_1] - f[x_0]}{(x_1 - x_0)} \tag{A.9}$$

which, on rearrangement of the terms, gives the first-order expansion

$$\begin{aligned} P_1(x) &= f[x_0] + (x - x_0)\,\frac{f[x_1] - f[x_0]}{(x_1 - x_0)} \\ &= f[x_0] + (x - x_0) f[x_0, x_1] \end{aligned} \tag{A.10}$$

where we have written $P_1(x)$ instead of $f(x)$ to indicate that the expansion is a polynomial approximation to the function $f(x)$.

To find the second-order expansion, we consider the second divided differences

$$f[x_0, x_1, x_2] \equiv \frac{f[x_2, x_1] - f[x_1, x_0]}{(x_2 - x_1)(x_1 - x_0)} \tag{A.11}$$

which corresponds to the slope of the slope, or the second derivative. This must be constant for a second-order function, so we have

$$f[x, x_0, x_1] = f[x_0, x_1, x_2] \tag{A.12}$$

which leads to the second-order expansion

$$P_2(x) = f[x_0] + (x - x_0) f[x_0, x_1] + (x - x_0)(x - x_1) f[x_0, x_1, x_2] \tag{A.13}$$

The general form for the nth-order expansion should again be obvious.

Remainders

The extrapolation formula for an nth-order expansion is only exact when the function itself is an nth-degree polynomial. Otherwise, the *remainder* at x after n terms $R_n(x)$, defined as the difference between the original function $f(x)$ and the expansion $P_n(x)$, is given by

$$\begin{aligned} R_n(x) &= f(x) - P_n(x) \\ &= (x - x_0)(x - x_1) \cdots (x - x_n) f[x, x_0, x_1, \ldots, x_n] \end{aligned} \tag{A.14}$$

Calculation of the remainder requires the value of the function $f(x)$ at x, which is generally not available. (If it were, we might not be doing this expansion.) However, it may be possible to make an estimate of $f_n(x)$, or to use a nearby value, and thus find an estimate of $R_n(x)$. An expression for the remainder can also be obtained in terms of the $(n + 1)$th derivative of the function.[1]

Uniform Spacing

The divided difference expressions have a particular convenient form when the intervals in x are uniform; that is, if $x_2 - x_1 = x_3 - x_2 = x_i - x_{i-1} = h$. The divided difference of the previous discussion can be written

$$f[x_0, x_1] = \frac{f[x_1] - f[x_0]}{(x_1 - x_0)} = \frac{\Delta f(x_0)}{h}$$

or

$$\Delta f(x_0) \equiv f(x_1) - f(x_0) \quad \text{and} \quad h = x_1 - x_2 \tag{A.15}$$

and higher-order differences become

$$\Delta^2 f(x_0) \equiv \Delta[\Delta f(x_0)] = \Delta f(x_1) - \Delta f(x_0), \text{ etc.} \tag{A.16}$$

If we define the relative distance along the interval by

$$\alpha = (x - x_0)/h \tag{A.17}$$

we can write for the nth-order expansion,

$$\begin{aligned} P_n(x) = f(x_0) &+ \alpha \Delta f(x_0) + \alpha(\alpha - 1)\Delta^2 f(x_0)/2! + \cdots \\ &+ \alpha(\alpha - 1) \cdots (\alpha - n - 1)\Delta^n f(x_0)/n! \end{aligned} \tag{A.18}$$

Equation (A.18) is a finite difference analog of the familiar Taylor expansion with the important difference that the factors multiplying the coefficients $\Delta^k f(x_0)/n!$ are not successive powers of the relative distance from the starting point, but rather the product of relative distances from successive points used in the expansion, because $(\alpha - 1) = (x - x_0 - h)/h = (x - x_1)/h$, and so forth.

Extrapolation

Equations (A.15) through (A.18) are perfectly general for fitting exactly n sequential equally spaced data points with a polynomial of degree $n - 1$. In principle, the

---

[1]See Hildebrand (1956) for a derivation.

TABLE A.1
Uniform differences for cos θ

θ (degrees)	y	Δ_1	Δ_2	Δ_3	Δ_4	Δ_5
0	1.0000	−0.0489	−0.0931	0.0139	0.0078	−0.0021
18	0.9511	−0.1420	−0.0792	0.0217	0.0056	
36	0.8090	−0.2212	−0.0575	0.0273		
54	0.5878	−0.2788	−0.0302			
72	0.3090	−0.3090				
90	−0.0000					

TABLE A.2
Extrapolation from 0 to 10° and from 0 to 75° in various orders

θ (degrees)	cos θ	Order 1	2	3	4	5
10	0.9848	0.9728	0.9843	0.9851	0.9848	0.9848
75	0.2588	0.7961	0.1819	0.2481	0.2589	0.2588

position of the first data point (x_0, y_0) can be anywhere, but for optimum interpolation, the values of x_0 and x_n should straddle the interpolation point x and be approximately equidistant from it.

The same formula can be used for extrapolating to values beyond the region of data, but the uncertainties in the validity of the approximation increase as x gets farther from the average of x_1 and x_n. The approximation is limited by both the degree of the interpolating polynomial and by uncertainties in the coefficients of the polynomial resulting from fluctuations in the data.

Example A.1. Table A.1 shows a uniform divided difference table for the cosine function for a range of the argument θ between 0 and 90°. Table A.2 shows values of cos θ for θ = 10 and 75° calculated from the divided difference table in orders 1 through 5. The interpolation starts at 0° so that only the top row of Table A.1 is used and thus, θ > 18°, the calculation is an extrapolation. The true value of cos θ is also listed. As we should expect, the large extrapolation to 75° is very poor in low order. Usually, an approximation can be improved by increasing the number of terms in the expansion. However, the better method would be to drop to a different line of the table; that is, to ensure that the calculation is an interpolation rather than an extrapolation.

A.2. BASIC CALCULUS: DIFFERENTIATION AND INTEGRATION

Let us review some basic principles of differential calculus before considering discrete methods that are applicable to computer calculations.

Differentiation

Let $f(x)$ be a function of the variable x. If x increases by an amount Δx, the function varies by an amount $\Delta f = f(x + \Delta x) - f(x)$. The ratio $\Delta f/\Delta x$ is a measure of the relative variation of $f(x)$ with x. In the limit, as Δx becomes infinitesimally small, the ratio $\Delta f/\Delta x$ for a continuous function $f(x)$ approaches an asymptotic value, the *derivative* df/dx of the function $f(x)$ with respect to x.

$$\frac{df(x)}{dx} \equiv \lim_{\Delta x \to 0} \frac{f(x + \Delta x) - f(x)}{\Delta x} \tag{A.19}$$

The derivative of $f(x)$ at $x = x_0$ is written $\frac{df(x_0)}{dx}$ and corresponds to the slope of the function evaluated at x_0 or the tangent to the curve at that point.

Example A.2 To find the derivative of $f(x) = x^n$, we can expand the function $f(x + \Delta x)$ to first order in a Taylor series.

Thus, with $n = 4$, we have $f(x) = x^4$ and $df/dx = 4r^3$.

$$\frac{d(x^n)}{dx} = \lim_{\Delta x \to 0} \frac{(x^n + nx^{n-1}\Delta x) - x^h}{\Delta x}$$

$$= \frac{nx^{n-1}\Delta x}{\Delta x} = nx^{n-1}$$

Example A.3 For $f(x) = \sin x$, we can write

$$\sin(x + \Delta x) = (\sin x)(\cos \Delta x) + (\sin \Delta x)(\cos x)$$

and again expand $f(x)$ to obtain

$$\frac{d(\sin x)}{dx} = \lim_{\Delta x \to 0} \frac{\sin(x + \Delta x) - \sin x}{\Delta x}$$

$$= \lim_{\Delta x \to 0} \frac{(\sin x)(\cos \Delta x) + (\sin \Delta x)(\cos x) - \sin x}{\Delta x}$$

$$= \frac{\sin x + (\Delta x)(\cos x) - \sin x}{\Delta x} = \cos x$$

Similarly, for $f(x) = \cos x$, we find $df/dx = -\sin x$.

SUMS AND PRODUCTS The derivative of a sum of functions is equal to the sum of the derivatives of the individual functions. Consider the function

$$f(x) = g(x) + h(x)$$

The derivative of this function is the sum of the derivatives of the individual terms.

$$\frac{df(x)}{dx} = \frac{dg(x)}{dx} + \frac{dh(x)}{dx}$$

The derivative of a product of functions, however, is not equal to the product of the derivatives. Consider the function

$$f(x) = g(x) \times h(x)$$

We can rewrite Equation (A.19) as

$$\lim_{\Delta x \to 0} f(x + \Delta x) = \lim_{\Delta x \to 0} \left[f(x) = \Delta x \frac{df(x)}{dx} \right] \tag{A.20}$$

and show that

$$\begin{aligned} \frac{d[g(x) \times h(x)]}{dx} &= \lim_{\Delta x \to 0} \frac{g(x + \Delta x)h(x + \Delta x) - g(x)h(x)}{\Delta x} \\ &= \lim_{\Delta x \to 0} \frac{1}{\Delta x} \left\{ \left[g(x) + \Delta x \frac{dg(x)}{dx} \right] \left[h(x) + \Delta x \frac{dh(x)}{dx} \right] - g(x)h(x) \right\} \\ &= g(x) \frac{dh(x)}{dx} + h(x) \frac{dg(x)}{dx} \end{aligned}$$

FUNCTIONS OF FUNCTIONS If the function $f(x)$ can be expressed as a function of a function $g(x)$ of x,

$$f(x) = f[g(x)]$$

the derivative of $f(x)$ with respect to x can be expressed in terms of the derivative of $g(x)$ with respect to x. If we expand the definition of Equation (A.19) for the derivative, we can make use of the relationship of Equation (A.20) to expand still further.

$$\begin{aligned} \frac{df(x)}{dx} &= \lim_{\Delta x \to 0} \frac{f\left[g(x) + \Delta x \frac{dg(x)}{dx} \right] - f[g(x)]}{\Delta x} \\ &= \lim_{\Delta x \to 0} \frac{f[g(x)] + \Delta x \frac{dg(x)}{dx} \frac{df(x)}{dg(x)} - f[g(x)]}{\Delta x} \\ &= \frac{df(x)}{dg(x)} \frac{dg(x)}{dx} \end{aligned} \tag{A.21}$$

Example A.4 If $f(x) = (a - bx^3)^2$, define $g(x) = a + bx^3$ so that $f(x) = [g(x)]^2$. The first factor in Equation (A.21) is the derivative of a square, and the second factor is the derivative of a cubic polynomial.

$$\frac{df(x)}{dg(x)} = 2g(x) = 2(a + bx^3) \qquad \frac{dg(x)}{dx} = 3bx^2$$

$$\frac{df(x)}{dx} = 2(a + bx^3)3bx^2 = 6bx^2(a + bx^3)$$

HIGHER-ORDER DERIVATIVES Higher-order derivatives are defined as derivatives of derivatives. For example, the second derivative of a function $f(x)$ is just the derivative of the first derivative.

$$\frac{d^2f(x)}{dx^2} \equiv \frac{d}{dx}\left[\frac{df(x)}{dx}\right]$$

For the nth-order derivative $d^n f(x)/dx^n$, we simply take the derivative n times in succession. For example, if $f(x) = x^4$ as in Example A.2, the second derivative is $12x^2$. Similarly, the fourth derivative of either $\sin x$ or $\cos x$ is equal to itself.

PARTIAL DERIVATIVES If the function $f(x, y)$ is dependent on two variables x and y, we must define derivatives of the function with respect to each of the independent variables. To determine the *partial derivative* of f with respect to x, $\partial f/\partial x$, we consider that y is a constant and proceed as we would for an ordinary derivative. Similarly, to determine the partial derivative $\partial f/\partial y$ we consider that x is constant.

$$\frac{\partial f(x, y)}{\partial x} \equiv \lim_{\Delta x \to 0} \frac{f(x + \Delta x, y) - f(x, y)}{\Delta x} = \frac{df(x)}{dx}$$

$$\frac{\partial f(x, y)}{\partial y} \equiv \lim_{\Delta y \to 0} \frac{f(x, y + \Delta y) - f(x, y)}{\Delta y} = \frac{df(y)}{dy}$$

Higher-order partial derivatives include not only higher-order derivatives with respect to one variable, but also cross-partial derivatives with respect to two or more variables simultaneously.

$$\frac{\partial^2 f(x, y)}{\partial x^2} \equiv \frac{\partial}{\partial x}\left[\frac{\partial f(x, y)}{\partial x}\right]$$

$$\frac{\partial^2 f(x, y)}{\partial x\, \partial y} \equiv \frac{\partial}{\partial x}\left[\frac{\partial f(x, y)}{\partial y}\right] = \frac{\partial}{\partial y}\left[\frac{\partial f(x, y)}{\partial x}\right] = \frac{\partial^2 f(x, y)}{\partial y\, \partial x}$$

MINIMA AND MAXIMA A function $f(x)$ is said to have a *local minimum* at $x = x_{min}$ if the values of $f(x_{min} \pm \Delta x)$ are larger than the value of $f(x_{min})$ for infinitesimal changes Δx about x_{min}. Similarly, the function has a *local maximum* if the values of $f(x_{max} \pm \Delta x)$ are smaller than $f(x_{max})$. At either a minimum or a maximum of a function, the derivative of the function is zero,

$$\frac{df(x_m)}{dx} = 0$$

corresponding to a tangent that is parallel to the x-axis.

The question of whether the function is a minimum or a maximum at x_m can be resolved by examining the second derivative. If the second derivative is positive, the curvature of the function is upward and $f(x_m)$ is a minimum. If the second derivative is negative, the $f(x_m)$ is a maximum.

FUNCTIONS OF MORE THAN ONE VARIABLE With functions of more than one variable, for example $f(x, y)$, we can still consider the function to have a minimum in parameter space, but we must be careful to assure that the function has a minimum simultaneously with respect to all parameters.

Integration

Integration is the inverse of differentiation. To find the integral $F(x)$ of the function $f(x)$,

$$F(x) = \int f(x)dx$$

we must find a function $F(x)$ such that $\frac{dF(x)}{dx} = f(x)$.

However, this definition is not unique. An undetermined constant must be added to the solution to allow for the fact that the derivative of a constant is zero.

Example A.5 Consider the integral of the function $f(x) = x^3$. We observe that $F(x) = x^4/4$ is a solution:

$$\frac{dF(x)}{dx} = \frac{d(x^4/4)}{dx} = x^3 = f(x)$$

However, $F(x) = x^4/4 + C$ is also a solution, where C is any quantity that is not a function of x. Thus, the solution to an *indefinite integral* must include an added constant.

A *definite integral* is the integral of a function between two specific values of the independent variable, and is written

$$I = \int_a^b f(x)dx$$

To find the definite integral of a function, we integrate it, calculate the value of the integral at $x = b$ and at $x = a$, and find the difference between the two values. This is equivalent to calculating the area under the function $f(x)$ between the two limits a and b.

Example A.6 Consider the integral of the function $f(x) = x^3$ between the limits $x = 1.0$ and $x = 2.0$.

$$I = \int_{1.0}^{2.0} f(x)dx = \int_{1.0}^{2.0} x^3 dx = x^4/4\Big|_{1.0}^{2.0} = [2^4 - 1^4]/4 = 15/4$$

Note that a definite integral is not a function of variable of integration x.

From the results of Example A.3,

$$\int \sin x\, dx = -\cos x + C \qquad \text{and} \qquad \int \cos x\, dx = \sin x + C$$

A.3. NUMERICAL DIFFERENTIATION AND INTEGRATION

With the interpolation expressions discussed in Section A.1, it is relatively straightforward to obtain expressions for derivatives and integrals in terms of expansions to order n.

Differentiation

We can differentiate Equation (A.18) to find approximations for the derivatives of the function $f(x)$. We obtain

$$\frac{dP_n(x)}{dx} = \frac{1}{h}\frac{dP_n(x)}{d\alpha} = [\Delta f(x_0) + (2\alpha - 1)\Delta^2 f(x_0)/2! + (3\alpha^2 - 6\alpha + 2)\Delta^3 f(x_0)/3! + \cdots]/h \tag{A.22}$$

and

$$\frac{d^2P_n(x)}{dx^2} = \frac{1}{h^2}\frac{d}{d\alpha}\left[\frac{dP_n(x)}{d\alpha}\right] = [\Delta^2 f(x_0) + (\alpha - 1)\Delta^3 f(x_0) + \cdots]/h^2 \tag{A.23}$$

We should note that the use of forward differences introduces an asymmetry in the calculation. For a general solution, we could replace the forward differences by central differences, which are taken symmetrically about a central starting point. For a particular problem, we can usually arrange the expansion to provide reasonable symmetry of the differences about the point of interest. Thus, we can replace Equations (A.22) and (A.23) by

$$\frac{dP_n(x)}{dx} = \Delta f(x_0)/h = \frac{f(x + h/2) - f(x - h/2)}{h} \tag{A.24}$$

and

$$\frac{d^2P_n(x)}{dx^2} = \Delta^2 f(x_0)/h^2 = \frac{f(x + h) - 2f(x) + f(x - h)}{h^2} \tag{A.25}$$

Integration

Integrating Equation (A.18) leads to expressions for calculating the numerical integral in various orders, depending on the number of terms in the polynomial approximation. There are various forms for each order, depending on how we choose the limits of integration. We quote three of the most useful forms with the remainder estimates.

First-order, endpoint trapezoidal

$$\int_{x_0}^{x_1} f(x)\,dx = \frac{h}{2}[f(x_0) + f(x_1)] - \frac{h^3}{12}f^{(2)}(\xi)$$

(first-order closed-end trapezoidal)

$$\int_{x_0}^{x_2} f(x)\,dx = 2hf(x_1) + \frac{h^3}{3}f^{(2)}(\xi) \quad \text{(first-order open end)}$$

$$\int_{x_0}^{x_2} f(x)\,dx = \frac{h}{3}[f(x_0) + 4f(x_1) + f(x_2)] - \frac{h^5}{90}f^{(4)}(\xi)$$

(second-order closed-end Simpson's rule)

The factors $f^{(n)}(\xi)$ in the remainder estimates represent the nth derivative of the function evaluated at some (unknown) value of x in the range of the integral.

Note the large reduction on the error estimate in going from either of the first-order approximations to the second-order approximation.

For an integral over an extended range of x, it is usually advisable to employ a series of first- or second-order integrals over sections of the function, rather than to attempt to fit a large region with a higher-order function. In fact, it can be shown that the gain in accuracy in going from a second- to a third-order numerical integral is relatively small, and, for the same number of calculations of the ordinate y_j, the second-order Simpson rule may be more accurate than the third-order form. This relation applies in general to even and odd orders, so that, to make a significant improvement in the numerical integration of a function, one should advance to the next higher *even* order.

Thus, to find the integral by Simpson's rule of $f(x)$ over an extended range between $x = x_0$ and $x = x_n$, we divide the region into n equal intervals in x, with $nh = (x_n - x_0)$, to obtain

$$\int_{x_0}^{x_n} f(x)\,dx = \frac{h}{3}[f(x_0) + 4f(x_1) + 2f(x_2) + 4f(x_3) + \cdots \\ + 4f(x_{n-1}) = f(x_n)] - \frac{nh^5}{180} f^{(4)}(\xi) \qquad \text{(A.26)}$$

where ξ is the value of x somewhere in the range of integration.

Program A.1 SIMPSON (Appendix E) calculates an extended integral by the second-order approximation of Equation (A.26). See Programs 11.1 and 11.2 for examples of the use of this routine.

The user supplies four arguments:

1. FUNCT: the name of the function to be integrated. The function must have one real argument. If other arguments are required, they must be made accessible to the function as global variables.
2. NINT: the number of *double* intervals. The interval is calculated as DX = (HILIM-LOLIM)\(2*NINT);
3. LOLIM and
4. HILIM: the integration limits.

A.4 CUBIC SPLINES

If we attempt to represent by an nth-degree polynomial a function that is tabulated at $n + 1$ points, we are apt to obtain disappointing results if n is large. The polynomial will necessarily coincide with the data points, but may exhibit large oscillations between points. In addition, if there are many data points, the calculations can become rather cumbersome. It is often better to make several low-order polynomial fits to separate regions of the function, and this procedure is usually satisfactory for simple interpolation in tables. However, if we want a smooth function, which passes through the data points, the results may not be satisfactory.

Suppose we have calculated a function at $n + 1$ points, and want to represent the function as a smooth curve on a graph. The nth-order polynomial is out—too wiggly. Breaking the curve up into small sections produces disjointed segments on the plot. It is unlikely that they will combine to form a smooth curve. What do we do now? Reach for our pencil and trusty drafting spline? No, we call up our spline fitting subroutine and let it join up the separate fits for us.

Spline fitting procedures have other uses besides plotting pretty curves on graphs, but the plotting function is of interest to us and is easily illustrated. Suppose we choose to make a series of cubic fits to successive groups of data points. What conditions do we need to produce a smooth curve that passes through the data points? We want the first and second derivatives, as well as the function itself, to be continuous at the data points. Suppose we consider a separate cubic polynomial for each interval on the graph, or a total of n polynomials for the $n + 1$ points. Then we write the polynomial equation, take derivatives, and, at each data point, equate the first and second derivatives of the left-side polynomial to those of the right-side polynomial.

Following the method discussed in Thompson (1984), we begin by writing the Taylor series for the cubic polynomial for interval i, expanded about the point x_i

$$y(x) = y(x_i) + (x - x_i)\frac{dy(x_i)}{dx} + (x - x_i)^2 \frac{d^2y(x_i)}{dx^3}\Big/2! + (x - x_i)^3 \frac{d^3y(x_i)}{dx^3}\Big/3! \tag{A.27}$$

where the function and derivatives are evaluated at x_i. This can be written in a more concise form as

$$y(x) = y_i + (x - x_i)y'_i + (x - x_i)^2 y''_i/2 + (x - x_i)^3(y''_{i+1} - y''_i)/6h \tag{A.28}$$

where y'_i and y''_i stand for the first and second derivatives evaluated at $x = x_i$ and the third derivative has been replaced by its divided difference form, which is exact for a cubic function. At $x = x_i$, we have $y = y_i$, as required. We can also set $x = x_{i+1} = x_i + h$ and solve the equation

$$y(x_{i+1}) = y_i + (x_{i+1} - x_i)y'_i + (x_{i+1} - x_i)^2 y''_i/2 + (x_{i+1} - x_i)^3(y''_{i+1} + y''_i)/6h \tag{A.29}$$

to obtain

$$y_{i+1} - y_i = hy'_i + h^2[2y''_i + y''_{i+1}]/6 \tag{A.30}$$

We repeat the calculation, using the equation for $y(x)$ in interval $i - 1$ [i.e., we replace i by $i - 1$ in Equation (A.29)],

$$y(x) = y_{i-1} + (x - x_{i-1})y'_{i-1} + (x - x_{i-1})^2 y''_{i-1}/2 + (x - x_{i-1})^3(y''_i - y'_{i-1})/6h \tag{A.31}$$

and again require that $y(x) = y(x_i)$ at the ith data point and obtain

$$y_i - y_{i-1} = hy'_{i-1} + h^2[2y''_{i-1} + y''_i]/6 \tag{A.32}$$

To establish the continuity conditions at the data points, we need the first derivative in the interval i,

$$y'(x) = y'_i + (x - x_i)y''_i + (x - x_i)^2(y''_{i+1} - y''_i)/2h \tag{A.33}$$

which we equate to the first derivative in the interval $i - 1$ at the position $x = x_i$, to obtain

$$y'_i - y'_{i-1} = h[y''_i + y''_{i-1}]/2 \tag{A.34}$$

Similarly, equating the derivatives at the boundary $x = x_{i+1}$ gives

$$y'_{i+1} - y'_i = h[y''_{i+1}y''_i]/2 \tag{A.35}$$

(Repeating the procedure with the second derivative leads to an identity, because our use of the divided difference form for the third derivative assures continuity of the second derivative across the boundaries.) Eliminating the first derivatives from Equations (A.30), (A.32), (A.34), and (A.35) gives us the spline equation

$$y''_{i-1} + 4y''_i + y''_{i+1} = D_i \tag{A.36}$$

with

$$D_i = y[y_{i+1} - 2y_i + y_{i-1}]/h^2 \tag{A.37}$$

Note that the D_i are proportional to the second differences of the tabulated data and are all known. We can write Equation (A.36) as a set of linear equations relating the unknown variables y'', beginning with $i = 2$ and ending with $i = n - 1$:

$$y''_1 + 4y''_2 + y''_3 = D_2 \tag{A.38a}$$

$$y''_2 + 4y''_3 + y''_4 = D_3 \tag{A.38b}$$

$$\vdots$$

$$y''_{n-3} + 4y''_{n-2} + y''_{n-1} = D_{n-2} \tag{A.38c}$$

$$y''_{n-2} + 4y''_{n-1} + y''_n = D_{n-1} \tag{A.38d}$$

These equations can be solved for the second derivatives y''_i, as long as we know the values of y''_1 and y''_n. One possibility is to set the second derivatives to 0 to obtain *natural splines*. Alternatively, we may use the true second derivatives, if they are known, or a numerical approximation.

For example, suppose we have only four points to consider. Then, if we know y''_1 and y''_4, we can solve the simultaneous Equations (A.38a) and (A.38b) for y_2 and y_3. Similarly, if we have a full set of n equations, we can rewrite Equation (A.38a) to express $y''_2 = (D_2 - y''_1 - y''_3)/4$, and substitute this expression into Equation (A.38b) to eliminate y''_2. Then, we repeat the procedure to eliminate y''_3 from the next equation. We continue this procedure until we reach the last equation, which will contain only terms in y''_1, y''_{n-1}, and y''_n. Because y''_1 and y''_n are known, we can solve this equation for y''_{n-1}, and then work back down the chain determining successively y''_{n-2}, y''_{n-3}, and so forth, until we reach Equation (A.38a) from which we determine the last unknown y''_2. Once we have found the values of the y''_i, we can find the y'_i from Equation (A.30) or (A.32), and use Equation (A.28) to interpolate in each interval.

The solution of Equation (A.38) is discussed in several textbooks. Essentially, one sets up recursion relations to build a table of the second derivatives y''. The method is illustrated by the computer routines SPLINEMAKE listed in Appendix E.

An interesting alternative method of solving the set of simultaneous equations, Equations (A.38), is to set them up in a spreadsheet program. Then, when the boundary values y''_i and y''_n are supplied, the program will readjust the variables until they stabilize at the solutions to Equations (A.38). Although this method is not very practical for graphical applications where we want to build the solution into our plotting program, it does provide a quick way of finding the second derivatives and an interesting illustration of the solution.

As with all techniques, a certain amount of care must be exercised in using spline routines. The choice of a second derivative at the boundary may have an important effect on the interpolation at the ends of the function, and a wrong choice, for example, can produce undesirable shapes at the edges of a plot. Then too, although the spline routine assures a smooth variation between the data points, with continuity of the function and first and second derivative across the points, it cannot guarantee that there will be no peculiar oscillation between the points.

Program A.2 SPLINE INTERPOLATION (Appendix E)
SPLINEMAKE numerically calculates a table of second derivatives for a spline interpolation by the method discussed in the previous paragraphs.
SPLINEINT performs the interpolation. For simplicity, we have chosen to store only the second derivatives and to calculate the first and third derivatives as needed in functions D1YDX1 and D3YDX3. If speed is important, the derivatives could be computed and stored in arrays.

A spline interpolation routine is especially useful for plotting curves on graphs. The routine has been used to produce many of the graphs in this book.

A.5 ROOTS OF NONLINEAR EQUATIONS

Finding roots of nonlinear equations is essentially the reverse of an interpolation problem. When we interpolate a function, our object is to find a value of the dependent variable y at a specific value of the independent variable x. When we are searching for the root of a function, we are trying to find the value of x at a particular value, usually 0, of y. However, interchanging the variables completely changes the nature of the problem. Interpolation involves straightforward application of well-defined equations that are independent of the form of the original function: Finding roots of nonlinear equations may require different equations for different problems and almost always requires some sort of a search and iteration procedure.

The diffraction of light by a single slit provides an interesting example of a nonlinear equation. It is well known that the position of the interference maxima and minima from double slits and diffraction gratings can be determined analytically from consideration of the phase difference between the rays that pass through each slit, but only the minima of the diffraction pattern of a single slit can be found in this way. To find the position of a maximum, with the exception of the central one, we must differentiate the expression for the intensity with respect to the phase α:

$$I = I_0 \left(\frac{\sin\alpha}{\alpha}\right)^2 \qquad \text{with} \qquad \alpha = \frac{\pi a}{\lambda}\sin\theta \tag{A.39}$$

In Equation (A.39), I_0 is the intensity of the light at the central maximum ($\theta = 0$), I is the intensity at angle θ, λ is the wavelength of the light, and a is the slit width. The position of the maximum is given by solving

$$\frac{dI}{d\alpha} = 2I_0\left(\frac{\sin\alpha}{\alpha^3}\right)(\alpha\cos\alpha - \sin\alpha) = 0 \tag{A.40}$$

to obtain the value α_r at the root of the equation

$$f(\alpha) = \alpha_r - \tan\alpha_r = 0 \tag{A.41}$$

The first root is at $\alpha_r = 0$. The other roots cannot be calculated analytically and must be found by an iterative method. An approximate solution can be obtained by rewriting Equation (A.41) as

$$\alpha_r = \tan\alpha_r \tag{A.42}$$

and plotting separately the left and right sides to find the intersection of the straight line and the tangent curves. There are several mathematical ways to solve the problem, but making a plot of the function is always a good starting procedure.

Trial-and-Error: The Half-Interval Method

With a personal computer, trial-and-error may be a suitable method for solving the occasional root finding problem. An orderly approach is advisable and the half-interval method is convenient. The procedure is to write a little program that requests a trial value of the root and calculates the function and displays its value. The initial trial value might be obtained from a graph, or perhaps by mapping the function for various values of the independent variable x, until a reasonable estimate of the root has been obtained. Then, a second trial x is submitted, which produces a value of y on the other side of the root. The half-interval method begins at this point. The procedure is to select a third trial value that is midway between the two that bracket the root. For the fourth trial value, we use the mean of the most recent value, and whichever of the two previous trials was on the other side of the root. The process continues until the root is found to the desired accuracy.

This rather primitive method of root finding could be improved with a little programming to let the program decide which root to choose, to calculate the mean, and perform the next trial. The program could proceed in a loop until the root had been found to a predefined degree of accuracy, or the calculation could be stopped manually. However, if we are willing to program that little bit of logic, slightly more effort will produce a much faster root-finding program.

Secant Methods

The gain in speed comes from using the slope of the function in the calculation. We begin with two trial estimates of the root, x_k and x_{k+1}, preferably, but not necessarily, on either side of the root. Then we write an expression for a linear interpolation between the two points. Equation (A.10) gives

$$f(x) = y_k + (x - x_k)\frac{(y_{k+1} - y_k)}{(x_{k+1} - x_k)} \tag{A.43}$$

where we have written $y_k = f(x_k)$ and so forth. Setting $f(x) = 0$ and solving for x gives us an approximation to the value of x at the root:

$$x = x_k - y_k\frac{x_{k+1} - x_k}{y_{k+1} - y_k} = \frac{x_k y_{k+1} - x_{k+1} y_k}{y_{k+1} - y_k} \tag{A.44}$$

For the next trial, we replace x_{k+1} or x_{k+2} by the value x found in Equation (A.44) and repeat the calculation. The process can be repeated until the root is approximated as closely as desired. This is the first-order secant method.

There are various ways of choosing which of the previous values of x (x_k or x_{k+1}) to keep for the next iteration. The simplest is to keep the most recent value and discard the older value. Another way is to choose whichever is closer to the root [i.e., gives a smaller value of $f(x)$]. A third is to start the process with two values that straddle the root (i.e., give opposite signs for y_1 and y_2) and to continue to choose values that straddle the root after each iteration. This is the Regulo-Falsi method.

Clearly any method will find the root most quickly if the starting values are close to the root, but, in principle, the secant methods will almost always find a root of the function, eventually. With some functions, such as those that are antisymmetric about the root, there is the possibility that the search by the Regulo-Falsi method, for example, will jump back and forth across the root and never approach it. Additionally, for functions with several roots, we may not always find the one we want. Problems may also arise if two roots are very close together.

Newton-Raphson Method

Instead of calculating the slope by finite differences, as in the secant method, we could use the tangent, or derivative of the function, if it can be calculated. Then, we can replace Equation (A.43) by

$$f(x) = y_k + (x - x_k)\frac{df(x_k)}{dx} \tag{A.45}$$

where x_k and y_k are the values of x and $f(x)$ after the kth iteration. We find the next estimate x_{k+1} for the root, as before, by setting $f(x)$ in Equation (A.45) to zero to obtain

$$x_{k+1} = x_k - y_k \div \frac{df(x_k)}{dx} \tag{A.46}$$

Example A.7 Table A.3 shows steps in an iterative calculation of the second and third roots of Equation (A.41) by the secant and Newton-Raphson methods. Starting values were chosen by examining a plot of tan x versus x.

Simultaneous Nonlinear Equations

In the examples of alternate fitting methods in Section 6.6, we obtained two pairs of coupled, nonlinear equations, Equations (6.24) and (6.27), which we wished to

TABLE A.3
Determination of the first two nonzero roots of α = tan α

	First root		Second root	
Trial	x	y	x	y
	(a) Newton's Method†			
0	4.40000	1.30368	7.70000	1.25713
1	4.53598	−1.07376	7.73028	−0.31270
2	4.50186	−0.17769	7.72545	−0.01188
3	4.49375	−0.00679	7.72525	−0.00002
4	4.49341	−0.00001	7.72525	−0.00000
5	4.49341	−0.00000		
	(b) Secant Method‡			
0	4.40000	1.30368	7.80000	−10.70682
0	4.50000	−0.13733	7.70000	1.25713
1	4.49047	0.05854	7.71051	0.78849
2	4.49332	0.00184	7.72819	−0.17931
3	4.49341	−000003	7.72491	0.02025
4	4.49341	0.00000	7.72524	0.00047
5			7.72525	−0.00000

†The calculation continues without assistance after the initial trial value has been selected

‡Two x, y pairs are required for each stage of the calculation. After the first trial, the most recently calculated x, y pair was used with whichever of the two previous pairs was closer to the root.

solve for the parameters a and b. We used the secant method to solve these equations.

Consider the two equations

$$f_a(u, v) = 0 \quad \text{and} \quad f_b(u, v) = 0 \tag{A.47}$$

which we wish to solve for u and v. We define the first *partial divided differences*,

$$\begin{aligned}
f_{au} &= f_a[u_0, v_0; u_1, v_0] \equiv \frac{f_a(u_1, v_0) - f_a(u_0, v_0)}{u_1 - u_0} \\
f_{av} &= f_a[u_0, v_0; u_0, v_1] \equiv \frac{f_a(u_0, v_1) - f_a(u_0, v_0)}{v_1 - v_0} \\
f_{bu} &= f_b[u_0, v_0; u_1, v_0] \equiv \frac{f_b(u_1, v_0) - f_b(u_0, v_0)}{u_1 - u_0} \\
f_{bv} &= f_b[u_0, v_0; u_0, v_1] \equiv \frac{f_b(u_0, v_1) - f_b(u_0, v_0)}{v_1 - v_0}
\end{aligned} \tag{A.48}$$

and, following Equation (A.43), write for a first-order expansion

$$\begin{aligned}
& f_a(u, v) = f_a(u_0, v_0) + (u - u_0)f_{au} + (v - v_0)f_{av} \\
\text{and} \quad & f_b(u, v) = f_b(u_0, v_0) + (u - u_0)f_{bu} + (v - v_0)f_{bv}
\end{aligned} \tag{A.49}$$

If we assume that f_a and f_b are linear in u and v, we can find a first approximation to the roots by setting $f_a(u, v)$ and $f_b(u, v)$ to zero in Equation (A.49) and solving the two coupled linear equations for u and v:

$$\begin{aligned} uf_{au} + vf_{av} - u_0 f_{au} - v_0 f_{av} + f_a(u_0, v_0) = 0 \\ uf_{bu} + vf_{bv} - u_0 f_{bu} - v_0 f_{bv} + f_b(u_0, v_0) = 0 \end{aligned} \tag{A.50}$$

Solution by the determinant method gives

and

$$\begin{aligned} u_2 = u = (Af_{bv} - Bf_{av})/D \\ v_2 = v = (Bf_{au} - Af_{bu})/D \end{aligned} \tag{A.51}$$

with

$$\begin{aligned} D &= f_{au}f_{bv} - f_{av}f_{bu} \\ A &= -u_0 f_{au} - v_0 f_{av} + f_a(u_0, v_0) \\ B &= -u_0 f_{bu} - v_0 f_{bv} + f_b(u_0, v_0) \end{aligned} \tag{A.52}$$

We then repeat the procedure with coordinate pairs (u_1, v_1) and (u_2, v_2), to obtain the next approximation, until the roots have been found to the desired degree of accuracy.

A.6 DATA SMOOTHING

The concept of smoothing is not one that meets with universal approval. The discussion that follows should be considered with one caveat: For rigorously valid least-squares fitting, smoothing is neither desirable nor permissible; however, there are cases where smoothing can be beneficial, and, therefore, the techniques are introduced.

Consider, for example, the discussion of Section 9.2 of the determination of the area under a peak from a least-squares fit to a histogram of the data. Least-squares fitting techniques applied to data that are distributed according to Poisson distributions, rather than Gaussian distributions, underestimate the area of a peak by an amount equal to the value of χ^2. We have seen that we can improve the result by decreasing the value of χ^2 at its minimum. Similarly, if the shape of the fitting function does not exactly simulate that of the parent distribution, a better fit to the data by decreasing χ^2 can yield an improved estimate of the area under a peak.

Another example that might benefit from application of a smoothing algorithm is the parameterization of data for use in a Monte Carlo or other program. In preparing experimental proposals, it is often necessary to estimate yields and distributions based on currently available data. Such data are often sparse and generally must be expressed in parametric form for ease and speed of use in the Monte Carlo simulation program. Smoothing can be useful to average out fluctuations and allow the data to be expressed with a few parameters by a least-squares fit or an interpolation procedure.

In other words, if rigorously valid results are not required, but rather an averaged estimate of the distribution, smoothing may help obtain more reliable estimates. The improvement in the estimate of one parameter must, of course, be accompanied by a decrease in information of some other parameter or parameters.

For example, an improved estimate of the area under a peak would be accompanied by an increased uncertainty in the estimates of the width and position of the peak.

Whatever smoothing or other manipulation is done must conserve the information pertaining to the desired parameters. The averaging techniques that we shall discuss, for example, conserve the area under a peak but not the width of the peak Similarly, this method would be useful for improving the estimate of the constant term of a polynomial but not the coefficients of the other terms.

Data smoothing is similar to the data "smearing" introduced in Chapter 5 to simulate measuring uncertainties in "measurements" generated by a Monte Carlo program. In the Monte Carlo program we used Gaussian smearing; that is, we allowed each event a Gaussian probability distribution about its mean.

In this section, we are dealing with binned data, and thus, for Gaussian smoothing, could consider a Gaussian integration that spreads each event over adjacent bins. Because our object here is to smooth the data, we are at liberty to choose the width of the smearing function to produce the desired degree of uniformity in the data, limited by the requirement that we do not damage the very variable we are trying to study.

The binomial distribution is a useful smoothing function. Suppose we want to smooth low statistics experimental data that follow a Gaussian peak in a way that preserves the area under the peak. Let us assume that the background slope is gentle enough that smoothing will not affect its determination drastically.

We can approximate the Gaussian peak with a binomial distribution with $p = ½$ (see Section 2.1):

$$y(x) = \frac{1}{\sigma\sqrt{2\pi}} e^{-1/2\left(\frac{x-\mu}{\sigma}\right)^2} \simeq \left(\frac{1}{2}\right)^n \frac{n!}{X!(n-X)!} \tag{A.53}$$

We can relate the widths σ and the means of the two distributions

$$\sigma_B^2 = np(1-p) = n/4 = \sigma^2 \quad \bar{X} = np = n/2 \quad \bar{x} = \mu \tag{A54}$$

to find the relationships among the parameters

$$n = 4\sigma^2 \quad X = x - \mu + n/2 = x - \mu + 2\sigma^2 \tag{A.55}$$

We can then express the binomial distribution of Equation (A.53) as

$$y(x) = \left(\frac{1}{2}\right)^n \frac{n!}{(n/2 + x - \mu)!(n/2 - x + \mu)!} \tag{A.56}$$

Let us smooth the data by averaging over adjacent channels with a binomial distribution spanning three channels:

$$y'(x) = 1/4y(x-1) + 1/2y(x) + 1/4y(x+1) \tag{A.57}$$

If we fold this averaging into the distribution of Equation (A.53), the result is also binomial:

$$y'(x) = \left(\frac{1}{2}\right)^{n+2} \frac{(n+2)!}{(n/2 + 1 + x - \mu)!(n/2 + 1 - x + \mu)!} \tag{A.58}$$

The new distribution has the same mean $\bar{x} = \mu$ but a larger width $\sigma'^2 = n'/4 = (n + 2)/4$ with the variance increased by ½:

$$\sigma'^2 = \sigma^2 + 1/2 \tag{A.59}$$

Similarly, we could smooth over five channels by using a formula similar to Equation (A.57) but with five terms with coefficients given by the binomial expansion

$$\begin{aligned} y''(x) = {} & 1/16y(x-2) + 1/4y(x-1) + 3/8y(x) \\ & + 1/4y(x+1) + 1/16y(x+2) \end{aligned} \tag{A.60}$$

A five-channel smoothing is identical to two successive smoothings over three channels and yields a variance that is increased accordingly, $\sigma''^2 = \sigma^2 + 1$. Any such smoothings over $2n + 1$ adjacent channels is equivalent to n smoothings over three channels.

If we apply the smoothing of Equation (A.57) to a Gaussian distribution, the resulting distribution will also be nearly Gaussian because the shapes of the binomial and Gaussian distributions are nearly alike. In fact, if we are applying the smoothing because the original shape is not Gaussian enough, the averaging may make the shape more nearly Gaussian. If we apply binomial smoothing to a distribution that is not Gaussian, we should be aware that we are distorting the shape of the peak and making it more Gaussian.

If the width of the original Gaussian is not too small ($\sigma > 1$), the increase of Equation (A.59) should not be drastic because the addition is in quadrature. For a width $\sigma = 2$, for example, the new width $\sigma' = 2$ is only 5% larger. If the original width is very small ($\sigma < 1$), the approximation of Equation (A.53) is not valid because the Gaussian and binomial distributions are only similar in the limits of large n. A Gaussian fit to the data without smoothing would not be valid either, however, because the parameters of the fit are only meaningful if $\sigma \geq 1$. Because the averaging itself is a binomial distribution, the result is still expected to be a better approximation to a Gaussian distribution than the original data. For a smoothing over three channels, a Gaussian fit requires $\sigma \geq \sqrt{1/2}$ for the original data.

APPENDIX B

MATRICES

B.1 DETERMINANTS

In applying the method of least squares to both linear and nonlinear functions, we required the solution of a set of n simultaneous equations in n unknowns a_i similar to the following:

$$\begin{aligned} y_1 &= a_1 X_{11} + a_2 X_{12} + a_3 X_{13} \\ y_2 &= a_1 X_{21} + a_2 X_{22} + a_3 X_{23} \\ y_3 &= a_1 X_{31} + a_2 X_{32} + a_3 X_{33} \end{aligned} \tag{B.1}$$

where the constants y_i and X_{ij} are known quantities calculated from the data.

The symmetry of the right-hand side suggests that we write elements of the equations in a two-dimensional array

$$\boldsymbol{\alpha} = \begin{bmatrix} X_{11} & X_{12} & X_{13} \\ X_{21} & X_{22} & X_{23} \\ X_{31} & X_{32} & X_{33} \end{bmatrix} \tag{B.2}$$

and separate the other terms and coefficients into one-dimensional arrays.

$$\mathbf{a} = \begin{bmatrix} a_1 \\ a_2 \\ a_3 \end{bmatrix} \qquad \text{and} \qquad \boldsymbol{\beta} = \begin{bmatrix} y_1 \\ y_2 \\ y_3 \end{bmatrix} \tag{B.3}$$

Such arrays are called matrices, and we can write Equations (B.1) in matrix form as

$$\boldsymbol{\beta} = \boldsymbol{\alpha} \cdot \mathbf{a} \tag{B.4}$$

Alternatively, because in our problems the matrix $\boldsymbol{\alpha}$ is always symmetric, that is, the element α_{ij} is equal to the element α_{ji}, we can write the matrices **a** and $\boldsymbol{\beta}$ as row matrices

$$a = [a_1 \quad a_2 \quad a_3] \qquad \text{and} \qquad b = [y_1 \quad y_2 \quad y_3] \tag{B.5}$$

and express Equation (B.1) as

$$\boldsymbol{\beta} = \mathbf{a} \cdot \boldsymbol{\alpha} \tag{B.6}$$

We shall be concerned primarily with linear one-dimensional matrices and with symmetric square two-dimensional matrices that have the same number of rows and columns and are mirror-symmetric about the diagonal. Consider a square matrix **A**:

$$\mathbf{A} = \begin{bmatrix} A_{11} & A_{12} & \cdots & A_{1k} & \cdots & A_{1n} \\ A_{21} & A_{22} & \cdots & A_{2k} & \cdots & A_{2n} \\ \vdots & \vdots & \vdots & \vdots & \vdots & \vdots \\ A_{j1} & A_{j2} & \cdots & A_{jk} & \cdots & A_{jn} \\ \vdots & \vdots & \vdots & \vdots & \vdots & \vdots \\ A_{n1} & A_{n2} & \cdots & A_{nk} & \cdots & A_{nn} \end{bmatrix} \tag{B.7}$$

The *degree* of the matrix **A** is the number n of rows and columns; the jkth *element* (or *component*) of the matrix is A_{jk}; the *diagonal terms* are A_{jj}. If the matrix is diagonally *symmetric,* $A_{jk} = A_{kj}$ and there are n^2 elements but only $n(n + 1)/2$ different elements.

Matrix Algebra

If **A** and **B** are two square symmetric matrices of degree n, then their sum **S** is a square symmetric matrix of degree n with elements that are the sums of the corresponding elements of the two matrices

$$\mathbf{A} + \mathbf{B} = \mathbf{S} \qquad S_{jk} = A_{jk} + B_{jk} \tag{B.8}$$

The product **P** of the matrices **A** and **B** is a square matrix of degree n, with elements determined in the following way:

$$\mathbf{AB} = \mathbf{P} \qquad P_{jk} = \sum_{m=1}^{n} (A_{jm} B_{mk}) \tag{B.9}$$

The elements of the jth row of **A** are multiplied by the elements of the kth column of **B** and the products are summed to obtain the jkth element of **P**. In general, the matrix **P** will not be symmetric.

If **a** is a linear one-dimensional matrix, the product of **A** and **a** is only well defined if the product is taken in a particular order. If **a** is a column matrix, it must be multiplied on the left by the square matrix to yield another column matrix **c**:

$$\begin{bmatrix} A_{11} & \cdots & \cdots & A_{1n} \\ \vdots & \cdots & \vdots & \vdots \\ A_{j1} & \cdots & A_{jk} & \cdots & A_{jn} \\ \vdots & \cdots & \vdots & \vdots \\ A_{n1} & \cdots & \cdots & A_{nn} \end{bmatrix} \begin{bmatrix} a_1 \\ \vdots \\ a_k \\ \vdots \\ a_n \end{bmatrix} = \begin{bmatrix} c_1 \\ \vdots \\ c_k \\ \vdots \\ c_n \end{bmatrix} \qquad c_j = \sum_{k=1}^{n} (A_{jk} a_k) \tag{B.10}$$

If **a** is a row matrix, it must multiply the square matrix on the left to yield another row matrix **r**.

$$[a_1 \ \cdots \ a_j \ \cdots \ a_n] \begin{bmatrix} A_{11} & \cdots & & \cdots & A_{1n} \\ \vdots & \cdots & & \cdots & \vdots \\ A_{j1} & \cdots & A_{jk} & \cdots & A_{jn} \\ \vdots & \cdots & & \cdots & \vdots \\ A_{n1} & \cdots & & \cdots & A_{nn} \end{bmatrix}$$

$$= [r_1 \ \cdots \ r_k \ \cdots \ r_n] \qquad r_j = \sum_{j=1}^{n} (a_j A_{jk}) \tag{B.11}$$

The product of two linear matrices depends on the order of multiplication. The product of a row matrix **a** times a column matrix **b** is a scalar. If the order is reversed, the result is a square matrix that is *diagonal;* that is, for which only the diagonal terms are nonzero:

$$[a_1 \ \cdots \ a_n] \begin{bmatrix} b_1 \\ \vdots \\ b_n \end{bmatrix} = \sum_{j=1}^{n} (a_j b_j)$$

$$\begin{bmatrix} b_1 \\ \vdots \\ b_n \end{bmatrix} [a_1 \ \cdots \ a_n] = \begin{bmatrix} a_1 b_1 & \cdots & 0 & \cdots & 0 \\ \vdots & \cdots & \vdots & \cdots & \vdots \\ 0 & \cdots & a_j b_j & \cdots & 0 \\ \vdots & \cdots & \vdots & \cdots & \vdots \\ 0 & \cdots & 0 & \cdots & a_n b_n \end{bmatrix} \tag{B.12}$$

Determinants

The *determinant* of a square matrix is defined in terms of its algebra. The *order* of the determinant of a square matrix is equal to the degree n of the matrix. In this section, we shall mainly use determinants of order 3 as examples, although, unless otherwise specified, the comments apply to matrices of all orders. Manipulation of the rows may be substituted for columns throughout.

1. The determinant of the *unity matrix* is 1 where the unity matrix is defined as the diagonal matrix with all diagonal elements equal to 1:

$$|\mathbf{1}| = \begin{vmatrix} 1 & 0 & 0 \\ 0 & 1 & 0 \\ 0 & 0 & 1 \end{vmatrix} = 1 \tag{B.13}$$

2. If a column matrix of degree n is added to one column of a square matrix of degree n, the determinant of the result is the sum of the determinant of the original square matrix plus that of another square matrix obtained by substituting the column matrix for the modified column:

$$\begin{vmatrix} A_{11}+a_1 & A_{12} & A_{13} \\ A_{21}+a_2 & A_{22} & A_{23} \\ A_{31}+a_3 & A_{32} & A_{33} \end{vmatrix} = \begin{vmatrix} A_{11} & A_{12} & A_{13} \\ A_{21} & A_{22} & A_{23} \\ A_{31} & A_{32} & A_{33} \end{vmatrix} + \begin{vmatrix} a_1 & A_{12} & A_{13} \\ a_2 & A_{22} & A_{23} \\ a_3 & A_{32} & A_{33} \end{vmatrix} \tag{B.14}$$

3. If one column of a square matrix is multiplied by a scalar, the determinant of the result is the product of the scalar and the determinant of the original matrix:

$$\begin{vmatrix} cA_{11} & A_{12} & A_{13} \\ cA_{21} & A_{22} & A_{23} \\ cA_{31} & A_{32} & A_{33} \end{vmatrix} = c\begin{vmatrix} A_{11} & A_{12} & A_{13} \\ A_{21} & A_{22} & A_{23} \\ A_{31} & A_{32} & A_{33} \end{vmatrix} \tag{B.15}$$

4. If two columns of a square matrix are interchanged, the determinant retains the same magnitude but changes sign:

$$\begin{vmatrix} A_{12} & A_{11} & A_{13} \\ A_{22} & A_{21} & A_{23} \\ A_{32} & A_{31} & A_{33} \end{vmatrix} = -\begin{vmatrix} A_{11} & A_{12} & A_{13} \\ A_{21} & A_{22} & A_{23} \\ A_{31} & A_{32} & A_{33} \end{vmatrix} \tag{B.16}$$

5. The *minor* A^{jk} of an element A_{jk} of a square matrix of degree n is defined as the determinant of the square matrix of degree $n-1$ formed by removing the jth row and the kth column:

$$\mathbf{A} = \begin{vmatrix} A_{11} & A_{12} & A_{13} \\ A_{21} & A_{22} & A_{23} \\ A_{31} & A_{32} & A_{33} \end{vmatrix} \qquad A^{21} = \begin{vmatrix} A_{12} & A_{13} \\ A_{32} & A_{33} \end{vmatrix} \tag{B.17}$$

6. The *cofactor* $\mathrm{cof}(A_{jk})$ of an element A_{jk} of a square matrix of degree n is defined as the product of the minor and a phase factor:

$$\mathrm{cof}(A_{jk}) \equiv (-1)^{j+k}A^{jk} \tag{B.18}$$

7. With the preceding definitions 5 and 6, the determinant of a square matrix of degree n can be expressed in terms of cofactors of minors:

$$|\mathbf{A}| = \sum_{k=1}^{n}[A_{jk}\,\mathrm{cof}(A_{jk})] = \sum_{k=1}^{n}[(-1)^{j+k}A_{jk}A^{jk}] \tag{B.19}$$

Equation (B.19) is an iterative definition, because the cofactor is itself a determinant. The determinant of a matrix of degree 1, however, is equal to the single element of that matrix. The determinant of a square matrix of degree 2 is encountered often enough to make its explicit formula useful:

$$\begin{vmatrix} a & b \\ c & d \end{vmatrix} = ad - bc \tag{B.20}$$

and we can evaluate the determinant of a third-order matrix with the help of Equation (B.19):

$$\begin{vmatrix} A_{11} & A_{12} & A_{13} \\ A_{21} & A_{22} & A_{23} \\ A_{31} & A_{32} & A_{33} \end{vmatrix} = A_{11}\begin{vmatrix} A_{22} & A_{23} \\ A_{32} & A_{33} \end{vmatrix} - A_{12}\begin{vmatrix} A_{21} & A_{23} \\ A_{31} & A_{33} \end{vmatrix} + A_{13}\begin{vmatrix} A_{21} & A_{22} \\ A_{31} & A_{32} \end{vmatrix} \tag{B.21}$$

Computation

Matrix computation is generally simpler if we can manipulate matrices into diagonal form in which only the diagonal elements A_{jj} are nonzero. The determinant of a diagonal matrix is equal to the product of all the diagonal elements and the *trace* is their sum:

$$\left|\mathbf{A}_{\text{diag}}\right| = \prod_{j=1}^{n} A_{jj} \tag{B.22}$$

If we combine rules 2, 3, and 4 of the algebra for determinants, we can show that the determinant of a matrix is unchanged if the elements of any column, multiplied by an arbitrary scalar, are added to the elements of any other column. The determinant of the sum is equal to the sum of the two determinants, but one of these determinants has two identical columns except for a scalar factor that may be extracted, and is therefore equal to 0:

$$\begin{vmatrix} A_{11}+cA_{12} & A_{12} & A_{13} \\ A_{21}+cA_{22} & A_{22} & A_{23} \\ A_{31}+cA_{32} & A_{32} & A_{33} \end{vmatrix} = \begin{vmatrix} A_{11} & A_{12} & A_{13} \\ A_{21} & A_{22} & A_{23} \\ A_{31} & A_{32} & A_{33} \end{vmatrix} + c\begin{vmatrix} A_{12} & A_{12} & A_{13} \\ A_{22} & A_{22} & A_{23} \\ A_{32} & A_{32} & A_{33} \end{vmatrix} = \left|\mathbf{A}\right| \tag{B.23}$$

Thus, it is possible to eliminate all elements except one from a row by successively subtracting one column, appropriately scaled, from each of the others. For example, if we perform the subtraction

$$A'_{jk} = A_{jk} - A_{1j}\frac{A_{j1}}{A_{11}} \tag{B.24}$$

on each row except the first, we eliminate all elements of the first column except A_{11} to obtain

$$\mathbf{A}' = \begin{vmatrix} A_{11} & A_{12} & A_{13} \\ 0 & A'_{22} & A'_{23} \\ 0 & A'_{32} & A'_{33} \end{vmatrix} \tag{B.25}$$

Similarly, if we subsequently start with element A'_{22} and subtract an appropriately scaled second row from the rest of the rows,

$$A''_{jk} = A'_{jk} - A'_{2k}A'_{jk}/A'_{22} \tag{B.26}$$

all the elements of the second column vanish except A'_{22}:

$$\mathbf{A}'' = \begin{vmatrix} A_{11} & 0 & A''_{13} \\ 0 & A'_{22} & A''_{23} \\ 0 & 0 & A''_{33} \end{vmatrix} \tag{B.27}$$

Note that A'_{22} is not the original value A_{22}, but is modified as a result of the first subtraction.

By successively subtracting rows (or columns) scaled to their diagonal elements, we can produce a matrix that is diagonal. In practice, it is sufficient to eliminate only half of the nondiagonal elements so that all elements on one side of a diagonal are 0:

$$\mathbf{A} = \begin{vmatrix} A_{11} & 0 & 0 \\ A_{21} & A_{22} & 0 \\ A_{31} & A_{32} & A_{33} \end{vmatrix} = \begin{vmatrix} A_{11} & A_{12} & A_{13} \\ 0 & A_{22} & A_{23} \\ 0 & 0 & A_{33} \end{vmatrix} = \begin{vmatrix} A_{11} & 0 & 0 \\ 0 & A_{22} & 0 \\ 0 & 0 & A_{33} \end{vmatrix} = A_{11}A_{22}A_{33} \tag{B.28}$$

B.2 SOLUTION OF SIMULTANEOUS EQUATIONS BY DETERMINANTS

Consider the following set of three equations in three coefficients a_1, a_2, and a_3. We shall consider the y_k and X_{jk} to be known quantities; that is, constants:

$$\begin{aligned} y_1 &= a_1X_{11} + a_2X_{12} + a_3X_{13} \\ y_2 &= a_1X_{21} + a_2X_{22} + a_3X_{23} \\ y_3 &= a_1X_{31} + a_2X_{32} + a_3X_{33} \end{aligned} \tag{B.29}$$

Let us consider the set of equations as if they were one matrix equation as in Equation (B.10):

$$\begin{bmatrix} y_1 \\ y_2 \\ y_3 \end{bmatrix} = \begin{bmatrix} X_{11} & X_{12} & X_{13} \\ X_{21} & X_{22} & X_{23} \\ X_{31} & X_{32} & X_{33} \end{bmatrix} \begin{bmatrix} a_1 \\ a_2 \\ a_3 \end{bmatrix} \tag{B.30}$$

with **a** and **y** represented by linear matrices and **X** represented by a square matrix. If we multiply the first equation of Equations (B.29) by the cofactor of X_{11} in the matrix of Equation (B.30), multiply the second equation by the cofactor of X_{21}, and multiply the third by the cofactor of X_{31}, then the sum of the three equations is an equation involving determinants according to Equation (B.18):

$$\begin{aligned} \begin{vmatrix} y_1 & X_{12} & X_{13} \\ y_2 & X_{22} & X_{23} \\ y_3 & X_{32} & X_{33} \end{vmatrix} &= a_1 \begin{vmatrix} X_{11} & X_{12} & X_{13} \\ X_{21} & X_{22} & X_{23} \\ X_{31} & X_{32} & X_{33} \end{vmatrix} + a_2 \begin{vmatrix} X_{12} & X_{12} & X_{13} \\ X_{22} & X_{22} & X_{23} \\ X_{32} & X_{32} & X_{33} \end{vmatrix} \\ &+ a_3 \begin{vmatrix} X_{13} & X_{12} & X_{13} \\ X_{23} & X_{22} & X_{23} \\ X_{33} & X_{32} & X_{33} \end{vmatrix} \end{aligned} \tag{B.31}$$

The determinants in the two rightmost terms of Equation (B.31) both vanish because they have two columns that are identical. Thus, the solution for the coefficient a_1 is the ratio of the two determinants:

$$a_1 = \frac{\begin{vmatrix} y_1 & X_{12} & X_{13} \\ y_2 & X_{22} & X_{23} \\ y_3 & X_{32} & X_{33} \end{vmatrix}}{\begin{vmatrix} X_{11} & X_{12} & X_{13} \\ X_{21} & X_{22} & X_{23} \\ X_{31} & X_{32} & X_{33} \end{vmatrix}} \tag{B.32}$$

The denominator is the determinant of the square matrix **X** of Equation (B.30) and the numerator is the determinant of a matrix that is formed by substituting the column matrix y for the first column of the **X** matrix.

Similarly, *Cramér's rule* gives the solution for the jth coefficient a_j of a set of n simultaneous equations as the ratio of two determinants:

$$y_k = \sum_{j=1}^{n} (a_j X_{kj}) \qquad k = 1, n$$

$$a_j = \frac{|\mathbf{X}'(j)|}{|\mathbf{X}|} \tag{B.33}$$

The denominator is the determinant of the **X** matrix. The numerator $|\mathbf{X}'(j)|$ is the determinant of the matrix formed by substituting the **y** matrix for the jth column.

A matrix is singular if its determinant is 0. If the **X** matrix is singular, there is no solution for Equation (B.33). For example, if two of the n simultaneous equations are identical, except for a scale factor, there are really only $n - 1$ independent simultaneous equations, and therefore no solution for the n unknowns. In this case, the **X** matrix has two identical rows and therefore a 0 determinant.

Solution by Matrix Equations

Let us consider Equation (B.33) as if it were a matrix equation as in Equation (B.30). If the **X** matrix is square, we can consider the **y** and **a** linear matrices as either column matrices as in Equation (B.10) or row matrices as in Equation (B.11):

$$[y_k] = [a_j][X_{kj}] \tag{B.34}$$

If we could multiply this matrix by another matrix $\mathbf{X}'$ such that the right-hand side becomes just the linear matrix **a**, then we will have our solution for the coefficients a_j directly. The multiplication of matrices is associative; that is,

$$\mathbf{A}(\mathbf{BC}) = (\mathbf{AB})\mathbf{C} \tag{B.35}$$

Therefore, we require a matrix $\mathbf{X}'$ such that if it is multiplied by the matrix **X**, the result is the unity matrix:

$$[X_{kj}][X'_{kj}] = \mathbf{1} \tag{B.36}$$

The matrix $\mathbf{X}'$ that satisfies Equation (B.36) is called the inverse matrix $\mathbf{X}^{-1}$ of **X**. Equation (B.34) multiplied from the right by $\mathbf{X}^{-1}$ gives the coefficients a_j explicitly, because any matrix is unchanged when multiplied by the unity matrix:

$$[y_k][X_{jk}]^{-1} = [a_j]\mathbf{1} = [a_j] \tag{B.37}$$

We can express Equation (B.37) in more conventional form to give the solution for each of the coefficients a_j:

$$a_j = \sum_{k=1}^{n} (y_k X_{kj}^{-1}) \tag{B.38}$$

Thus, the solution for the n unknowns with n simultaneous equations is reduced to evaluating the elements of the inverse matrix $\mathbf{X}^{-1}$.

B.3 MATRIX INVERSION

The adjoint $\mathbf{A}^{\dagger}$ of a matrix $\mathbf{A}$ is defined as the matrix obtained by substituting for each element A_{jk} the cofactor of the transposed element A_{kj}:

$$A_{jk}^{\dagger} = \mathrm{cof}(A_{kj}) \tag{B.39}$$

For a square symmetric matrix, the transposition makes no difference.

The inverse matrix $\mathbf{A}^{-1}$ defined in Equation (B.36) may be evaluated by dividing the adjoint matrix $\mathbf{A}^{\dagger}$ by the determinant of $\mathbf{A}$:

$$A_{jk}^{-1} = \frac{A_{jk}^{\dagger}}{|\mathbf{A}|} \tag{B.40}$$

To show that this equality holds, we multiply both sides of Equation (B.40) by $|\mathbf{A}|\mathbf{A}$.

$$|\mathbf{A}|\mathbf{A}\mathbf{A}^{-1} = |\mathbf{A}|\mathbf{1} = \mathbf{A}\mathbf{A}^{\dagger} \tag{B.41}$$

Diagonal terms of the matrices in Equation (B.41) are equivalent to the formula of Equation (B.19) for evaluating the determinant:

$$|\mathbf{A}| = \sum_{k=1}^{n} (A_{jk}A_{jk}^{\dagger}) = \sum_{k=1}^{n} [A_{jk}\mathrm{cof}(\mathrm{A}_{jk})] \tag{B.42}$$

Off-diagonal elements can be shown to vanish like those of the determinants of Equation (B.31). If the matrix $\mathbf{A}$ is singular (that is, if $|\mathbf{A}| = 0$), the inverse matrix $\mathbf{A}^{-1}$ does not exist and there is no solution to the matrix equation of Equation (B.34).

Gauss-Jordan Elimination

The formula of Equation (B.40) is generally too cumbersome for use in computing the inverse of a matrix. Instead, the Gauss-Jordan method of elimination is used to invert a matrix by building up the inverse matrix from a unity matrix while reducing the original matrix to unity.

Consider the inverse matrix $\mathbf{A}^{-1}$ as the ratio of the unity matrix divided by the original matrix, $\mathbf{A}^{-1} = \mathbf{1/A}$. If we manipulate the numerator and denominator of this ratio in the same manner (multiplying rows or columns by the same constant factor and adding the same rows scaled to the same constants), the ratio remains unchanged. If we perform the proper manipulation, we can change the denominator into the unity matrix; the numerator must then become equal to the inverse matrix $\mathbf{A}^{-1}$.

Let us write the 3×3 matrix $\mathbf{A}$ and the 3×3 unity matrix side by side and manipulate both to reduce the matrix $\mathbf{A}$ to the unity matrix. We start by using the formula of Equation (B.24) to eliminate the two off-diagonal elements of the first column:

$$\begin{bmatrix} A_{11} & A_{12} & A_{13} \\ A_{21} & A_{22} & A_{23} \\ A_{31} & A_{32} & A_{33} \end{bmatrix} \qquad \begin{bmatrix} 1 & 0 & 0 \\ 0 & 1 & 0 \\ 0 & 0 & 1 \end{bmatrix}$$

$$\begin{bmatrix} A_{11} & A_{12} & A_{13} \\ 0 & A_{22} - A_{12}\dfrac{A_{21}}{A_{11}} & A_{23} - A_{13}\dfrac{A_{21}}{A_{11}} \\ 0 & A_{32} - A_{12}\dfrac{A_{31}}{A_{11}} & A_{33} - A_{13}\dfrac{A_{31}}{A_{11}} \end{bmatrix} \quad \begin{bmatrix} 1 & 0 & 0 \\ -\dfrac{A_{21}}{A_{11}} & 1 & 0 \\ -\dfrac{A_{21}}{A_{11}} & 0 & 1 \end{bmatrix} \tag{B.43}$$

Now, we divide the first row by A_{11} to get a diagonal element of

$$\begin{bmatrix} 1 & \dfrac{A_{12}}{A_{11}} & \dfrac{A_{13}}{A_{11}} \\ 0 & A_{22} - A_{12}\dfrac{A_{21}}{A_{11}} & A_{23} - A_{13}\dfrac{A_{21}}{A_{11}} \\ 0 & A_{32} - A_{12}\dfrac{A_{31}}{A_{11}} & A_{33} - A_{13}\dfrac{A_{31}}{A_{11}} \end{bmatrix} \begin{bmatrix} \dfrac{1}{A_{11}} & 0 & 0 \\ -\dfrac{A_{21}}{A_{11}} & 1 & 0 \\ -\dfrac{A_{21}}{A_{11}} & 0 & 1 \end{bmatrix} \tag{B.44}$$

The left matrix now has the proper first column. Let us relabel the matrices $\mathbf{B}$ (on the left) and $\mathbf{B}'$ (on the right) and perform the corresponding manipulations to obtain zeros in place of B_{12} and B_{32}, and then divide the second row by B_{22}:

$$\begin{bmatrix} 1 & 0 & B_{13} - B_{23}\dfrac{B_{12}}{B_{22}} \\ 0 & 1 & \dfrac{B_{23}}{B_{22}} \\ 0 & 0 & B_{33} - B_{23}\dfrac{B_{32}}{B_{22}} \end{bmatrix} \begin{bmatrix} B'_{11} - B'_{21}\dfrac{B_{12}}{B_{22}} & -\dfrac{B_{12}}{B_{22}} & 0 \\ \dfrac{B'_{21}}{B_{22}} & \dfrac{1}{B_{22}} & 0 \\ B'_{31} - B'_{21}\dfrac{B_{32}}{B_{22}} & -\dfrac{B_{32}}{B_{22}} & 1 \end{bmatrix} \tag{B.45}$$

After similar manipulation of the third column, the matrix on the left becomes the unity matrix and that on the right, therefore, must be the inverse matrix.

For computational purposes, even this method is somewhat inefficient in that two matrices must be manipulated throughout. Note, however, that at each stage of the reduction, there are only n (or three) useful columns of information in the two matrices. As each column is eliminated from the left matrix, the corresponding column is accumulated on the right.

Therefore, we can combine the manipulation into the range of a single matrix. We start with the matrix $\mathbf{A}$ and use the formula of Equation (B.24) as for Equation

(B.43), but instead of applying this formula to the first column, we divide the first column by $-A_{11}$ to get the first column on the right of Equation (B.43); the diagonal element must be divided twice to become $1/A_{11}$. Divide the rest of the first row by A_{11} to get the composite of the two matrices of Equation (B.44):

$$\begin{bmatrix} \dfrac{1}{A_{11}} & \dfrac{A_{12}}{A_{11}} & \dfrac{A_{13}}{A_{11}} \\ -\dfrac{A_{21}}{A_{11}} & A_{22} - A_{12}\dfrac{A_{21}}{A_{13}} & A_{23} - A_{13}\dfrac{A_{21}}{A_{11}} \\ -\dfrac{A_{21}}{A_{11}} & A_{32} - A_{12}\dfrac{A_{31}}{A_{11}} & A_{33} - A_{13}\dfrac{A_{31}}{A_{11}} \end{bmatrix} \tag{B.46}$$

A corresponding manipulation of the second column yields a matrix with the first two columns identical to those of the right side of Equation (B.45) whereas the last column is identical to that of the left side of Equation (B.45). Thus the inverse matrix is accumulated in the space vacated by the original matrix.

Computer Routine PROGRAM B.1 MATRIX (WEBSITE) includes two routines, MATINV and LINEARBYSQUARE. MATINV inverts a square matrix and calculates its determinant, substituting the inverted matrix into the same array as the original matrix.[1] Input variables are ARRAY, the matrix to be inverted, and NORDER, the order of its determinant.

The initial program loop iterates through the n columns of the matrix, reorganizing the matrix to get the largest element in the diagonal in order to reduce rounding errors and improve computational precision. The inversion procedure discussed above is then carried out and the determinant DET of the matrix is calculated from the diagonalized matrix. After inversion, the inverted matrix is stored back in ARRAY and the variable DET, the value of the determinant of the original matrix, is returned.

LINEARBYSQUARE multiplies a linear matrix (on the right) by a square matrix (on the left). For example, see Equation (B.30).

[1]The subroutine MATINV follows the procedure of the subroutine MINV of the IBM System/360 Scientific Subroutine Package.

APPENDIX C

GRAPHS AND TABLES

The tables and graphs in this appendix are provided for easy reference. Computer routines for calculating several of the distributions and probability functions are listed in Appendix E. Routines are also available on the website for calculating probabilities.

C.1 GAUSSIAN PROBABILITY DISTRIBUTION

The probability density function $p_G(x; \mu, \sigma)$ for the Gaussian or normal error distribution is given by

$$p_G(x; \mu, \sigma) = \frac{1}{\sigma\sqrt{2\pi}} \exp\left[-\frac{1}{2}\left(\frac{x-\mu}{\sigma}\right)^2\right]$$

If measurements of a quantity x are distributed in this manner around a mean μ with standard deviation σ, the probability $dP_G(x; \mu, \sigma)$ for observing a value of x, within an infinitesimally small interval dx, in a random sample measurement is given by

$$dP_G(x; \mu, \sigma) = p_G(x; \mu, \sigma)\, dx$$

Values of the probability density function $p_G(x; \mu, \sigma)$ are tabulated in Table C.1 as a function of the dimensionless deviation

$$z = |x - \mu|/\sigma$$

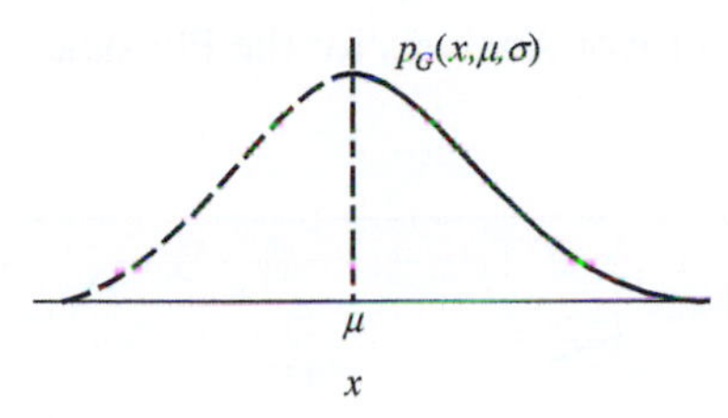

TABLE C.1

Gaussian probability density distribution. The Gaussian or normal error distribution $p_G(x; \mu, \sigma)$ versus $z = |x - \mu|/\sigma$

z	0.00	0.01	0.02	0.03	0.04	0.05	0.06	0.07	0.08	0.09
0.0	0.39894	0.39892	0.39886	0.39876	0.39862	0.39844	0.39822	0.39797	0.39767	0.39733
0.1	0.39695	0.39654	0.39608	0.39559	0.39505	0.39448	0.39387	0.39322	0.39253	0.39181
0.2	0.39104	0.39024	0.38940	0.38853	0.38762	0.38667	0.38568	0.38466	0.38361	0.38251
0.3	0.38139	0.38023	0.37903	0.37780	0.37654	0.37524	0.37391	0.37255	0.37115	0.36973
0.4	0.36827	0.36678	0.36526	0.36371	0.36213	0.36053	0.35889	0.35723	0.35553	0.35381
0.5	0.35207	0.35029	0.34849	0.34667	0.34482	0.34294	0.34105	0.33912	0.33718	0.33521
0.6	0.33322	0.33121	0.32918	0.32713	0.32506	0.32297	0.32086	0.31874	0.31659	0.31443
0.7	0.31225	0.31006	0.30785	0.30563	0.30339	0.30114	0.29887	0.29659	0.29431	0.29200
0.8	0.28969	0.28737	0.28504	0.28269	0.28034	0.27799	0.27562	0.27324	0.27086	0.26848
0.9	0.26609	0.26369	0.26129	0.25888	0.25647	0.25406	0.25164	0.24923	0.24681	0.24439
1.0	0.24197	0.23995	0.23713	0.23471	0.23230	0.22988	0.22747	0.22506	0.22266	0.22025
1.1	0.21785	0.21546	0.21307	0.21069	0.20831	0.20594	0.20357	0.20122	0.19887	0.19652
1.2	0.19419	0.19186	0.18955	0.18724	0.18494	0.18265	0.18038	0.17811	0.17585	0.17361
1.3	0.17137	0.16915	0.16694	0.16475	0.16256	0.16039	0.15823	0.15609	0.15395	0.15184
1.4	0.14973	0.14764	0.14557	0.14351	0.14147	0.13944	0.13742	0.13543	0.13344	0.13148
1.5	0.12952	0.12759	0.12567	0.12377	0.12189	0.12002	0.11816	0.11633	0.11451	0.11271
1.6	0.11093	0.10916	0.10741	0.10568	0.10397	0.10227	0.10059	0.09893	0.09729	0.09567
1.7	0.09406	0.09247	0.09090	0.08934	0.08780	0.08629	0.08478	0.08330	0.08184	0.08039
1.8	0.07896	0.07755	0.07615	0.07477	0.07342	0.07207	0.07075	0.06944	0.06815	0.06688
1.9	0.06562	0.06439	0.06316	0.06196	0.06077	0.05960	0.05845	0.05731	0.05619	0.05509
2.0	0.05400	0.05293	0.05187	0.05083	0.04981	0.04880	0.04781	0.04683	0.04587	0.04492
2.1	0.04399	0.04307	0.04217	0.04129	0.04041	0.03956	0.03871	0.03788	0.03707	0.03627
2.2	0.03548	0.03471	0.03395	0.03320	0.03247	0.03175	0.03104	0.03034	0.02966	0.02899
2.3	0.02833	0.02769	0.02705	0.02643	0.02582	0.02522	0.02464	0.02406	0.02350	0.02294
2.4	0.02240	0.02187	0.02135	0.02083	0.02033	0.01984	0.01936	0.01889	0.01843	0.01798
2.5	0.01753	0.01710	0.01667	0.01626	0.01585	0.01545	0.01506	0.01468	0.01431	0.01394
2.6	0.01359	0.01324	0.01290	0.01256	0.01224	0.01192	0.01160	0.01130	0.01100	0.01071
2.7	0.01042	0.01015	0.00987	0.00961	0.00935	0.00910	0.00885	0.00861	0.00837	0.00814
2.8	0.00792	0.00770	0.00749	0.00728	0.00707	0.00688	0.00668	0.00649	0.00631	0.00613
2.9	0.00595	0.00578	0.00562	0.00546	0.00530	0.00514	0.00500	0.00485	0.00471	0.00457

	0.00	0.10	0.20	0.30	0.40
3.0	0.0044318	0.0032668	0.0023841	0.0017226	0.0012322
3.5	0.00087269	0.00061191	0.00042479	0.00029195	0.00019866
4.0	0.00013383	0.000089264	0.000058945	0.000038536	0.000024943
4.5	0.000015984	0.000010141	0.0000063701	0.0000039615	0.0000024391
5.0	0.0000014868	0.00000089730	0.00000053614	0.00000031716	0.00000018575
5.5	0.00000010771	0.00000006183	0.00000003514	0.00000001978	0.00000001102

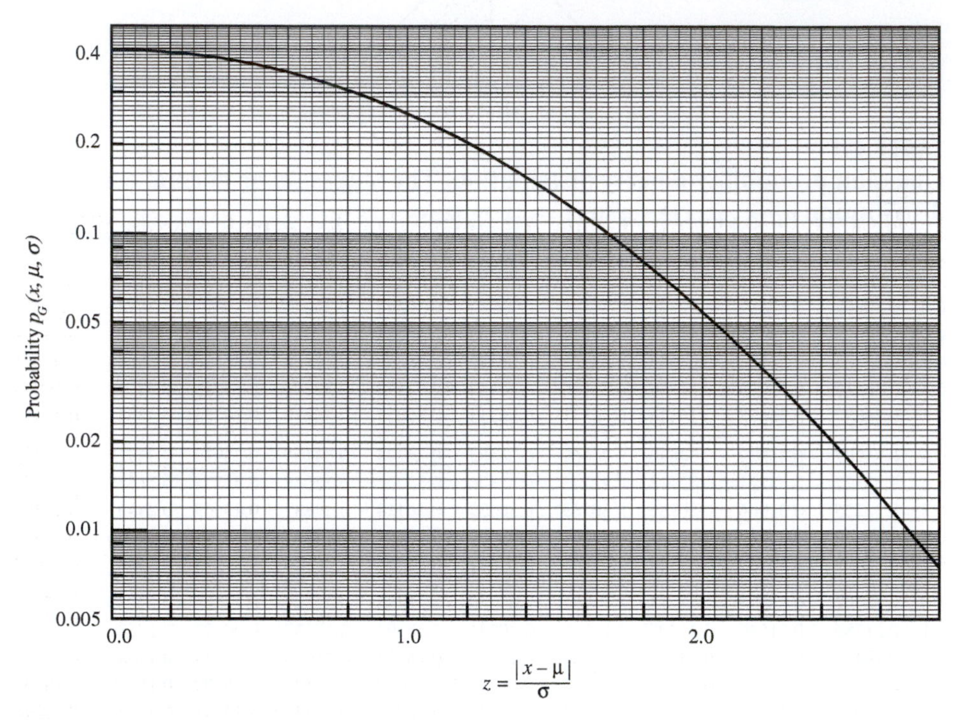

FIGURE C.1
Gaussian probability density distribution, $p_G(x; \mu, \sigma)$ versus $z = |x - \mu|/\sigma$

for z ranging from 0.0 to 3.0 in increments of 0.01 and up to 5.9 in increments of 0.1. This function is graphed on a semi-logarithmic scale as a function of z in Figure C.1.

C.2 INTEGRAL OF GAUSSIAN DISTRIBUTION

The integral $P_G(x; \mu, \sigma)$ of the probability density function $p_G(x; \mu, \sigma)$ for the Gaussian or normal error distribution is given by

$$P_G(x; \mu, \sigma) = \frac{1}{\sigma\sqrt{2\pi}} \int_{\mu - z\sigma}^{\mu + z\sigma} \exp\left[-\frac{1}{2}\left(\frac{x - \mu}{\sigma}\right)^2\right] dx$$

with

$$z = \frac{|x - \mu|}{\sigma}$$

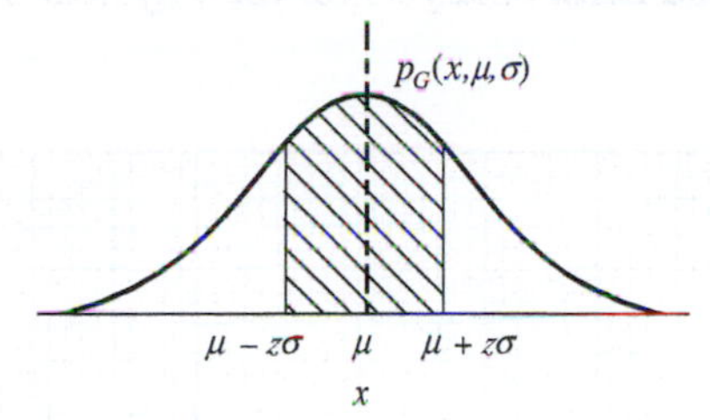

TABLE C.2

Integral of Gaussian distribution. The integral of the Gaussian probability density distribution. $P_G(x; \mu, \sigma)$ versus $z = |x - \mu|/\sigma$

z	0.00	0.01	0.02	0.03	0.04	0.05	0.06	0.07	0.08	0.09
0.0	0.0	0.00798	0.01596	0.02393	0.03191	0.03988	0.04784	0.05581	0.06376	0.07171
0.1	0.07966	0.08759	0.09552	0.10343	0.11134	0.11924	0.12712	0.13499	0.14285	0.15069
0.2	0.15852	0.16633	0.17413	0.18191	0.18967	0.19741	0.20514	0.21284	0.22052	0.22818
0.3	0.23582	0.24344	0.25103	0.25860	0.26614	0.27366	0.28115	0.28862	0.29605	0.30346
0.4	0.31084	0.31819	0.32551	0.33280	0.34006	0.34729	0.35448	0.36164	0.36877	0.37587
0.5	0.38292	0.38995	0.39694	0.40389	0.41080	0.41768	0.42452	0.43132	0.43809	0.44481
0.6	0.45149	0.45814	0.46474	0.47131	0.47783	0.48431	0.49075	0.49714	0.50350	0.50981
0.7	0.51607	0.52230	0.52847	0.53461	0.54070	0.54674	0.55274	0.55870	0.56461	0.57047
0.8	0.57629	0.58206	0.58778	0.59346	0.59909	0.60467	0.61021	0.61570	0.62114	0.62653
0.9	0.63188	0.63718	0.64243	0.64763	0.65278	0.65789	0.66294	0.66795	0.67291	0.67783
1.0	0.68269	0.68750	0.69227	0.69699	0.70166	0.70628	0.71085	0.71538	0.71985	0.72428
1.1	0.72866	0.73300	0.73728	0.74152	0.74571	0.74985	0.75395	0.75799	0.76199	0.76595
1.2	0.76985	0.77371	0.77753	0.78130	0.78502	0.78869	0.79232	0.79591	0.79945	0.80294
1.3	0.80639	0.80980	0.81316	0.81647	0.81975	0.82298	0.82616	0.82930	0.83240	0.83546
1.4	0.83848	0.84145	0.84438	0.84727	0.85012	0.85293	0.85570	0.85843	0.86112	0.86377
1.5	0.86638	0.86895	0.87148	0.87397	0.87643	0.87885	0.88123	0.88358	0.88588	0.88816
1.6	0.89039	0.89259	0.89476	0.89689	0.89898	0.90105	0.90308	0.90507	0.90703	0.90896
1.7	0.91086	0.91272	0.91456	0.91636	0.91813	0.91987	0.92158	0.92326	0.92491	0.92654
1.8	0.92813	0.92969	0.93123	0.93274	0.93422	0.93568	0.93711	0.93851	0.93988	0.94123
1.9	0.94256	0.94386	0.94513	0.94638	0.94761	0.94882	0.95000	0.95115	0.95229	0.95340
2.0	0.95449	0.95556	0.95661	0.95764	0.95864	0.95963	0.96059	0.96154	0.96247	0.96338
2.1	0.96426	0.96513	0.96599	0.96682	0.96764	0.96844	0.96922	0.96999	0.97074	0.97147
2.2	0.97219	0.97289	0.97358	0.97425	0.97490	0.97555	0.97617	0.97679	0.97739	0.97797
2.3	0.97855	0.97911	0.97965	0.98019	0.98071	0.98122	0.98172	0.98221	0.98268	0.98315
2.4	0.98360	0.98404	0.98448	0.98490	0.98531	0.98571	0.98610	0.98648	0.98686	0.98722
2.5	0.98758	0.98792	0.98826	0.98859	0.98891	0.98922	0.98953	0.98983	0.99012	0.99040
2.6	0.99067	0.99094	0.99120	0.99146	0.99171	0.99195	0.99218	0.99241	0.99264	0.99285
2.7	0.99306	0.99327	0.99347	0.99366	0.99385	0.99404	0.99422	0.99439	0.99456	0.99473
2.8	0.99489	0.99504	0.99520	0.99534	0.99549	0.99563	0.99576	0.99589	0.99602	0.99615
2.9	0.99627	0.99638	0.99650	0.99661	0.99672	0.99682	0.99692	0.99702	0.99712	0.99721

	0.00	0.10	0.20	0.30	0.40
3.0	0.9973002	0.9980648	0.9986257	0.99903315	0.99932614
3.5	0.99953474	0.99968178	0.99978440	0.99985530	0.999903805
4.0	0.999936656	0.999958684	0.999973308	0.999982920	0.999989174
4.5	0.9999932043	0.9999957748	0.9999973982	0.9999984132	0.99999904149
5.0	0.99999942657	0.99999966024	0.99999980061	0.99999988410	0.99999993327
5.5	0.99999996193	0.99999997847	0.99999998793	0.99999999328	0.99999999627

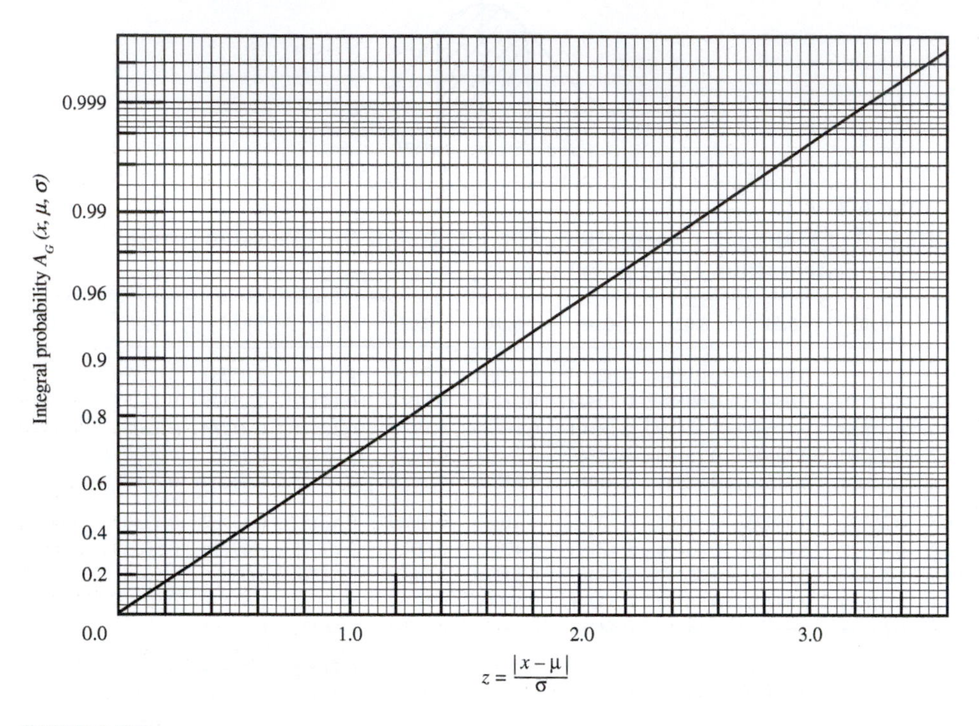

FIGURE C.2
Integral of the Gaussian probability density distribution, $p_G(x; \mu, \sigma)$ versus $z = |x - \mu|/\sigma$

If measurements of the quantity x are distributed according to the Gaussian distribution around a mean μ with standard deviation σ, $P_G(x; \mu, \sigma)$ is equal to the probability for observing a value of x in a random sample measurement that is between $\mu - z\sigma$ and $\mu + z\sigma$; that is, it is the probability that $|x - \mu| < z\sigma$.

Values of the integral $P_G(x; \mu, \sigma)$ are tabulated in Table C.2 as a function of z, for z ranging from 0.0 to 3.0 in increments of 0.01 and up to 5.9 in increments of 0.1. This function is graphed on a probability scale as a function of z in Figure C.2.

A related function is the *error function* erf Z:

$$\text{erf } Z = \frac{1}{\sqrt{\pi}} \int_{-Z}^{Z} e^{-z^2} dz = P_G(z\sqrt{2}; 0, 1)$$

The function that is tabulated and graphed is the shaded area between the limits $\mu \pm z\sigma$ as indicated.

C.3 LINEAR-CORRELATION COEFFICIENT

The probability distribution $p_r(r, \nu)$ for the linear-correlation coefficient r for ν degrees of freedom is given by

$$p_r(r; \nu) = \frac{1}{\sqrt{\pi}} \frac{\Gamma[(\nu+1)/2]}{\Gamma(\nu/2)} (1 - r^2)^{(\nu-2)/2}$$

The probability of observing a value of the correlation coefficient larger than r for a random sample of N observations with ν degrees of freedom is the integral of this probability $P_c(r; N)$:

$$P_c(r; N) = \frac{1}{\sqrt{\pi}} \frac{\Gamma[(\nu+1)/2]}{\Gamma(\nu/2)} \int_{|r|}^{1} (1 - x^2)^{(\nu-2)/2} dx \qquad \nu = N - 2$$

If two variables of a parent population are uncorrelated, the probability that a random sample of N observations will yield a correlation coefficient for those two variables greater in magnitude than $|r|$ is given by $P_c(r; N)$.

Values of the coefficient $|r|$corresponding to various values of the probability $P_c(r; N)$ are tabulated in Table C.3 for N ranging from 3 to 100, and values of $P_c(r; N)$ ranging from 0.001 to 0.5. The functional dependence of r corresponding to representative values of $P_c(r, N)$ is graphed on a semi-logarithmic scale as a smooth variation with the number of observations N in Figure C.3.

The function that is tabulated and graphed is the shaded area under the tails of the probability curve for values larger than $|r|$ as indicated.

C.4 χ^2 DISTRIBUTION

The probability density distribution $p_\chi(\chi^2; \nu)$ for χ^2 is given by

$$p_\chi(\chi^2; \nu) = \frac{1}{2^{\nu/2}\Gamma(\nu/2)} (\chi^2)^{(\nu-2)/2} e^{-\chi^2/2}$$

The probability of observing a value of χ^2 that is larger than a particular value for a random sample of N observations with ν degrees of freedom is the integral of this probability $P_\chi(\chi^2; \nu)$:

$$P_\chi(\chi^2; \nu) = \frac{1}{2^{\nu/2}\Gamma(\nu/2)} \int_{\chi^2}^{\infty} (x^2)^{(\nu-2)/2} e^{-x^2/2} d(x^2)$$

Values of the reduced chi-square $\chi_\nu^2 = \chi^2/\nu$ corresponding to various values of the integral probability $P_\chi(\chi^2; \nu)$ of exceeding χ^2 in a measurement with ν degrees of freedom are tabulated in Table C.4 for ν ranging from 1 to 200. The functional dependence of $P_\chi(\chi^2; \nu)$ corresponding to representative values of ν is graphed in Figure C.4 as a smooth variation with the reduced chi-square χ_ν^2.

The function that is tabulated and graphed is the shaded area under the tail of the probability curve for values larger than χ^2 as indicated.

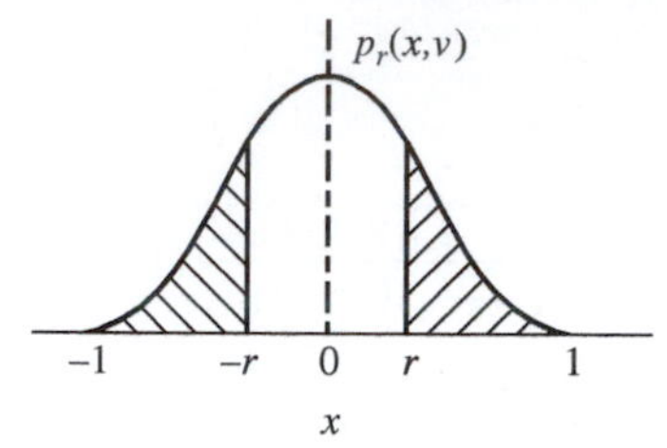

TABLE C.3

Linear-correlation coefficient. The linear-correlation coefficient r versus the number of observations N and the corresponding probability $P_c(r; N)$ of exceeding r in a random sample of observations taken from an uncorrelated parent population ($\rho = 0$)

	P								
N	**0.50**	**0.20**	**0.10**	**0.050**	**0.020**	**0.010**	**0.005**	**0.002**	**0.001**
3	0.707	0.951	0.988	0.997	1.000	1.000	1.000	1.000	1.000
4	0.500	0.800	0.900	0.950	0.980	0.990	0.995	0.998	0.999
5	0.404	0.687	0.805	0.878	0.934	0.959	0.974	0.986	0.991
6	0.347	0.608	0.729	0.811	0.882	0.917	0.942	0.963	0.974
7	0.309	0.551	0.669	0.754	0.833	0.875	0.906	0.935	0.951
8	0.281	0.507	0.621	0.707	0.789	0.834	0.870	0.905	0.925
9	0.260	0.472	0.582	0.666	0.750	0.798	0.836	0.875	0.898
10	0.242	0.443	0.549	0.632	0.715	0.765	0.805	0.847	0.872
11	0.228	0.419	0.521	0.602	0.685	0.735	0.776	0.820	0.847
12	0.216	0.398	0.497	0.576	0.658	0.708	0.750	0.795	0.823
13	0.206	0.380	0.476	0.553	0.634	0.684	0.726	0.772	0.801
14	0.197	0.365	0.458	0.532	0.612	0.661	0.703	0.750	0.780
15	0.189	0.351	0.441	0.514	0.592	0.641	0.683	0.730	0.760
16	0.182	0.338	0.426	0.497	0.574	0.623	0.664	0.711	0.742
17	0.176	0.327	0.412	0.482	0.558	0.606	0.647	0.694	0.725
18	0.170	0.317	0.400	0.468	0.543	0.590	0.631	0.678	0.708
19	0.165	0.308	0.389	0.456	0.529	0.575	0.616	0.662	0.693
20	0.160	0.299	0.378	0.444	0.516	0.561	0.602	0.648	0.679
22	0.152	0.284	0.360	0.423	0.492	0.537	0.576	0.622	0.652
24	0.145	0.271	0.344	0.404	0.472	0.515	0.554	0.599	0.629
26	0.138	0.260	0.330	0.388	0.453	0.496	0.534	0.578	0.607
28	0.133	0.250	0.317	0.374	0.437	0.479	0.515	0.559	0.588
30	0.128	0.241	0.306	0.361	0.423	0.463	0.499	0.541	0.570
32	0.124	0.233	0.296	0.349	0.409	0.449	0.484	0.526	0.554
34	0.120	0.225	0.287	0.339	0.397	0.436	0.470	0.511	0.539
36	0.116	0.219	0.279	0.329	0.386	0.424	0.458	0.498	0.525
38	0.113	0.213	0.271	0.320	0.376	0.413	0.446	0.486	0.513
40	0.110	0.207	0.264	0.312	0.367	0.403	0.435	0.474	0.501
42	0.107	0.202	0.257	0.304	0.358	0.393	0.425	0.463	0.490
44	0.104	0.197	0.251	0.297	0.350	0.384	0.416	0.453	0.479
46	0.102	0.192	0.246	0.291	0.342	0.376	0.407	0.444	0.469
48	0.100	0.188	0.240	0.285	0.335	0.368	0.399	0.435	0.460
50	0.098	0.184	0.235	0.279	0.328	0.361	0.391	0.427	0.451
60	0.089	0.168	0.214	0.254	0.300	0.330	0.358	0.391	0.414
70	0.082	0.155	0.198	0.235	0.278	0.306	0.332	0.363	0.385
80	0.077	0.145	0.185	0.220	0.260	0.286	0.311	0.340	0.361
90	0.072	0.136	0.174	0.207	0.245	0.270	0.293	0.322	0.341
100	0.068	0.129	0.165	0.197	0.232	0.256	0.279	0.305	0.324

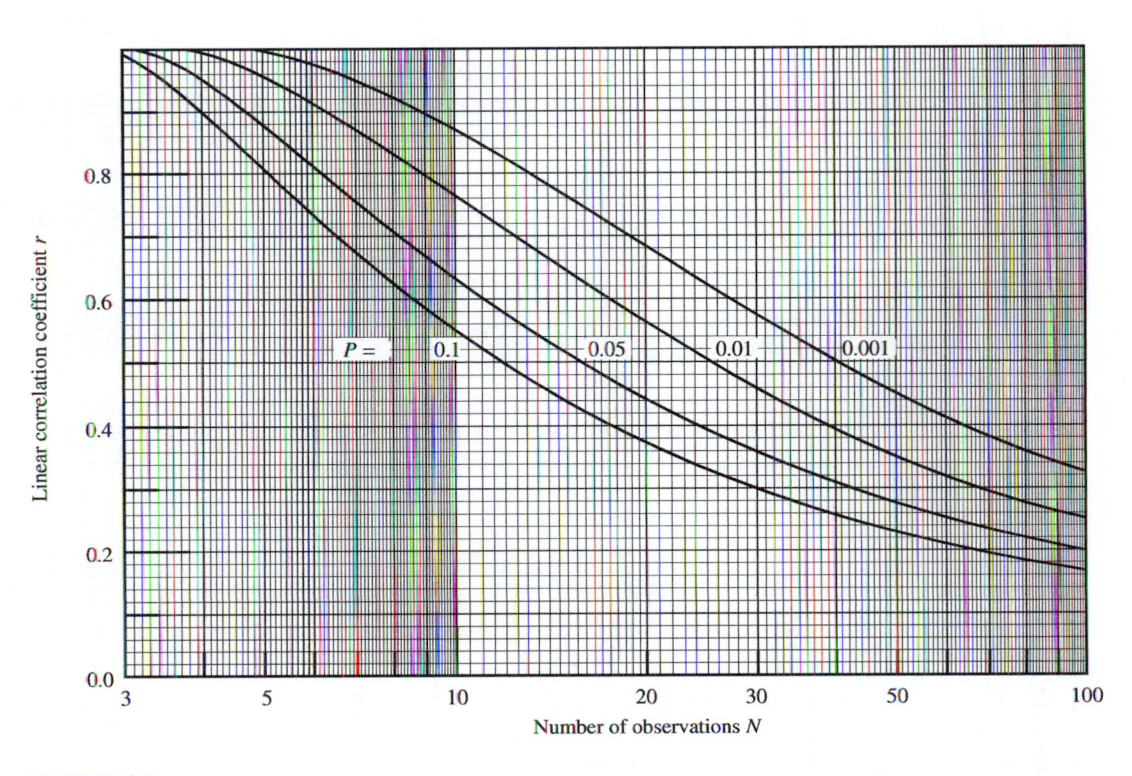

FIGURE C.3
The linear-correlation coefficient τ versus the number of observations N and the corresponding probability $P_c(r; N)$ that the variables are not correlated.

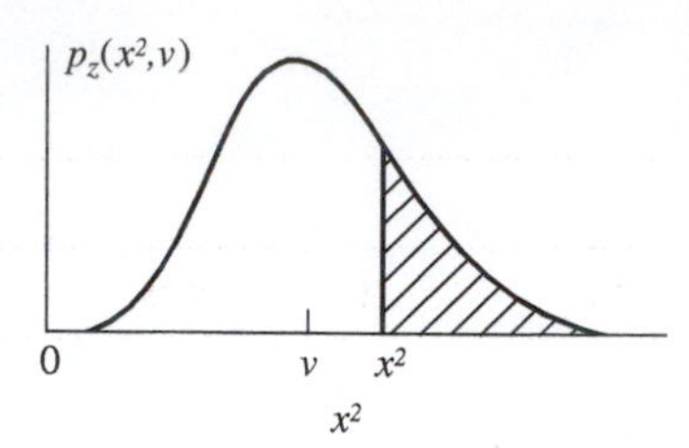

TABLE C.4

χ^2 distribution. Values of the reduced chi-square $\chi^2_\nu = \chi^2/\nu$ corresponding to the probability $P_\chi(\chi^2; \nu)$ of exceeding χ^2 versus the number of degrees of freedom ν

	P							
ν	**0.99**	**0.98**	**0.95**	**0.90**	**0.80**	**0.70**	**0.60**	**0.50**
1	0.00016	0.00063	0.00393	0.0158	0.0642	0.148	0.275	0.455
2	0.0100	0.0202	0.0515	0.105	0.223	0.357	0.511	0.693
3	0.0383	0.0617	0.117	0.195	0.335	0.475	0.623	0.789
4	0.0742	0.107	0.178	0.266	0.412	0.549	0.688	0.839
5	0.111	0.150	0.229	0.322	0.469	0.600	0.731	0.870
6	0.145	0.189	0.273	0.367	0.512	0.638	0.762	0.891
7	0.177	0.223	0.310	0.405	0.546	0.667	0.785	0.907
8	0.206	0.254	0.342	0.436	0.574	0.691	0.803	0.918
9	0.232	0.281	0.369	0.463	0.598	0.710	0.817	0.927
10	0.256	0.306	0.394	0.487	0.618	0.727	0.830	0.934
11	0.278	0.328	0.416	0.507	0.635	0.741	0.840	0.940
12	0.298	0.348	0.436	0.525	0.651	0.753	0.848	0.945
13	0.316	0.367	0.453	0.542	0.664	0.764	0.856	0.949
14	0.333	0.383	0.469	0.556	0.676	0.773	0.863	0.953
15	0.349	0.399	0.484	0.570	0.687	0.781	0.869	0.956
16	0.363	0.413	0.498	0.582	0.697	0.789	0.874	0.959
17	0.377	0.427	0.510	0.593	0.706	0.796	0.879	0.961
18	0.390	0.439	0.522	0.604	0.714	0.802	0.883	0.963
19	0.402	0.451	0.532	0.613	0.722	0.808	0.887	0.965
20	0.413	0.462	0.543	0.622	0.729	0.813	0.890	0.967
22	0.434	0.482	0.561	0.638	0.742	0.823	0.897	0.970
24	0.452	0.500	0.577	0.652	0.753	0.831	0.902	0.972
26	0.469	0.516	0.592	0.665	0.762	0.838	0.907	0.974
28	0.484	0.530	0.605	0.676	0.771	0.845	0.911	0.976
30	0.498	0.544	0.616	0.687	0.779	0.850	0.915	0.978
32	0.511	0.556	0.627	0.696	0.786	0.855	0.918	0.979
34	0.523	0.567	0.637	0.704	0.792	0.860	0.921	0.980
36	0.534	0.577	0.646	0.712	0.798	0.864	0.924	0.982
38	0.545	0.587	0.655	0.720	0.804	0.868	0.926	0.983
40	0.554	0.596	0.663	0.726	0.809	0.872	0.928	0.983
42	0.563	0.604	0.670	0.733	0.813	0.875	0.930	0.984
44	0.572	0.612	0.677	0.738	0.818	0.878	0.932	0.985
46	0.580	0.620	0.683	0.744	0.822	0.881	0.934	0.986
48	0.587	0.627	0.690	0.749	0.825	0.884	0.936	0.986
50	0.594	0.633	0.695	0.754	0.829	0.886	0.937	0.987
60	0.625	0.662	0.720	0.774	0.844	0.897	0.944	0.989
70	0.649	0.684	0.739	0.790	0.856	0.905	0.949	0.990
80	0.669	0.703	0.755	0.803	0.865	0.911	0.952	0.992
90	0.686	0.718	0.768	0.814	0.873	0.917	0.955	0.993
100	0.701	0.731	0.779	0.824	0.879	0.921	0.958	0.993
120	0.724	0.753	0.798	0.839	0.890	0.928	0.962	0.994
140	0.743	0.770	0.812	0.850	0.898	0.934	0.965	0.995
160	0.758	0.784	0.823	0.860	0.905	0.938	0.968	0.996
180	0.771	0.796	0.833	0.868	0.910	0.942	0.970	0.996
200	0.782	0.806	0.841	0.874	0.915	0.945	0.972	0.997

TABLE C.4
(continued)

	P							
ν	**0.40**	**0.30**	**0.20**	**0.10**	**0.05**	**0.02**	**0.01**	**0.001**
1	0.708	1.074	1.642	2.706	3.841	5.412	6.635	10.827
2	0.916	1.204	1.609	2.303	2.996	3.912	4.605	6.908
3	0.982	1.222	1.547	2.084	2.605	3.279	3.780	5.423
4	1.011	1.220	1.497	1.945	2.372	2.917	3.319	4.617
5	1.026	1.213	1.458	1.847	2.214	2.678	3.017	4.102
6	1.035	1.205	1.426	1.774	2.099	2.506	2.802	3.743
7	1.040	1.198	1.400	1.717	2.010	2.375	2.639	3.475
8	1.044	1.191	1.379	1.670	1.938	2.271	2.511	3.266
9	1.046	1.184	1.360	1.632	1.880	2.187	2.407	3.097
10	1.047	1.178	1.344	1.599	1.831	2.116	2.321	2.959
11	1.048	1.173	1.330	1.570	1.789	2.056	2.248	2.842
12	1.049	1.168	1.318	1.546	1.752	2.004	2.185	2.742
13	1.049	1.163	1.307	1.524	1.720	1.959	2.130	2.656
14	1.049	1.159	1.296	1.505	1.692	1.919	2.082	2.580
15	1.049	1.155	1.287	1.487	1.666	1.884	2.039	2.513
16	1.049	1.151	1.279	1.471	1.644	1.852	2.000	2.453
17	1.048	1.148	1.271	1.457	1.623	1.823	1.965	2.399
18	1.048	1.145	1.264	1.444	1.604	1.797	1.934	2.351
19	1.048	1.142	1.258	1.432	1.586	1.773	1.905	2.307
20	1.048	1.139	1.252	1.421	1.571	1.751	1.878	2.266
22	1.047	1.134	1.241	1.401	1.542	1.712	1.831	2.194
24	1.046	1.129	1.231	1.383	1.517	1.678	1.791	2.132
26	1.045	1.125	1.223	1.368	1.496	1.648	1.755	2.079
28	1.045	1.121	1.215	1.354	1.476	1.622	1.724	2.032
30	1.044	1.118	1.208	1.342	1.459	1.599	1.696	1.990
32	1.043	1.115	1.202	1.331	1.444	1.578	1.671	1.953
34	1.042	1.112	1.196	1.321	1.429	1.559	1.649	1.919
36	1.042	1.109	1.191	1.311	1.417	1.541	1.628	1.888
38	1.041	1.106	1.186	1.303	1.405	1.525	1.610	1.861
40	1.041	1.104	1.182	1.295	1.394	1.511	1.592	1.835
42	1.040	1.102	1.178	1.288	1.384	1.497	1.576	1.812
44	1.039	1.100	1.174	1.281	1.375	1.485	1.562	1.790
46	1.039	1.098	1.170	1.275	1.366	1.473	1.548	1.770
48	1.038	1.096	1.167	1.269	1.358	1.462	1.535	1.751
50	1.038	1.094	1.163	1.263	1.350	1.452	1.523	1.733
60	1.036	1.087	1.150	1.240	1.318	1.410	1.473	1.660
70	1.034	1.081	1.139	1.222	1.293	1.377	1.435	1.605
80	1.032	1.076	1.130	1.207	1.273	1.351	1.404	1.560
90	1.031	1.072	1.123	1.195	1.257	1.329	1.379	1.525
100	1.029	1.069	1.117	1.185	1.243	1.311	1.358	1.494
120	1.027	1.063	1.107	1.169	1.221	1.283	1.325	1.446
140	1.026	1.059	1.099	1.156	1.204	1.261	1.299	1.410
160	1.024	1.055	1.093	1.146	1.191	1.243	1.278	1.381
180	1.023	1.052	1.087	1.137	1.179	1.228	1.261	1.358
200	1.022	1.050	1.083	1.130	1.170	1.216	1.247	1.338

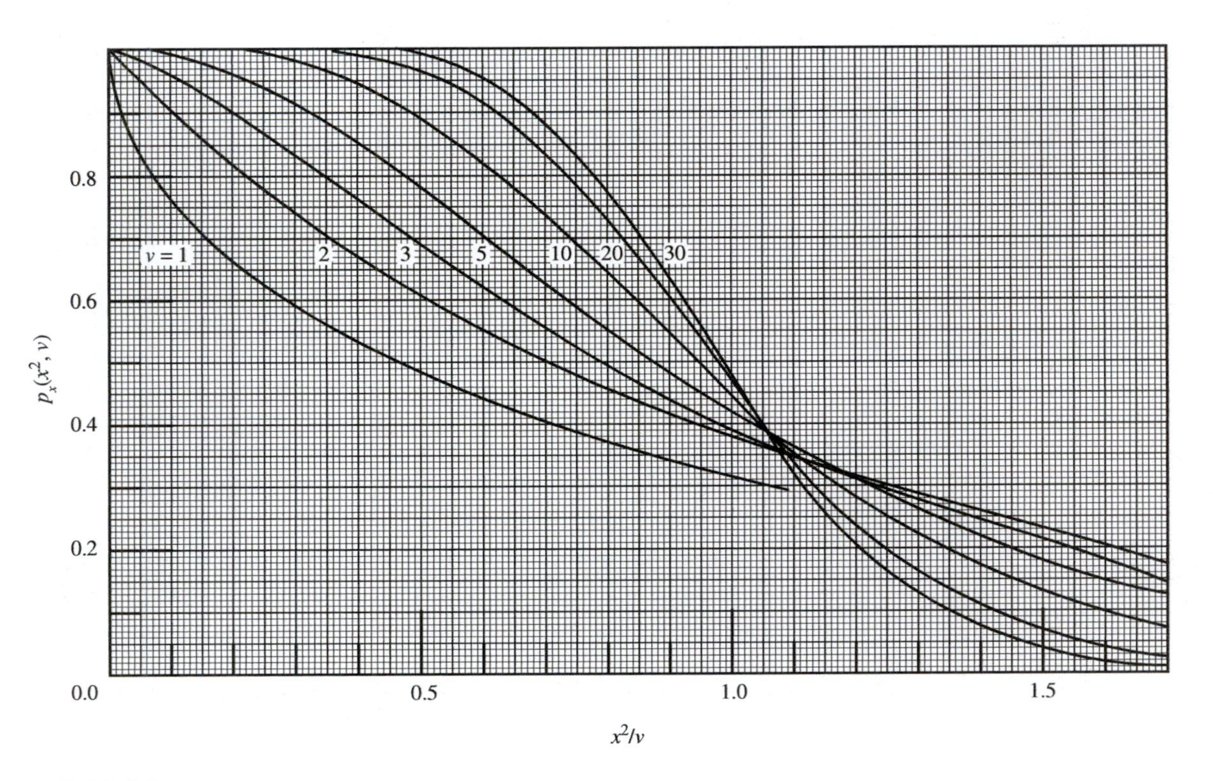

FIGURE C.4
The probability $P_\chi(\chi^2; \nu)$ of exceeding χ^2 versus the reduced chi-square $\chi^2_\nu = \chi^2/\nu$ and the number of degrees of freedom ν.

C.5 *F* DISTRIBUTION

The probability distribution for F is given by

$$p_f(f, \nu_1, \nu_2) = \frac{\Gamma[(\nu_1 + \nu_2)/2]}{\Gamma(\nu_1/2)\Gamma(\nu_2/2)} \left(\frac{\nu_1}{\nu_2}\right)^{\nu_1/2} \frac{f^{(\nu_1-2)/2}}{(1 + f\nu_1/\nu_2)^{1/2(\nu_1+\nu_2)}}$$

The probability of observing a value of F that is larger than a particular value for a random sample with ν_1 and ν_2 degrees of freedom is the integral of this probability:

$$P_F(F; \nu_1, \nu_2) = \int_F^\infty p_f(f; \nu_1, \nu_2)\, df$$

Values of F corresponding to various values of the integral probability $P_F(F; \nu_1, \nu_2)$ of exceeding F in a measurement are tabulated in Table C.5 for $\nu_1 = 1$ and graphed in Figure C.5 as a smooth variation with the probability. Values of F corresponding to various values of ν_1 and ν_2 ranging from 1 to ∞ are listed in Table C.6 and graphed in Figure C.6 for $P_F(F; \nu_1, \nu_2) = 0.05$ and in Table C.7 and Figure C.7 for $P_F(F; \nu_1, \nu_2) = 0.01$. These values were adapted by permission from Dixon and Massey (1969).

The function that is tabulated and graphed is the shaded area under the tail of the probability curve for values larger than F as indicated.

C.6 STUDENT'S *t* DISTRIBUTION

The probability distribution for Student's t is given by[1]

$$f(t, \nu) = \frac{1}{\sqrt{(\nu\pi)}} \frac{\Gamma[(\nu + 1)/2]}{\Gamma(\nu/2)} \left(1 + \frac{t^2}{\nu}\right)^{-(\nu+1)/2}$$

Student's t distribution describes, as a function of the number of degrees of freedom ν, the distribution of the parameter $t = |x - \bar{x}|/s_\mu$, where t is the number of standard deviations s_μ of the sample distribution by which x differs from $\bar{x}$. This distribution takes account of the fact that the sample standard deviation s_μ is an *estimate* of the parent standard error σ_μ and, as such, will vary for different samples drawn from the same parent distribution, just as the sample means vary. If $\bar{x}$ represents the mean of N numbers and x is not derived from the data, then $\nu = N - 1$. If both x and $\bar{x}$ are means, s_μ must be the joint standard deviation of both x and $\bar{x}$, and ν must be the total number of degrees of freedom. In the limit of large numbers of degrees of freedom, Student's t and Gaussian probability distributions agree; for small ν, that is, low-statistics experiments, the Gaussian distribution overestimates the probability and Student's t is preferred.

Table C.8 lists probabilities obtained by integrating Student's t distribution from $x = \bar{x} - ts_\mu$ to $x = \bar{x} + ts_\mu$ where $t = |\bar{x} - x|/s_\mu$. The integrals are listed as functions of t and of the number of degrees of freedom ν. The values corresponding to Gaussian probability (which are independent of ν) are listed in the last column.

---

[1]"Review of Particle Physics" *The European Physical Journal C*, vol. 15 (2000), p. 193.

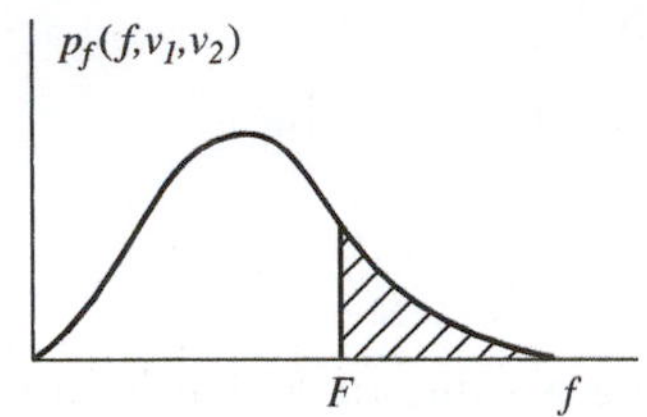

TABLE C.5

F distribution, $\nu = 1$. Values of F corresponding to the probability $P_F(F;1, \nu_2)$ of exceeding F (with $\nu_1 = 1$ degrees of freedom) versus the larger number of degrees of freedom ν_2.

Degrees of freedom	Probability (P) of exceeding F							
ν_2	0.50	0.25	0.10	0.05	0.025	0.01	0.005	0.001
1	1.000	5.83	39.90	161.00	648.00	4050.00	16200.00	406000.0
2	0.667	2.57	8.53	18.50	38.50	98.50	198.00	998.0
3	0.585	2.02	5.54	10.10	17.40	34.10	55.60	167.0
4	0.549	1.81	4.54	7.71	12.20	21.20	31.30	74.1
5	0.528	1.69	4.06	6.61	10.00	16.30	22.80	47.2
6	0.515	1.62	3.78	5.99	8.81	13.70	18.60	35.5
7	0.506	1.57	3.59	5.59	8.07	12.20	16.20	29.2
8	0.499	1.54	3.46	5.32	7.57	11.30	14.70	25.4
9	0.494	1.51	3.36	5.12	7.21	10.60	13.60	22.9
10	0.490	1.49	3.28	4.96	6.94	10.00	12.80	21.0
11	0.486	1.47	3.23	4.84	6.72	9.65	12.20	19.7
12	0.484	1.46	3.18	4.75	6.55	9.33	11.80	18.6
15	0.478	1.43	3.07	4.54	6.20	8.68	10.80	16.6
20	0.472	1.40	2.97	4.35	5.87	8.10	9.94	14.8
24	0.469	1.39	2.93	4.26	5.72	7.82	9.55	14.0
30	0.466	1.38	2.88	4.17	5.57	7.56	9.18	13.3
40	0.463	1.36	2.84	4.08	5.42	7.31	8.83	12.6
60	0.461	1.35	2.79	4.00	5.29	7.08	8.49	12.0
120	0.458	1.34	2.75	3.92	5.15	6.85	8.18	11.4
∞	0.455	1.32	2.71	3.84	5.02	6.63	7.88	10.8

Note: For larger values of the probability P, the value of F is approximately $F \simeq [1.25(1 - P)]^2$.

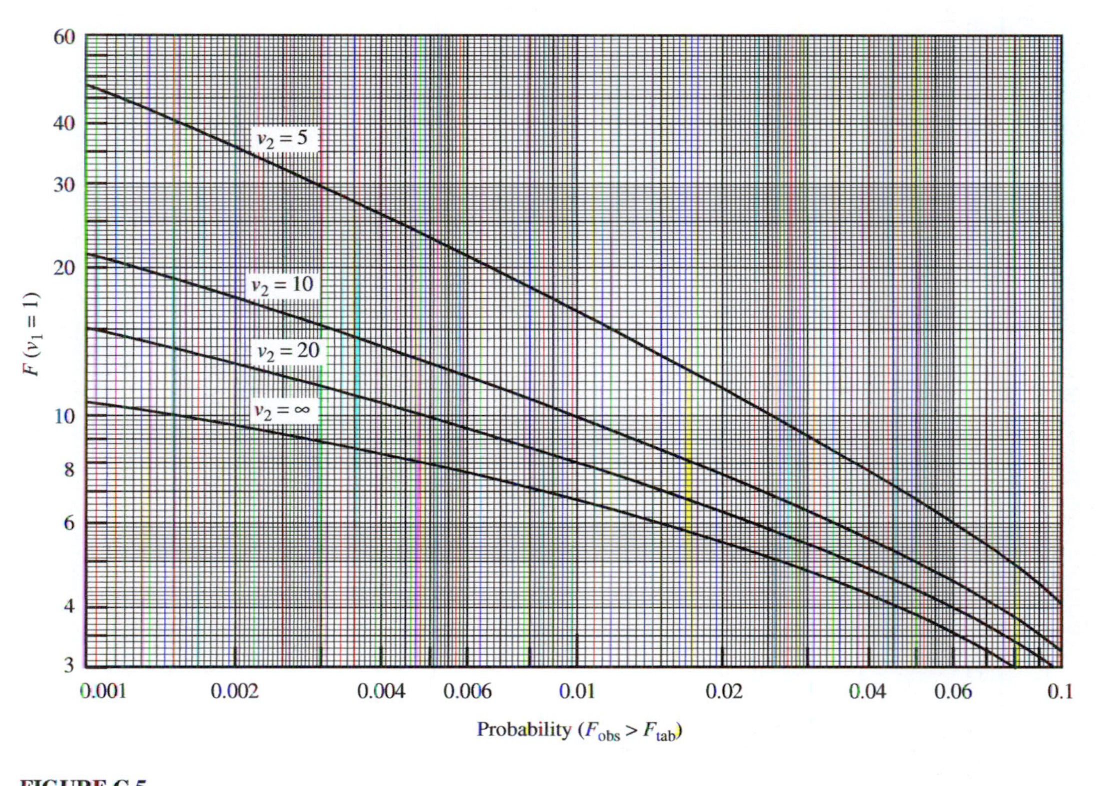

FIGURE C.5
The probability $P_F(F; 1, \nu_2)$ of exceeding F versus F and ν_2 for $\nu_1 = 1$.

TABLE C.6

F distribution, 5%. Values of F corresponding to the probability $P_F(F; \nu_1, \nu_2) = 0.05$ of exceeding F for ν_1 versus ν_2 degrees of freedom

Degrees of freedom	Degrees of freedom ν_1							
ν_2	2	4	6	8	10	15	20	100
1	200.00	225.00	234.00	239.00	242.00	246.00	248.00	253.00
2	19.00	19.20	19.30	19.40	19.40	19.40	19.40	19.50
3	9.55	9.12	8.94	8.85	8.79	8.70	8.66	8.55
4	6.94	6.39	6.16	6.04	5.96	5.86	5.80	5.66
5	5.79	5.19	4.95	4.82	4.73	4.62	4.56	4.41
6	5.14	4.53	4.28	4.15	4.60	3.94	3.87	3.71
7	4.74	4.12	3.87	3.73	3.64	3.51	3.44	3.27
8	4.46	3.84	3.58	3.44	3.35	3.22	3.15	2.97
9	4.26	3.63	3.37	3.23	3.14	3.01	2.94	2.76
10	4.10	3.48	3.22	3.07	2.98	2.85	2.77	2.59
11	3.98	3.36	3.09	2.95	2.85	2.72	2.65	2.46
12	3.89	3.26	3.00	2.85	2.75	2.62	2.54	2.35
15	3.68	3.06	2.79	2.64	2.54	2.40	2.33	2.12
20	3.49	2.87	2.60	2.45	2.35	2.20	2.12	1.91
24	3.40	2.78	2.51	2.36	2.25	2.11	2.03	1.80
30	3.32	2.69	2.42	2.27	2.16	2.01	1.93	1.70
40	3.23	2.61	2.34	2.18	1.08	1.92	1.84	1.59
60	3.15	2.53	2.25	2.10	1.99	1.84	1.75	1.48
120	3.07	2.45	2.18	2.02	1.91	1.75	1.66	1.37
∞	3.00	2.37	2.10	1.94	1.83	1.67	1.57	1.24

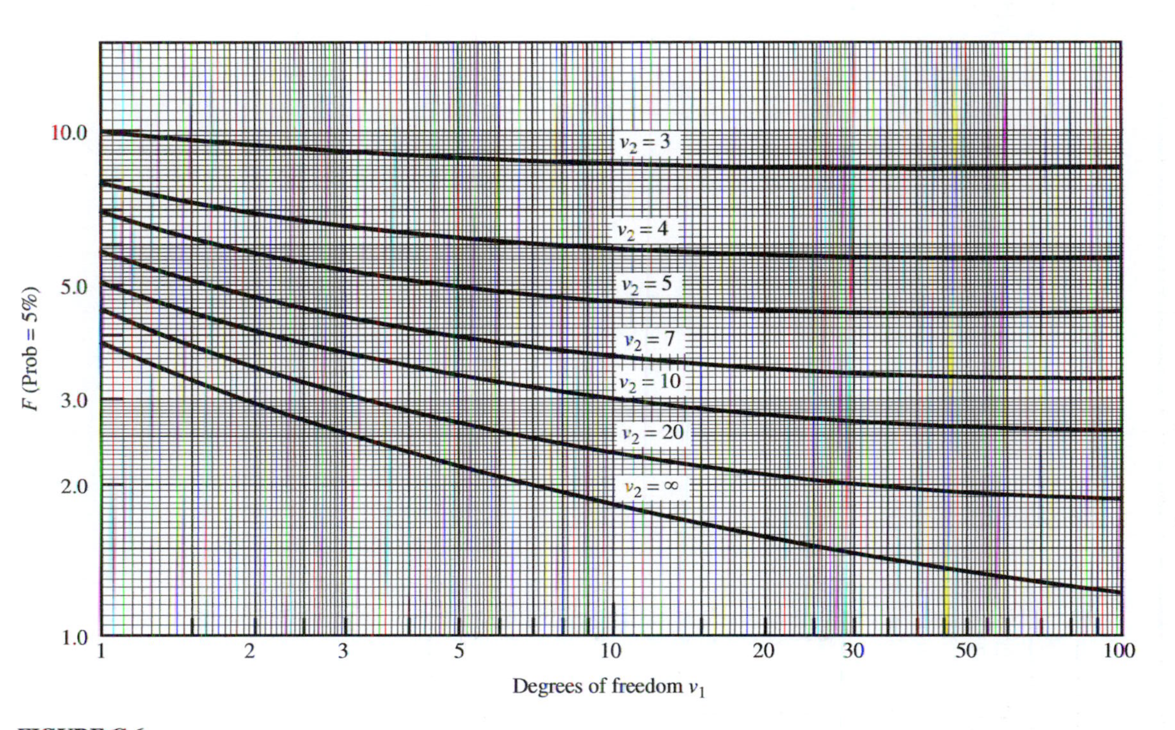

FIGURE C.6
Test values of $F(\nu_1, \nu_2)$ versus the numbers of degrees of freedom ν_1 and ν_2 for a probability $P_F(F; \nu_1, \nu_2) = 0.05$ of exceeding F.

TABLE C.7

F distribution, 1%. Values of F corresponding to the probability $P_F(F; \nu_1, \nu_2) = 0.01$ of exceeding F for ν_1 versus ν_2 degrees of freedom

Degrees of freedom	Degrees of freedom ν_1							
ν_2	2	4	6	8	10	15	20	100
1	5000.00	5620.00	5860.00	5980.00	6060.00	6160.00	6210.00	6330.00
2	99.00	99.20	99.30	99.40	99.40	99.40	99.40	99.50
3	30.80	28.70	27.90	27.50	27.20	26.90	26.70	26.20
4	18.00	16.00	15.20	14.80	14.50	14.20	14.00	13.60
5	13.30	11.40	10.70	10.30	10.10	9.72	9.55	9.13
6	10.90	9.15	8.47	8.10	7.87	7.56	7.40	6.99
7	9.55	7.85	7.19	6.84	6.62	6.31	6.16	5.75
8	8.65	7.01	6.37	6.03	5.81	5.52	5.36	4.96
9	8.02	6.42	5.80	5.47	5.26	4.96	4.81	4.42
10	7.56	5.99	5.39	5.06	4.85	4.56	4.41	4.01
11	7.21	5.67	5.07	4.74	4.54	4.25	4.10	3.71
12	6.93	5.41	4.82	4.50	4.30	4.01	3.86	3.47
15	6.36	4.89	4.32	4.00	3.80	3.52	3.37	2.98
20	5.85	4.43	3.87	3.56	3.37	3.09	2.94	2.54
24	5.61	4.22	3.67	3.36	3.17	2.89	2.74	2.33
30	5.39	4.02	3.47	3.17	2.98	2.70	2.55	2.13
40	5.18	3.83	3.29	2.99	2.80	2.52	2.37	1.94
60	4.98	3.65	3.12	2.82	2.63	2.35	2.20	1.75
120	4.79	3.48	2.96	2.66	2.47	2.19	2.03	1.56
∞	4.61	3.32	2.80	2.51	2.32	2.04	1.88	1.36

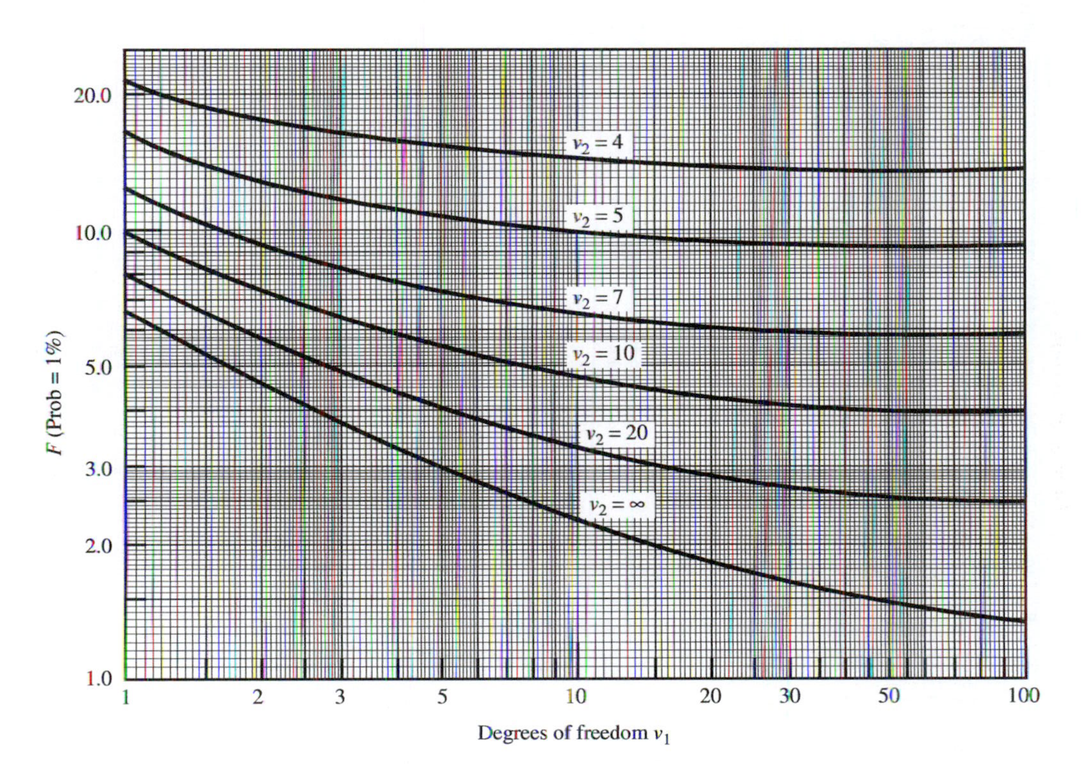

FIGURE C.7
Test values of $F(v_1, v_2)$ versus the numbers of degrees of freedom v_1 and v_2 for a probability $F_F(F; v_1, v_2) = 0.01$ of exceeding F.

TABLE C.8

$P_T(x; \mu, \sigma)$ versus $t = |x - \mu|/\sigma$; Integral of Student's t distribution between $x = \bar{x} - ts_\mu$ and $\bar{x} + ts_\mu$, expressed in percent.

	$\nu = N - 1$															
t	2	3	4	5	6	8	10	12	16	20	25	30	35	40	50	Gaussian probability
0.6	39.1	40.9	41.9	42.5	43.0	43.5	43.8	44.0	44.3	44.5	44.6	44.7	44.8	44.8	44.9	45.1
0.7	44.4	46.6	47.8	48.5	49.0	49.6	50.0	50.3	50.6	50.8	51.0	51.1	51.1	51.2	51.3	51.6
0.8	49.3	51.8	53.2	54.0	54.6	55.3	55.8	56.1	56.5	56.7	56.9	57.0	57.1	57.2	57.3	57.6
0.9	53.7	56.6	58.1	59.1	59.7	60.6	61.1	61.4	61.9	62.1	62.3	62.5	62.6	62.7	62.8	63.2
1.0	57.8	60.9	62.6	63.7	64.4	65.3	65.9	66.3	66.8	67.1	67.3	67.5	67.6	67.7	67.8	68.3
1.1	61.4	64.8	66.7	67.9	68.7	69.7	70.3	70.7	71.2	71.6	71.8	72.0	72.1	72.2	72.3	72.9
1.2	64.7	68.4	70.4	71.6	72.5	73.6	74.2	74.7	75.2	75.6	75.9	76.0	76.2	76.3	76.4	77.0
1.3	67.7	71.6	73.7	75.0	75.9	77.0	77.7	78.2	78.8	79.2	79.5	79.7	79.8	79.9	80.0	80.6
1.4	70.4	74.4	76.6	78.0	78.9	80.1	80.8	81.3	81.9	82.3	82.6	82.8	83.0	83.1	83.2	83.8
1.5	72.8	77.0	79.2	80.6	81.6	82.8	83.6	84.1	84.7	85.1	85.4	85.6	85.7	85.9	86.0	86.6
1.6	75.0	79.2	81.5	83.0	83.9	85.2	85.9	86.4	87.1	87.5	87.8	88.0	88.1	88.3	88.4	89.0
1.7	76.9	81.3	83.6	85.0	86.0	87.3	88.0	88.5	89.2	89.5	89.9	90.1	90.2	90.3	90.5	91.1
1.8	78.7	83.1	85.4	86.8	87.8	89.1	89.8	90.3	90.9	91.3	91.6	91.8	92.0	92.1	92.2	92.8
1.9	80.2	84.7	87.0	88.4	89.4	90.6	91.3	91.8	92.4	92.8	93.1	93.3	93.4	93.5	93.7	94.3
2.0	81.7	86.1	88.4	89.8	90.8	92.0	92.7	93.1	93.7	94.1	94.4	94.5	94.7	94.8	94.9	95.4
2.1	83.0	87.4	89.7	91.0	92.0	93.1	93.8	94.3	94.8	95.1	95.4	95.6	95.7	95.8	95.9	96.4
2.2	84.1	88.5	90.8	92.1	93.0	94.1	94.8	95.2	95.7	96.0	96.3	96.4	96.6	96.6	96.8	97.2
2.3	85.2	89.5	91.7	93.0	93.9	95.0	95.6	96.0	96.5	96.8	97.0	97.1	97.3	97.3	97.4	97.9
2.4	86.2	90.4	92.6	93.9	94.7	95.7	96.3	96.7	97.1	97.4	97.6	97.7	97.8	97.9	98.0	98.4
2.5	87.1	91.3	93.3	94.6	95.4	96.3	96.9	97.2	97.6	97.9	98.1	98.2	98.3	98.3	98.4	98.8
2.6	87.9	92.0	94.0	95.2	95.9	96.8	97.4	97.7	98.1	98.3	98.5	98.6	98.6	98.7	98.8	99.1
2.7	88.6	92.6	94.6	95.7	96.5	97.3	97.8	98.1	98.4	98.6	98.8	98.9	98.9	99.0	99.1	99.3
2.8	89.3	93.2	95.1	96.2	96.9	97.7	98.1	98.4	98.7	98.9	99.0	99.1	99.2	99.2	99.3	99.5
2.9	89.9	93.8	95.6	96.6	97.3	98.0	98.4	98.7	99.0	99.1	99.2	99.3	99.4	99.4	99.5	99.6
3.0	90.5	94.3	96.0	97.0	97.6	98.3	98.7	98.9	99.2	99.3	99.4	99.5	99.5	99.5	99.6	99.7
3.2	91.5	95.1	96.7	97.6	98.2	98.7	99.1	99.2	99.4	99.6	99.6	99.7	99.7	99.7	99.8	99.9
3.4	92.4	95.8	97.3	98.1	98.6	99.1	99.3	99.5	99.6	99.7	99.8	99.8	99.8	99.9	99.9	99.9
3.6	93.1	96.3	97.7	98.5	98.9	99.3	99.5	99.6	99.8	99.8	99.9	99.9	99.9	99.9	99.9	100.0
3.8	93.7	96.8	98.1	98.8	99.1	99.5	99.7	99.8	99.8	99.9	99.9	99.9	99.9	100.0	100.0	100.0
4.0	94.3	97.2	98.4	99.0	99.3	99.6	99.8	99.8	99.9	99.9	100.0	100.0	100.0	100.0	100.0	100.0

Note: The Gaussian probability for each value of t is listed in the last column.

APPENDIX D

HISTOGRAMS AND GRAPHS

Graphs of experimental data and of theoretical predictions have always been important tools for scientists, in both the actual performance of research and in presentations of results. In recent years we have seen a proliferation of graphics displays as fast inexpensive computers and printers have facilitated the display-making process. Scientists have benefited from the new techniques and equipment, with many excellent commercial programs available for creating high-quality scientific graphics suitable for publication.

In science, the object is to present results in a straightforward manner so that relevant points are illustrated clearly and without bias. Graphs with suppressed zeros, which are common in advertisements, are not often seen in scientific papers. Bar graphs tend to be simple histograms rather than the multibar, brightly colored displays of magazines and newspapers. In fact, although the use of color is growing, especially in direct publication on the Internet, few scientific preprints and papers are printed in color, although discrete use can clarify graphical presentations significantly. Error bars, which are rare indeed in advertisements, are essential in a scientific presentation. Exaggerated perspective and distorted scales have very limited use in scientific work whereas semilogarithmic plots that are often used in science are not often seen in business publications.

It is often convenient to have graphics routines that are part of a simulation or an analysis program, rather than to use a separate graphing program. For example, in a Monte Carlo simulation, it is essential to be able to produce histograms and data graphs quickly at each stage of the study. Generations of scientists have made simple histograms on monitors or printers to make preliminary studies of their data. We provide some simple routines of this type in the source files associated with this book.

More elegant and detailed graphs can be created by using the graphics features of particular programming languages, and those provided by data analysis programs and spreadsheets. Such programs can produce high-quality graphs and charts suitable for presentations and publications. Many of the graphs in this book, such as those in Chapter 2, were created by programs written in Fortran and Pascal. Others, such as those in Chapter 11, were created in Origin, a very powerful data analysis program with strong graphing features.

D.1 MAKING A GRAPH

Whether a scientific graph is produced by hand or by computer, there are several basic principles that should be followed. The graph should be large enough to be read and understood easily, with appropriately proportioned abscissa and ordinate. Axes should be labeled with large, clean letters, and the axes scales should be clearly indicated. If more than a single function is displayed, or if both data and curves are displayed, a box, or legend, may be superimposed on the graph to indicate the meaning of different symbols. In scientific journals, a description of the graph is generally included as text below the abscissa label. In internal papers and preprints, these descriptions are often collected in a separate section of the paper. For visual presentation, some descriptive material may be included in a box on the graph, but it is important that text be large enough to be clearly legible. One should avoid scattering too much material over any graph, which gives a busy appearance. A properly made graph should not require many words of explanation.

It is generally advisable to plot the independent variable as the abscissa and the dependent variable as the ordinate. However, if the independent data have a high degree of uncertainty while the corresponding measurements of the dependent data can be made with high precision, then it might be wise to interchange the two axes to simplify least-squares fitting.

Reasonable, convenient values and intervals should be chosen for the scale marks on the two axes. For example, if abscissa values range from 0 to 400, it might be reasonable to divide the x-axis into eight parts and thus to mark the abscissa with major, labeled ticks at 0, 100, 200, 300, and 400, with minor ticks half-way between. Dividing the axis into six parts and putting ticks at 66.7, 133.3, and so forth, would make it very difficult for a reader to interpret.

In general, error bars should be included for ordinate variables except for simple histograms where the text clearly specifies that the uncertainties are statistical and therefore given by the square root of the value of the coordinate. Unless otherwise noted, error bars generally indicate the standard deviation. Error bars usually are not necessary for abscissa variables. However, if appropriate, they may be drawn to indicate the resolution of the measurement or setting, or they may simply indicate the range of the variable over which data have been collected or grouped, as in the case of the width of a histogram bin. The text must explain the meaning of such error bars. If no error bar is shown for the abscissa, then it is useful to draw a circle or other symbol at each data point to indicate the position of the central values of each coordinate pair.

D.2 GRAPHICAL ESTIMATION OF PARAMETERS

A graph of y versus x often provides a convenient way of estimating parameters of the relation $y = y(x)$. The simplest example is the straight line

$$y = A + Bx \tag{D.1}$$

where the slope and the intercept can be estimated by making a graph and drawing a straight line that relates y to x. Clearly the better way to handle this problem is by a least-squares fitting technique, but the graphical method can be useful in both research and instructional laboratories for obtaining quick preliminary estimates of experimental results.

If we wish to find from the graph the uncertainty in our estimate of the slope, then we should attempt to draw two lines through the data, corresponding to estimates of the largest and smallest *reasonable* slopes, s_1 and s_2. We should take account in the uncertainties in the data points, if they are available, and, because we are trying to estimate the uncertainty as a standard deviation, we should attempt to draw these two lines to bracket about two-thirds of the data points—not all the points. Making this estimate is often difficult and subjective, especially if there are few points and they exhibit a lot of scatter. The mean slope s is just the average of our two slopes,

$$s = (s_1 + s_2)/2 \tag{D.2}$$

and an approximate estimate of the uncertainty is the magnitude of half the difference

$$\sigma = |s_1 - s_2|/2 \tag{D.3}$$

To gain practice in determining parameters from a graph, it is a worthwhile exercise to estimate the parameters from the graph and to compare those estimates with the results of a least-squares fit to the data. We should note that the two lines selected to give a reasonable estimate of the uncertainty in the slope may not be the same two lines we might draw to obtain a reasonable estimate of the uncertainty in the intercept. Figure D.1 displays the data of Figure 1.1b, with lines bracketing the points to show (a) reasonable ranges for estimating the intercept, and (b) reasonable ranges for estimating the slope. These lines were actually calculated from the results of a least-squares fit of the equation $Y = A + Bx$ to the data, which yielded the parameters A and B and their uncertainties σ_A and σ_B. We calculated the two lines in Figure D.1a from the equations $Y = (A \pm \sigma_A) + Bx$ and those in Figure D.1b from the equations $Y = A + (B \pm \sigma_B)x$. We note that these lines are just particular examples of an infinite number of such lines corresponding to all combinations of the slope and intercept within one standard deviation ranges, and in any given graph, a decision must be made on which lines to draw. In particular, allowing the lines to intersect at the intercept as in Figure D.1b may not give the best solution, although it can be a good starting point.

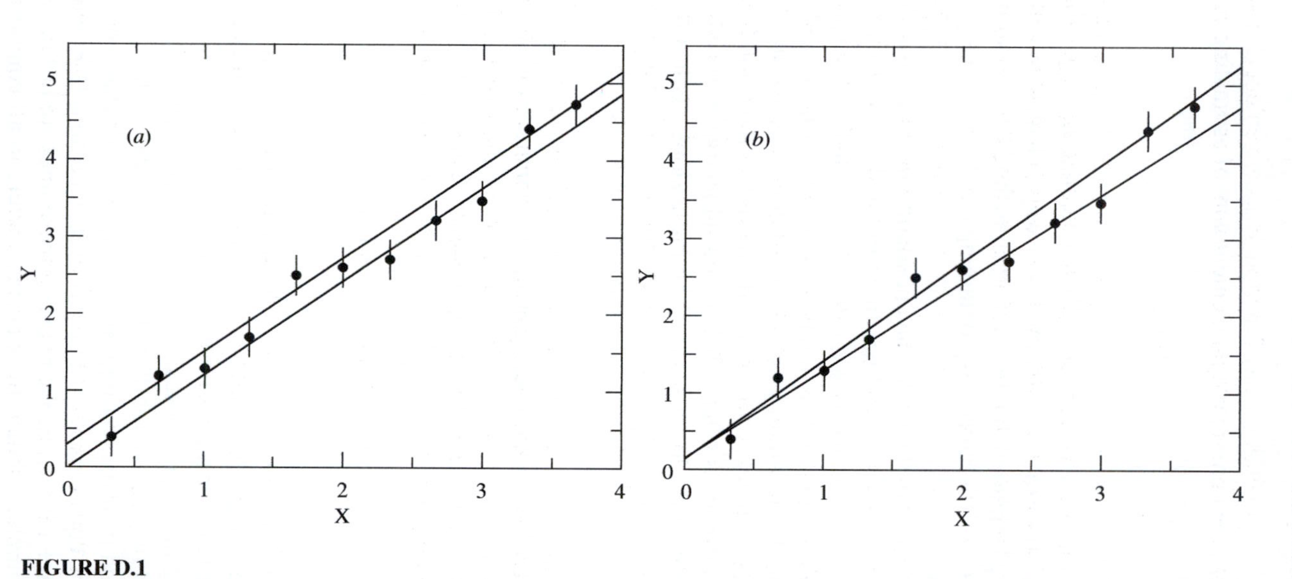

FIGURE D.1
Results of a least-squares fit of a straight line to the data of Figure 1.1b plotted to show separately the range of ± 1 standard deviation (*a*) in the intercept, and (*b*) in the slope. Units on the axes are arbitrary.

Semilogarithmic Graphs

When dealing with an exponential decay function, it is convenient to display the activity as a function of time on a semilogarithmic graph. That is, if the relation is

$$y(t) = y_0 e^{-at} \tag{D.4}$$

we plot a graph of $\log(y)$ versus x. Fortunately semilogarithmic graph paper is readily available so that it is not necessary actually to calculate any logarithms to make this plot. We merely have to select paper with the appropriate number of powers of 10 for our plot, label the axes, and plot y versus x on the graph. Such a graph is illustrated in Figure 8.1 for Example 8.1.

Semilogarithmic graph paper comes in various *cycles,* corresponding to the number of *decades* or powers of 10 that can be plotted on a single sheet. Thus, for example, on three-cycle paper we can plot y values that range from 1 to 1000 (or from 0.01 to 10.0, etc.). Note that we can never plot y values that are zero or negative on semilogarithmic paper. This is a problem when dealing with subtracted distributions, such as the counting experiment of Example 8.1, where, if we wish to plot the number of counts remaining after we have subtracted the average background from cosmic rays, we discover that, at large times, some bins have negative net counts. Those points, of course, cannot be displayed on a semilogarithmic graph. A full, least-squares fit to the total, unsubtracted data sample is clearly the right way to solve this problem, but if we are to attempt a graphical solution, we should be aware of this limitation.

We can determine from our data the parameter a in Equation (D.4) by finding the slope of the straight line on the semilogarithmic graph just as we found the slope on ordinary graph paper for a simple linear plot. Note that when calculating the slope we must compute the logarithms of the y values. Thus, if the two ends of the straight line have coordinates (x_1, y_1) and (x_2, y_2), the slope is given by

$$s = \frac{\ln(y_2) - \ln(y_1)}{x_2 - x_1} = \frac{\ln(y_2/y_1)}{x_2 - x_1} \tag{D.5}$$

The uncertainty in the slope can again be determined by drawing two straight lines that bracket the mean slope, although the logarithmic form of the plot decreases the accuracy in this determination.

Full-Logarithmic Graphs

If we wish to display a power relation of the form $y = Ax^n$, we may make a plot of y versus x on full-logarithmic paper or *log-log* paper. The result will be a straight line with slope n and we can obtain the slope, and therefore the exponent n, from the graph. This technique could be used, for example, to check the $1/r^2$ law for radiation intensity as a function of distance, by plotting a graph of intensity versus distance on log-log paper.

In Section 7.4 we discuss variable transformation as a method of converting a nonlinear fitting problem to a linear problem, and the distortions that may be introduced into the uncertainties in the process. Plotting on semilogarithmic or full-logarithmic paper is equivalent to such a variable change and we should attempt to compensate for these distortions, if necessary.

D.3 HISTOGRAMS AND FREQUENCY PLOTS

If we wish to display the frequency distribution of a measured variable x, then a histogram is generally the simplest and clearest form of presentation. For example, we may have observed particles emitted in the decay of an unstable state and wish to present the number detected in successive time intervals as in Example 2.4. Alternatively, we may have measured secondary particles in a scattering experiment and wish to display the distribution of their energies. In such cases, we can display the frequency distribution of the individual measurements, or *events,* as a histogram of $f(x)$ versus x, where $f(x_i)$ is the number of events that have values of x between x_i and $x_i + \Delta x$, and Δx is the histogram interval or bin width.

An alternate procedure for displaying binned data, which is especially useful for distributions with large numbers of bins, or for data with nonstatistical uncertainties, is to make a regular graph of frequency versus the measured variable, a *frequency plot,* with the data points indicated by crosses and uncertainties by error bars. This procedure is especially convenient when there are many bins or when error bars must be displayed, as illustrated in Figure 8.1.

A convenient procedure for finding the frequency distribution of (or *binning*) a continuous variable x is to label a bin with a tick mark at the lower limit x_1 of the bin and to count within a bin those events for which $x_i \leq x < x_i + \Delta x$. This is suitable for most, but not all, data sets. Choice of the bin width depends on a number of factors. In the ideal situation with a large quantity of high-precision data, the bin width could be chosen to be very small. However, in real experiments, the number of events may not be very large and each x coordinate will have some uncertainty. As a general rule, the bin width should not be less than the uncertainty in the measured variable x and one should be very wary of any data structure that is narrower than the uncertainty in x. If the number of events is relatively small, then even wider binning may be necessary. With such data, the competition between statistical significance and resolution of narrow effects in the histogram may become important. A histogram with less than ten events in its highest bin is not generally very informative, considering that the uncertainty in that bin will be over 30%.

A problem arises when the bin width of a histogram is close to or equal to the least count of the data. This can happen when the data are integral numbers or with data that have been collected by a digital device. The previous suggestions that the histogram bins be labeled with the lower limit at the left of the bin may not be reasonable for such data, and it may be better to place tick marks at the center of the bins.

EXAMPLE D.1 A student in an introductory physics laboratory attempts to measure the value of the acceleration of gravity by timing a ball that she drops 50 times from a height of 3 m. She uses an electronic timer with a least count of 0.01 s. The

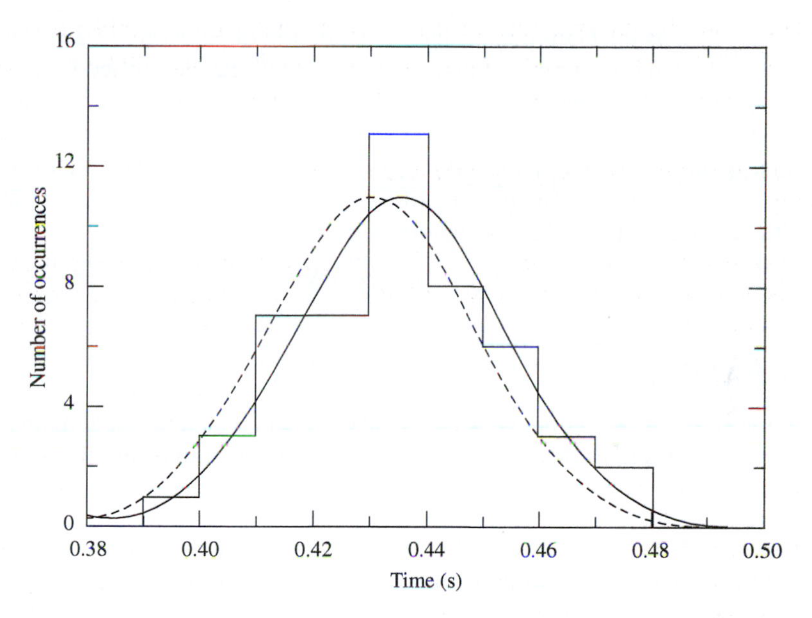

FIGURE D.2
Histogram of measured times plotted with the bin width equal to the least count of a digital clock. The numbers on the abscissa correspond to the lower time limit of the bin. The dashed Gaussian curve was calculated from the mean and standard deviation of the measurements. The solid curve was calculated with the mean increased by half the bin width to correct for the truncation of the data.

timer starts when the ball is released and stops when it hits the floor. Uncertainties in the measurements come mainly from variations in the starting and stopping times.

The student's measurements have been plotted in the histogram of Figure D.2 where the bin width is equal to the least count (0.01ts). We assume that the digital clock truncates the measured times so each time measurement corresponds to the left-hand edge of a bin and the actual value of the time is somewhere within the bin limits. Thus, in this case it is appropriate to indicate the lower value of the bin limit at the left-hand edge of the bin.

The dashed Gaussian curve was calculated from the mean ($\bar{t}$ = 0.431 s) and standard deviation (s = 0.0184) of the measurements. The curve clearly is shifted to the left relative to the data. The discrepancy is caused by the fact that we neglected to correct for the truncation of the data by the digital clock. To correct the mean we must add to it half the width of a bin to obtain $\bar{t} = 0.431 + 0.005 = 0.436$ s. The Gaussian curve, calculated from the corrected mean, is shown as a solid line.

Normalized Curves on Histograms

When superimposing a theoretical curve on a data histogram, we often want to scale the area of the curve to that of the histogram, or to *normalize* the curve. The required scale factor can be determined in the following way. We assume that the curve has been calculated from a probability distribution function that is normalized to unit area, such as the Gaussian probability function of Equation (2.23). The area of one event on the histogram is equal to the bin width Δx multiplied by a unit interval on the ordinate. Thus, the total area of the histogram is equal to the bin width multiplied by the total number of events ($A = N\Delta x$). To scale the curve to the area of the histogram, we multiply the values $p(x_i)$, calculated from the equation of the probability distribution, by the product of the number of events on the plot and the bin width, so that the plotted curve becomes

$$y(x_i) = p(x_i) \times N\Delta x \tag{D.6}$$

D.4 GRAPHICS ROUTINES

We include source files on the website for routines which can be used to make simple graph and histograms. Most of the sample computer routines in this book make calls to these routines.

Program D.1 QUIKSCRP (website) accepts data that define graphs and histograms and writes a script file that can be read and interpreted by the executable program QDISPLAY.EXE (website) to produce displays on the monitor. Details of the calling procedures can be seen in the routine PLOTIT in the program unit \CHAPT-6\FITUTIL (website) called from the program \CHAPT-6\FITLINE (APPENDIX E). For this program QuikScrp writes an output file FITLINE.SCR.

Program D.2 QUIKHIST (website) collects data and presents a character-based histogram on the monitor. Printed output is also available. PROGRAM 5.2: \CHAPT-5\POISDCAY illustrates use of this program.

Program D.3 QDISPLAY.EXE (website) is an executable program that reads a script file written by QUIKSCRP and interprets the file to create a graphics display on the monitor. The command line instruction for running QDISPLAY with the script file produced by the program FITLINE is QDISPLAY FITLINE.

APPENDIX E

COMPUTER ROUTINES IN FORTRAN

This appendix lists several routines that illustrate the material of the text. The routines are listed in Fortran 77, an old, but quite readable version of that ever-popular programming language. All routines have been tested; however, most of them requires subsidiary routines and drivers that are not listed. Complete programs and routines are available on the Web in C++ as well as in Fortran. Readers are urged to log onto the website at www.mhhe.com/bevington to download these programs.

We have tried to keep the routines simple, trading efficiency for clarity where necessary. To make explicit which modules are required to form a complete program, and to avoid the need for command strings to link the object programs into an executable program, we have chosen to use the INCLUDE statement to present the compiler with a single source file from which to compile a single object module incorporating all required routines. We also use the INCLUDE statement to copy blocks of COMMON and other variable-defining statements into routines.

Because readers may not be familiar with Fortran, we list a few basic principles that should help in understanding the instructions and following their logic. This list includes only a selection of language elements that appear in the sample programs.

STATEMENTS

The format of Fortran statements was defined in terms of the 80-column Hollerith card:

column 1: blank or with a "C" to indicate that the information on the line is a "Comment";

columns 2–5: statement label (a number);

column 6: reserved for a single digit number to indicate a continuation of the statement from the previous line;

columns 7–72: program statements;

columns 73–80: not used.

Although it is not necessary to follow rigorously this scheme with a modern interactive compiler on a personal computer (for example, "tabs" can be used), the general order must be followed.

PROGRAM FLOW

Program flow can be controlled by IF statements, by IF THEN statements (with ELSEIF and ENDIF), by DO AND DOWHILE statements that may refer to a termination label (all statements labels are numerical) or to the DO terminator; ENDDo, and by GOTO statements. Excessive use of the GOTO statement can lead to very confusing programs. In order to facilitate following the program flow, we have indented groups of instructions that are accessed through a control statement, such as IF THEN, or DO.

Examples

```
DO 100I = 1 TO 20      DO I = 1 TO 20      X = 1
  X = I                  X = I             DO WHILE X .LE. 20
  Y(I) = SQRT(X)         Y = SQRT(X)         Y = SQRT(X)
100 CONTINUE           ENDDo                 X = X + 1
                                           ENDDo
```

VARIABLE DEFINITIONS

Fortran does not require the rigorous variable typing of newer languages. As default typing, variables with names beginning with I, J, K, L, M, or N are defined as INTEGER; variable names beginning with other letters are identified as REAL. However, we have attempted to identify most of the variables in the routines and in some instances have violated the default typing for program clarity.

Examples

```
INTEGER S1, S3, N/10/
REAL X, T, TPRIME, SIGMAT/1.0/
LOGICAL NEXTVAR/.FALSE./
```

Note that the variables N, SIGMAT and, NEXTVAR in the preceding examples have been initialized to the values 10, 1.0, and FALSE, respectively. The DATA statement also can be used to initialize variables. DATA SQRTPI/1.7724539/

Other types include:

```
CHARACTER
COMPLEX
DOUBLE PRECISION
```

Variables defined in named COMMON statements are available to any routine that includes the statement. Local variables can be defined in DIMENSION STATEMENTS. Array sizes may be defined in PARAMETER statements or directly in a COMMON or DIMENSION statement.

Examples

```
PARAMETER(MAXPARAM 10)
COMMON/FITVARS/ NPTS, M, NFREE, MARRAY(MAXPARAM),
ZARRAY(200)
DIMENSION NPLAN(30).
```

Fortran has several types of subprograms that can be called from another routine: SUBROUTINE and FUNCTION are the most common. Data types defined in a subprogram must be consistent with the definitions in the calling routine. A function name must specify its own data type.

Examples

```
CALL SETRANDOMDEVIATESEED(S1, S2, S3)
TPRIME = GAUSSSMEAR(T,SIGMAT)
REAL FUNCTION GAUSSSMEAR(X,DX)
SUBROUTINE SETRANDOMDEVIATESEED(SA,SB,SC)
```

The INCLUDE statement copies the specified file into the body of the program.

Example

```
INCLUDE '\CHAPT-5\MONTEINC.FOR'
```

As well as comment statements that begin with a "C" in column 1, comments may appear in statement lines, preceded by the exclamation point (!).

E.1 Routines from Chapter 5

```
C PROGRAM 5.1: \CHAPT-5\HOTROD.FOR
C SIMULATED VARIATION OF TEMPERATURE ALONG A METAL ROD
C 10 CM ROD-TEMPERATURE IS ZERO AT ONE END, 100 DEGREES C AT OTHER.
C USES MONTELIB
      PROGRAM HOTROD
      INTEGER S1, S2, S3, N/10/        !--- GENERATE 10 POINTS AT 1 CM INTERVALS
      REAL  X, T, TPRIME, SIGMAT/1.0/ !--- WITH AN UNCERTAINTY OF +-1 DEGREE
      REAL GAUSSSMEAR
      S1 = 1171
      S2 = 343
      S3 = 1322
      CALL SETRANDOMDEVIATESEED(S1, S2, S3)
      PRINT *,'      HOT ROD TEST DATA,   SIGMA=',  SIGMAT
      X = -0.5
      DO 100 I = 1, N
         X     = X + 1.0           !--- POSITION ALONG ROD
         T     = 10.0*X            !--- CALCULATE MEAN TEMPERATURE AT POINT
         TPRIME = GAUSSSMEAR(T,SIGMAT)         !--- SMEAR IT
         PRINT *,I, X, T, TPRIME
  100 CONTINUE
      CALL EXIT
      END
      INCLUDE C:\CHAPT-5\MONTELIB.FOR

C PROGRAM 5.2: \CHAPT-5\POISDCAY.FOR
C SIMULATED DECAY OF AN UNSTABLE STATE.
C USES QUIKHIST, MONTELIB
      PROGRAM POISDCAY         !--- GENERATE A 200-EVENT POISSON  HISTOGRAM
      REAL LO/0/, INT/1/, HI/22/
      INTEGER NEVENTS/400/, POISSONDEVIATE
      REAL MU/8.4/
      INTEGER S1, S2, S3, I, K
      REAL   X
      S1 = 1171
      S2 = 343
      S3 = 1322
      CALL SETRANDOMDEVIATESEED(S1, S2, S3)
      CALL HISTINIT(' ')            !---OUTPUT FILE NAME OR ' ' FOR MONITOR OUTPUT
      CALL HISTSETUP(1,LO,INT,HI,'POISSON - COUNTS/10 SEC')
      K=POISSONDEVIATE(MU,.TRUE.)          !--- INITIALIZE - MAKE THE TABLE
      DO 100 I = 1, NEVENTS
         K = POISSONDEVIATE(MU,.FALSE.)
         X = K
         CALL HISTOGRAM(1,X)
  100 CONTINUE
      CALL HISTDISPLAYALL(.FALSE.) !DUMMY ARG-COMPAT. WITH QUIKSCRP
      CALL EXIT
      END
      INCLUDE \CHAPT-5\MONTELIB.FOR
      INCLUDE \APPEND-D\QUIKHIST.FOR !REPLACE WITH QUIKSCRP FOR GRAPHICS
```

```
C PROGRAM 5.3: \CHAPT-5\MONTELIB.FOR
C MONTE CARLO LIBRARY ROUTINES
      SUBROUTINE SETRANDOMDEVIATESEED(SA,SB,SC)
      INCLUDE '\CHAPT-5\MONTEINC.FOR'
      INTEGER SA, SB, SC
      SEED1 = SA
      SEED2 = SB
      SEED3 = SC
      RETURN
      END

      SUBROUTINE GETRANDOMDEVIATESEED(SA,SB,SC)
      INCLUDE '\CHAPT-5\MONTEINC.FOR'
      INTEGER SA, SB, SC
      SA = SEED1
      SB = SEED2
      SC = SEED3
      RETURN
      END

      REAL FUNCTION RANDOMDEVIATE()          !--- WICHMANN AND HILL
      INCLUDE '\CHAPT-5\MONTEINC.FOR'
      REAL  TEMP
      SEED1 = 171*MOD(SEED1,177) - 2*(SEED1 / 177)
      IF (SEED1 .LT. 0 ) SEED1 = SEED1 + 30269
      SEED2 = 172*MOD(SEED2,176) - 35*(SEED2 / 176)
      IF (SEED2 .LT. 0 ) SEED2 = SEED2 + 30307
      SEED3 = 170*MOD(SEED3,178) - 63*(SEED3 / 178)
      IF (SEED3 .LT. 0 ) SEED3 = SEED3 + 30323
      TEMP = SEED1/30269. + SEED2/30307. + SEED3/30323.
      RANDOMDEVIATE = TEMP-AINT(TEMP)
      RETURN
      END

C -FIND A RANDOM VARIABLE DRAWN FROM THE GAUSSIAN DISTRIBUTION-
      REAL FUNCTION RANDOMGAUSSDEVIATE()         !--- BOX-MUELLER
      INCLUDE '\CHAPT-5\MONTEINC.FOR'
      LOGICAL  NEXTVAR/.FALSE./
      REAL R, F, Z1, Z2, X1RANGAUSS, RANDOMDEVIATE
      IF (NEXTVAR) THEN
         NEXTVAR = .FALSE.
         RANDOMGAUSSDEVIATE = X2RANGAUSS
      ELSE
 100     Z1 = -1 + 2*RANDOMDEVIATE()
         Z2 = -1 + 2*RANDOMDEVIATE()
         R  = Z1*Z1 + Z2*Z2
         IF (R .GE. 1) GOTO 100
         F = SQRT(-2*ALOG(R)/R)
         X1RANGAUSS = Z1*F
         X2RANGAUSS = Z2*F
         RANDOMGAUSSDEVIATE = X1RANGAUSS
```

```
      NEXTVAR = .TRUE.
   ENDIF
   RETURN
   END

   REAL FUNCTION GAUSSSMEAR(X,DX)
   REAL X, DX
   REAL RANDOMGAUSSDEVIATE
   GAUSSSMEAR = X + RANDOMGAUSSDEVIATE() * DX
   RETURN
   END

C -RECURSION METHOD FOR POISSON PROBABILITY (P(N,M). TO FIND P(N,M) MUST
C    ALL WITH SUCCESSIVE ARGUMENTS J=0,1,..N. MAX MU=85, NO LIMIT ON X
   REAL FUNCTION POISSONRECUR(J, M)
   INCLUDE '\CHAPT-5\MONTEINC.FOR'
   INTEGER J
   REAL M
   IF (J.EQ.0 ) THEN
      POISS = EXP(-M)
   ELSE
      POISS = (POISS*M)/J          !--- POISS =  (M^J)EXP(-MU/J)
   ENDIF
   POISSONRECUR = POISS
   RETURN
   END

C -FIND A RANDOM VARIABLE DRAWN FROM THE POISSON DISTRIBUTION
   INTEGER FUNCTION POISSONDEVIATE(MU, INIT)
   INCLUDE '\CHAPT-5\MONTEINC.FOR'
   INTEGER   I, X, N
   REAL      MU, P , R, POISSONRECUR
   LOGICAL   INIT
   IF (INIT ) THEN                        ! --- MAKE TABLE OF SUMS ---
      N = AINT(MU + 8* SQRT(MU))          ! ---IE., 8*SIGMA
      IF (N .GT. MAXBINS ) THEN
         PRINT *, 'OVERFLOW ERROR IN ROUTINE POISSON DEVIATE'
         CALL EXIT
      ENDIF
      PTABLE(0) = POISSONRECUR(0,MU)
      DO 100 I = 1, N-1
         P = POISSONRECUR(I,MU)
         PTABLE(I) = PTABLE(I-1)+P
100   CONTINUE
      PTABLE(N) = 1                       ! --- ASSURE UNIT PROBABILITY ---
   ELSE                                   ! --- GENERATE AN EVENT ---
      X = -1
      R = RANDOMDEVIATE()
200   X = 1 + X
      IF (PTABLE(X) .LE. R) GOTO 200 !— REPEAT UNTIL PTABLE(X) >= X
      POISSONDEVIATE = X
```

```
      ENDIF
      RETURN
      END
```

```
PROGRAM 5.4: \CHAPT-5\KDECAY.FOR            (WEBSITE)
C ILLUSTRATION OF EXAMPLE 5.7
```

```
C PROGRAM 5.5: \CHAPT-5\MONTEINC.FOR
C COMMON FOR MONTE CARLO LIBRARY
      COMMON/MC/ SEED1, SEED2, SEED3, X2RANGAUSS, POISS, PTABLE
      PARAMETER (MAXBINS = 100)
      INTEGER SEED1, SEED2, SEED3
      REAL X2RANGAUSS, PTABLE(0:MAXBINS)
      REAL*8 POISS
C---------------END MONTEINC  -------------------------
```

E.2 Routines from Chapter 6

```
C PROGRAM 6.1: \CHAPT-6 FITLINE.FOR
C LEAST-SQUARES FIT TO A STRAIGHT LINE BY METHOD OF DETERMINANTS
C USES  FITUTIL
      PROGRAM FITLINE
C  -------------------------MAIN  ROUTINE----------------------
      INCLUDE '\CHAPT-6 FITVARS.FOR'
      CHARACTER*40  TITLE
      CHARACTER*1 VORG, READCHAR
      INTEGER I
      REAL DET, CHI2, CALCCHISQ
      M = 2                                          !--- FIND 2 PARAMETERS
      PRINT *, '(V)OLTS OR (G)EIGER? '
      VORG = READCHAR()
      IF (VORG .EQ. 'V') THEN
         CALL FETCHDATA('\CHAPT-6\VOLTS.DAT' ,TITLE)       !--- EXAMPLE 6.1
      ELSEIF ((VORG .EQ. 'G') .OR. (VORG .EQ. 'G')) THEN
         CALL FETCHDATA('\CHAPT-6\GEIGER.DAT',TITLE)       !--- EXAMPLE 6.2
         DO 100 I = 1 , NPTS
         X(I)  = 1/X(I)**2                                 !--- FITTING 1/R^2
 100     CONTINUE
      ENDIF
      CALL LINEFIT(DET)
      CALL CALCULATEY            !---  FILL ARRAY YCALC FOR CALCCHISQ AND PLOTIT
      CHI2 = CALCCHISQ()
      CALL OUTPUT(.FALSE. , 'CON', CHI2, TITLE)  !--- FALSE FOR NO ERROR MATRIX
      IF (VORG .EQ. 'V') THEN
         CALL PLOTIT('FITLINE.SCR',.FALSE.,.FALSE., !---SCRPT FILE,LOG?,SPLINE?
     1   'C', ABS(X(2)-X(1))/20,            !--- DATA CIRCLE, RAD OF CIRCLE
     2   0.0, 0.0, 100.0, 3.0,              !---  X1,Y1,X2,Y2
```

```
     3      5, 6,                          !--- # X-DIV, # Y-DIV
     4      'X (CM)', 'POTENTIAL DIFF(ERENCE (VOLTS)')  !--- AXIS LABELS
         ELSEIF (VORG .EQ. 'G') THEN
            CALL PLOTIT('FITLINE.SCR',.FALSE. ,.FALSE.,
     1      'C', ABS(X(2)-X(1))/50, 0.0, 0.0, 30.0, 1000.0, 6, 5,
     2      'SQUARED INVERSE DISTANCE (1/M^2)', 'NUMBER OF COUNTS PER SEC')
         ENDIF
         READ *
         CALL CLOSEGRAPHICS
         END

         SUBROUTINE CALCULATEY          !--- FILLS ARRAY YCALC
         INCLUDE '\CHAPT-6\FITVARS.FOR'
         INTEGER I
         DO 100 I= 1 , NPTS
            YCALC(I) = A(1) + A(2)*X(I)
   100   CONTINUE
         RETURN
         END

         REAL FUNCTION CALCCHISQ() !--- ASSUMES ARRAY YCALC HAS BEEN FILLED
         INCLUDE '\CHAPT-6\FITVARS.FOR'
         INTEGER  I
         REAL    CHI2
         CHI2=0.
         DO 100 I = 1 , NPTS
            CHI2 = CHI2 + ( (Y(I)-YCALC(I))/SIGY(I))**2
   100   CONTINUE
         CALCCHISQ = CHI2
         RETURN
         END

         SUBROUTINE LINEFIT(DET)
         INCLUDE '\CHAPT-6\FITVARS.FOR'
         REAL DET
         INTEGER I
         REAL SUMWT, SUMX, SUMY, SUMX2, SUMY2, SUMXY, WEIGHT
         SUMWT = 0
         SUM    = 0
         SUMY   = 0
         SUMX2  = 0
         SUMY2  = 0
         SUMXY  = 0
C --------- ACCUMULATE WEIGHTED SUMS -----------
         DO 100  I= 1 , NPTS
            WEIGHT  = 1/SIGY(I)**2
            SUMWT   = SUMWT  + WEIGHT
            SUMX    = SUMX   + WEIGHT * X(I)
```

```
      SUMY   = SUMY  + WEIGHT * Y(I)
      SUMX2  = SUMX2 + WEIGHT * X(I)**2
      SUMY2  = SUMY2 + WEIGHT * Y(I)**2
      SUMXY  = SUMXY + WEIGHT * X(I)*Y(I)
 100 CONTINUE
C ---CALCULATE THE PARAMETERS - CUT OUT IF DETERMINANT IS NOT > O ---
      DET = SUMWT * SUMX2 - SUMX * SUMX
      IF (DET .GT. O ) THEN
         A(1)    = (SUMX2*SUMY  - SUMX*SUMXY)/DET
         A(2)    = (SUMXY*SUMWT - SUMX*SUMY) /DET
         SIGA(1) = SQRT(SUMX2/DET)
         SIGA(2) = SQRT(SUMWT/DET)
      ELSE
         CALL ERRORABORT('DETERMINANT < OR = O IN LINEFIT')
      ENDIF
      RETURN
      END
      INCLUDE '\CHAPT-6\FITUTIL.FOR' ! FITUTIL INCLUDES QUIKSCRP.FOR
```

```
C PROGRAM 6.2: \CHAPT-6\FITVARS.FOR         (WEBSITE)
C INCLUDE FILE OF CONSTANTS, VARIABLES AND ARRAYS FOR LEAST-SQUARES FITS
C ALL GLOBAL TYPES, CONSTANTS AND VARIABLES ARE DECLARED HERE.
C THE ARRAY LIMITS MAXDATA AND MAXPARAM CAN BE SET AS REQUIRED
                                              FOR PARTICULAR PROBLEMS.

C PROGRAM 6.3: \CHAPT-6\FITUTIL.FOR         (WEBSITE)
C GENERAL UTILITY ROUTINES
```

E.3 Routines from Chapter 7

```
C PROGRAM 7.1: \CHAPT-7\MULTREGR.FOR
C LEAST-SQUARES FIT TO A POWER SERIES AND TO LEGENDRE POLYNOMIALS.
C USES  FITFUNC7, MAKEAB7, MATRIX, FITUTIL

      PROGRAM MULTREGR
C     M = NUM OF PARAMETERS, NPTS=NUMBER OF DATA PAIRS,
C     DATA AND UNCERTAINTIES ARE IN ARRAYS  X, Y, DY.
      INCLUDE '\CHAPT-6\FITVARS.FOR'
      COMMON /FITVARS7/PAE
      CHARACTER * 1 PAE
      REAL    DET, CHI2, CALCCHISQ
      INTEGER  I
      LOGICAL  SPL
      CHARACTER*1 READCHAR
      CHARACTER*40 TITLE
      PRINT *,   '(P)OWER SERIES, (A)LL LEGENDRE TERMS TO L = 4,'
```

```
      PRINT *,    'OR (E)VEN LEGENDRE TERMS(L = 0,2,4).'
      PRINT *,    'TYPE P,A OR E '
      PAE = READCHAR()
 1000 FORMAT(A1)
      IF (PAE .EQ. 'P') THEN
         CALL FETCHDATA ('\CHAPT-7\THERMCOU.DAT', TITLE)
         PRINT *,    'TYPE NUMBER OF PARAMETERS '
         READ *, M
      ELSEIF (PAE .EQ. 'A') THEN
         CALL FETCHDATA('\CHAPT-7\LEGENDRE.DAT',TITLE)
         M = 5
      ELSEIF (PAE .EQ. 'E' ) THEN
         CALL FETCHDATA('\CHAPT-7\LEGENDRE.DAT',TITLE)
         M = 3
      ENDIF  !--- PAE
      CALL MAKEBETA                       !--- SET UP THE LINEAR BETA MATRIX
      CALL MAKEALPHA                      !--- SET UP THE SQUARE ALPHA MATRIX
      CALL MATINV(M, ALPHA, DET)          !--- INVERT ALPH TO GET EPSILON MATRIX
      CALL LINEARBYSQUARE(M,BETA,ALPHA,A) !--- BETA X EPS = PARAMETER MATRIX
      CALL CALCULATEY
      CHI2 = CALCCHISQ()
      DO 100 I = 1, M
         SIGA(I) = SQRT(ALPHA(I,I))
  100 CONTINUE
      CALL OUTPUT(.TRUE., 'CON', CHI2, TITLE) !--- TRUE TO PRINT ERROR MATRIX
      IF (M .GT. 2 ) THEN
         SPL = .TRUE.                          !--- PLOT A CURVE
      ELSE
         SPL = .FALSE.                         !--- PLOT A LINE
      ENDIF
      IF (PAE .EQ. 'P') THEN
         CALL PLOTIT('MULTREGR.SCR', .FALSE., SPL,     !--- FILE,LOG?,SPLINE
     1   'C', (X(2)-X(1))/12,       !--- DATA CIRCLES, RADIUS OF DATA CIR
     2   -10., -2., 110., 4.,              !--- X1,Y1, X2,Y2
     3    6, 6,                            !— X,Y GRID MARKS
     4   'TEMPERATURE (DEGREES CELSIUS)','VOLTAGE (MV)')
      ELSE IF ((PAE .EQ. 'A') .OR. (PAE .EQ. 'E')) THEN
         CALL PLOTIT('MULTREGR.SCR', .FALSE., .TRUE.,
     1   'C', (X(2)-X(1))/10,   0., 0.,180., 1500., 6, 6,
     2   'THETA(DEGREES)', 'NUMBER OF COUNTS')
      ENDIF                                      !--- PAE
      CALL CLOSEGRAPHICS
      END
      INCLUDE '\CHAPT-7\FITFUNC7.FOR'
      INCLUDE '\CHAPT-7\MAKEAB7.FOR'
      INCLUDE '\CHAPT-6\FITUTIL.FOR'
      INCLUDE '\APPEND-B\MATRIX.FOR'

C PROGRAM 7.2: \CHAPT-7\FITFUNC7.FOR
C FITTING FUNCTIONS FOR CHAPTER 7 EXAMPLES.
```

```
      REAL FUNCTION POWERFUNC(K, XX)
      INTEGER K
      REAL    XX
      REAL    YY
      INTEGER I
      YY = 1
      IF (K .GT. 1 ) THEN
         DO 100 I= 2, K
         YY = XX * YY
 100  CONTINUE
      ENDIF
      POWERFUNC = YY
      RETURN
      END

      REAL FUNCTION LEGFUNC(K, XX)
C DEFINE SEPARATE TERMS IN A SERIES, Y = AO*LO(X) + A1*L1(X) + ..
C NOTE K = 1 CORRESPONDS TO ZEROTH ORDER.
C VAR   PAE : CHAR   'P'-POWER SERIES,
C 'A'-ALL LEGENDRE TERMS TO ORDER M,
C 'E'-EVEN LEGENDRE TERMS)
C
      COMMON /FITVARS7/PAE
      CHARACTER *1 PAE
      INTEGER K
      REAL    XX
      INTEGER KK, I
      REAL  C, PI/3.14159/, LEGPOLY(11) !---  I.E., 0TH THRU 10TH ORDER
      IF (PAE .EQ. 'E') KK = 2*K - 1
      IF (PAE .EQ.'A')  KK = K
      C = COS(PI*XX/180)
      LEGPOLY(1) = 1 !---  FOR BETTER EFFICIENCY, COULD CALC ONCE AND SAVE
      IF (KK .GT. 1 ) THEN
         LEGPOLY(2) = C
         IF (KK .GT. 2 ) THEN
         DO 100 I = 3, KK
           LEGPOLY(I)=((2*I-1)*C*LEGPOLY(I-1)-(I-1)*LEGPOLY(I-2))/I
 100     CONTINUE
      ENDIF            !---  KK > 2
      ENDIF               !---  KK > 1
      LEGFUNC = LEGPOLY(KK)
      RETURN
      END

      REAL FUNCTION FUNCT(K, XX)
      INTEGER K
      REAL XX
      REAL LEGFUNC, POWERFUNC
      COMMON /FITVARS7/PAE
      CHARACTER * 1 PAE
```

```
      IF ((PAE .EQ. 'A') .OR. (PAE.EQ.'E')) FUNCT = LEGFUNC(K,XX)
      IF  (PAE .EQ. 'P')  FUNCT = POWERFUNC(K,XX)
      RETURN
      END

      SUBROUTINE CALCULATEY
      INTEGER  I, K
      REAL     YY, FUNCT
      INCLUDE '\CHAPT-6\FITVARS.FOR'
      DO 100 I=1, NPTS
         YY = 0
         DO 200 K = 1, M
            YY = YY + A(K) * FUNCT(K,X(I))
200      CONTINUE
         YCALC(I) = YY
100   CONTINUE
      RETURN
      END

      REAL FUNCTION CALCCHISQ() !---  ASSUMES ARRAY YCALC HAS BEEN FILLED
      INTEGER I
      REAL    CHI2
      INCLUDE '\CHAPT-6\FITVARS.FOR'
      CHI2=0.
      DO 100 I = 1, NPTS
         CHI2 = CHI2 + ( (Y(I)-YCALC(I)) / SIGY(I))**2
100   CONTINUE
      CALCCHISQ = CHI2
      RETURN
      END

C PROGRAM 7.3: \CHAPT-7\MAKEAB7.FOR
C ROUTINES TO SET UP THE BETA AND ALPHA MATRICES FOR LINEAR REGRESSION
C USES  MATRIX, FITFUNC7
C
      SUBROUTINE MAKEBETA              !---  MAKE THE BETA MATRICES
      INTEGER  I, K
      REAL FUNCT
      INCLUDE 'C:\CHAPT-6\FITVARS.FOR'
      DO 100  K=1, M
         BETA(K)=0
         DO 200 I=1, NPTS
         BETA(K)=BETA(K) + Y(I)*FUNCT(K, X(I))/SIGY(I)**2
200   CONTINUE
100   CONTINUE
      RETURN
      END

      SUBROUTINE MAKEALPHA             !---  MAKE THE ALPHA MATRICES
      INTEGER I,J,K
      REAL FUNCT
```

```
        INCLUDE 'C:\CHAPT-6\FITVARS.FOR'
        DO 100 J=1, M
           DO 200 K=1, M
              ALPHA(J,K)=0
                 DO 300 I=1, NPTS
                    ALPHA(J,K) = ALPHA(J,K)+FUNCT(J, X(I))*FUNCT(K, X(I))/SIGY(I)**2
                 300 CONTINUE
           200   CONTINUE
        100   CONTINUE
        RETURN
        END
```

E.4 Routines from Chapter 8

```
C PROGRAM 8.0: \CHAPT-8\NONLINFT.FOR
C MAIN CALLING ROUTINE FOR NON-LINEAR FITTING METHODS
C USES GRIDSEAR, GRADSEAR, EXPNDFIT, MARQFIT, FITFUNC8, MAKEAB8,
C       NUMDERIV, MATRIX, FITUTIL
        PROGRAM NONLINFT
        INTEGER  TRIAL, J, METHOD
        REAL STEPDOWN, LAMBDA, CHISQR, CALCCHISQ
        CHARACTER*40 TITLE
        REAL STEPSCALE(4)/0.49999, 0.99999, 0.001, 0.001/
        INCLUDE '\CHAPT-6\FITVARS.FOR'
        PRINT *,'    (1)GRID SEARCH,     (2)GRADIENT SEARCH'
        PRINT *,'    (3)CHISQ EXPANSION, (4)FUNCTION EXPANSION'
        PRINT *, 'TYPE 1, 2, 3, OR 4 --- '
        READ *, METHOD
        CHICUT    = 0.01
        STEPDOWN  = 0.1         !--- STEP DOWN THE GRADIENT IN GRADLS
        LAMBDA    = 0.001       !--- FOR MARQUARDT METHOD ONLY
        STEPSIZE  = STEPSCALE(METHOD)  !--- SCALES DELTAA(J)
        CALL FETCHDATA('\CHAPT-8\RADIODK.HST',TITLE)
        CALL FETCHPARAMETERS      !--- USES NPTS, MUST FOLLOW FETCHDATA
        TRIAL     = 0
        CHISQR    = CALCCHISQ()
        CHIOLD = CHISQR + CHICUT + 1
        DO WHILE (ABS(CHIOLD - CHISQR) .GE. CHICUT)
           CHIOLD = CHISQR
           PRINT 1000, TRIAL, CHISQR
 1000 FORMAT(' TRIAL #', I4, ' CHISQ =', F10.1)
           PRINT 1100, (A(J), J = 1,M)
 1100 FORMAT(6F12.4)
           PRINT *
           GOTO (110, 120, 130, 140) , METHOD
 110       CALL GRIDLS(CHISQR)
           GOTO 150
 120       CALL GRADLS(CHISQR, STEPDOWN)
           GOTO 150
 130       CALL CHIFIT(CHISQR)
           GOTO 150
```

```
140     CALL MARQUARDT(CHISQR, CHICUT, LAMBDA)
150     TRIAL = TRIAL +1
      ENDDO
151   CALL CALCULATEY
      IF ((METHOD .EQ. 1) .OR. (METHOD .EQ. 2)) THEN
         DO 200 J = 1, M
            SIGA(J) = SIGPARAB(J)                  !--- DCHI2 = 1
200      CONTINUE
         CALL OUTPUT(.FALSE., 'CON' ,CHISQR, TITLE) !--- NO ERROR MATRIX
      ELSEIF ((METHOD .EQ. 3) .OR. (METHOD .EQ. 4)) THEN
         IF (METHOD .EQ.  4  ) THEN
            CALL MARQUARDT(CHISQR,CHICUT,0)        !--- GET ERROR MATRI
         ENDIF
         DO 300 J = 1, M
            SIGA(J) = SIGMATRX(J)                  !--- ERROR MATRIX
300      CONTINUE
         CALL OUTPUT(.TRUE., 'CON', CHISQR, TITLE) !--- WITH ERROR MATRIX
      ENDIF
      CALL PLOTIT('NONLIN.SCR', .TRUE., .TRUE., !--- SCRPT FILE, LOG?, SPLINE?
     1 'C', (X(2)-X(1))/5,      !--- DATA CIRCLES, RADIUS OF CIRCLES
     2 0., 1., 900., 1000.,     !--- RANGES-X1,Y1,X2,Y2
     3 6, 6,                  !--- NUM X-AXIS DIV, NUM Y-AXIS DIV
     4 'TIME (SEC)', 'NUMBER OF COUNTS') !--- AXIS LABELS
      CALL CLOSEGRAPHICS
      END

C SAMPLE FITTING FUNCTION FOR NON-LINEAR FITS
C EXAMPLE IS SUM OF 2 EXPONENTIALS ON A CONSTANT BACKGROUND
      REAL FUNCTION  EXPF(A,X)
      REAL A,X
      REAL YY, ARG
      ARG = ABS(X/A)
      IF (ARG .GT. 60 ) THEN
         YY = 0
      ELSE
         YY = EXP(-ARG)
      ENDIF
      EXPF = YY
      RETURN
      END

      FUNCTION  YFUNCTION(XX) !---  REAL
      REAL YFUNCTION, XX, EXPF
      INCLUDE '\CHAPT-6\FITVARS.FOR'
      YFUNCTION = A(1) + A(2)*EXPF(A(4),XX) + A(3)*EXPF(A(5),XX)
      RETURN
      END

      INCLUDE '\CHAPT-8\GRIDSEAR.FOR'     !---  1-GRID SEARCH METHOD
      INCLUDE '\CHAPT-8\GRADSEAR.FOR'     !---  2-GRADIENT SEARCH METHOD
      INCLUDE '\CHAPT-8\EXPNDFIT.FOR'     !---  3-FUNCTION EXPANSION METHOD
```

```
      INCLUDE '\CHAPT-8\MARQFIT.FOR'   !--- 4-MARQUARDT METHOD
      INCLUDE '\CHAPT-6\FITUTIL.FOR'
      INCLUDE '\CHAPT-8\FITFUNC8.FOR'     !--- USED BY ALL METHODS
      INCLUDE '\CHAPT-8\MAKEAB8.FOR'      !--- USED BY METHODS 4 AND 5
      INCLUDE '\CHAPT-8\NUMDERIV.FOR'     !--- USED BY METHODS 4 AND 5
      INCLUDE '\APPEND-B\MATRIX.FOR'      !--- USED BY METHODS 4 AND 5

C  PROGRAM 8.1: \CHAPT-8\GRIDSEAR.FOR
C NON-LINEAR FIT BY THE GRID-SEARCH METHOD
C USES FITFUNC8, FITUTIL
      SUBROUTINE GRIDLS(CHISQR)
      REAL CHISQR
      REAL CALCCHISQ
      REAL SAVE,DELTA, DELTA1,DEL1,DEL2,AA,BB,CC,DISC,ALPH,X1,X2
      INTEGER J
      INCLUDE '\CHAPT-6\FITVARS.FOR'
      CHISQ2 = CALCCHISQ()
C  -FIND LOCAL MINIMUM FOR EACH PARAMETER-
      DO 100 J = 1, M
         DELTA   = DELTAA(J)
         A(J)    = A(J) + DELTA
         CHISQ3  = CALCCHISQ()
         IF (CHISQ3 .GT. CHISQ2 ) THEN
            DELTA  = -DELTA             !--- STARTED IN WRONG DIRECTION
            A(J)   =  A(J) + DELTA
            SAVE   =  CHISQ2            !--- INTERCHANGE 2 AND 3 SO 3 IS LOWER
            CHISQ2 =  CHISQ3
            CHISQ3 =  SAVE
         ENDIF                         !--- IF (CHISQ3 ...
C  -INCREMENT OR DECREMENT A(J) UNTIL CHI SQUARED INCREASES-
  110 CONTINUE
      CHISQ1 = CHISQ2              !--- MOVE BACK TO PREPARE FOR QUAD FIT
      CHISQ2 = CHISQ3
      A(J)   = A(J) + DELTA
      CHISQ3 = CALCCHISQ()
      IF (CHISQ3 .LE.  CHISQ2) GOTO 110
C  -FIND MINIMUM OF PARABOLA DEFINED BY LAST THREE POINTS-
      DEL1 = CHISQ2 - CHISQ1
      DEL2 = CHISQ3 - 2*CHISQ2 + CHISQ1
      DELTA1 = DELTA * (DEL1/DEL2 + 1.5)
      A(J) = A(J)  - DELTA1
      CHISQ2 = CALCCHISQ()            !--- AT NEW LOCAL MINIMUM
C  -ADJUST DELTA FOR CHANGE OF 2 FROM CHISQ AT MINIMUM-
      AA = DEL2/2                     !--- CHISQ = AA*A(J)**2 + BB*A(J) + CC
      BB = DEL1 - DEL2/2
      CC = CHISQ1-CHISQ2
      DISC = BB**2 -4*AA*(CC-2)       !--- CHISQR DIFF(ERENCE) = 2
      IF (DISC .GT. 0 ) THEN      !--- IF NOT, THEN PROBABLY NOT PARABOLIC YET
         DISC =SQRT(DISC)
         ALPH = (-BB - DISC)/(2*AA)
```

```
        X1 = ALPH*DELTA + A(1) - 2*DELTA   !--- A(J) AT CHISQ MINIMUM+2
        DISC = BB**2 - 4*AA*CC
        IF (DISC.GT.0 ) THEN
           DISC = SQRT(DISC)
        ELSE
           DISC = 0                         !--- ELIM ROUNDING ERR
        ENDIF
        ALPH = (-BB - DISC)/(2*AA)
        X2 = ALPH*DELTA + A(1) - 2*DELTA !--- A(J) AT CHISQ MINIMUM
        DELTA = X1 - X2
        DELTAA(J) = DELTA
      ENDIF                                 !--- IF (DISC .GT. 0 ...
  100 CONTINUE                              !--- DO J
      CHISQR = CHISQ2
      RETURN
      END

C PROGRAM 8.2: \CHAPT-8\GRADSEAR.FOR
C NON-LINEAR LEAST-SQUARES FIT BY GRADIENT SEARCH METHOD
C USES FITFUNC8, FITUTIL
      SUBROUTINE CALCGRAD
      INTEGER J
      REAL  SUM, DELTA, FRACT/0.001/, CALCCHISQ
      INCLUDE '\CHAPT-6\FITVARS.FOR'
      SUM = 0
      DO 100 J = 1, M
        CHISQ2 = CALCCHISQ()
        DELTA  = FRACT * DELTAA(J) !---  DIFF(ERENTIAL ELEMENT FOR GRADENT
        A(J)   = A(J) + DELTA
        CHISQ1  = CALCCHISQ()
        A(J)   = A(J) - DELTA
        GRAD(J) = CHISQ2 - CHISQ1           !--- 2*DELTA*GRAD
        SUM    = SUM + GRAD(J)**2
  100 CONTINUE
      DO 200  J = 1, M
        GRAD(J) =  DELTAA(J)*GRAD(J)/SQRT(SUM)    !--- STEP * GRAD
  200 CONTINUE
      RETURN
      END

      SUBROUTINE GRADLS(CHISQR, STEPDOWN)
      REAL CHISQR, STEPDOWN
      REAL STEPSUM, STEP1, CALCCHISQ
      INTEGER J
      INCLUDE '\CHAPT-6\FITVARS.FOR'
      CALL CALCGRAD              !--- CALCULATE THE GRADIENT
C -EVALUATE CHISQR AT NEW POINT AND MAKE SURE CHISQR DECREASES-
      CHISQ3 = CHISQ2 + 1
      DO WHILE (CHISQ3 .GT. CHISQ2)
        DO J = 1, M
          A(J) = A(J) + STEPDOWN * GRAD(J) ! SLIDE DOWN
```

```
          ENDDo
          CHISQ3 = CALCCHISQ()
          IF (CHISQ3 .GE. CHISQ2 ) THEN
             DO J = 1, M      ! MUST HAVE OVERSHOT MINIMUM
                A(J) = A(J) - STEPDOWN * GRAD(J) ! RESTORE
             ENDDo
             STEPDOWN = STEPDOWN/2            ! DECREASE STEPSIZE
          ENDIF
       ENDDo
       STEPSUM = 0
C -INCREMENT PARAMETERS UNTIL CHISQR STARTS TO INCREASE-
       DO WHILE (CHISQ3 .LT. CHISQ2)
          STEPSUM = STEPSUM + STEPDOWN   ! COUNTS TOTAL INCREMENT
          CHISQ1 = CHISQ2
          CHISQ2 = CHISQ3
          DO J = 1, M
             A(J) = A(J) + STEPDOWN * GRAD(J)
          ENDDo
          CHISQ3 = CALCCHISQ()
       ENDDo              !DOWHILE
C -FIND MINIMUM OF PARABOLA DEFINED BY LAST THREE POINTS-
       STEP1=STEPDOWN*((CHISQ3-CHISQ2)/(CHISQ1-2*CHISQ2+CHISQ3)+0.5)
       DO J = 1, M
          A(J) = A(J) - STEP1 * GRAD(J)    ! MOVE TO MINIMUM
       ENDDo
       CHISQR = CALCCHISQ()
       STEPDOWN = STEPSUM               ! START WITH THIS NEXT TIME
       RETURN
       END

C PROGRAM 8.3: \CHAPT-8\EXPNDFIT.FOR
C NON-LINEAR LEAST-SQUARES FIT BY EXPANSION OF THE FITTING FUNCTION
C USES  FITFUNC8, MAKEAB8, MATRIX
       SUBROUTINE CHIFIT(CHISQR)
       INTEGER J
       REAL DET, CALCCHISQ
       INCLUDE '\CHAPT-6\FITVARS.FOR'
       CALL MAKEBETA
       CALL MAKEALPHA
       CALL MATINV(M, ALPHA, DET)      !---  INVERT MATRIX
       CALL LINEARBYSQUARE(M,BETA,ALPHA,DA)  !---  EVALULATE PARAM
INCREMENTS
       DO 100  J = 1, M
          A(J) = A(J) + DA(J)              !---  INCREMENT TO NEXT SOLUTION.
  100  CONTINUE
       PRINT *,'A',(A(J),J=1,M)
       CHISQR =  CALCCHISQ()
       RETURN
       END
```

```
C PROGRAM 8.4: \CHAPT-8\MARQFIT.FOR
C NON-LINEAR FIT BY THE GRADIENT-EXPANSION (MARQUARDT) METHOD
C USES  FITFUNC9, MAKEAB8, MATRIX
      SUBROUTINE MARQUARDT(CHISQR, XICUT, LAMBDA)
      INTEGER J
      REAL CHISQR, XICUT, LAMBDA
      REAL DET, CALCCHISQ
      INCLUDE '\CHAPT-6\FITVARS.FOR'
      DO
         CALL MAKEBETA
         CALL MAKEALPHA
         DO 100 J = 1, M
            ALPHA(J,J) = (1 + LAMBDA) * ALPHA(J,J)
 100     CONTINUE
         CALL MATINV(M, ALPHA, DET)          !---  INVERT MATRIX
         IF (LAMBDA .LE. 0 ) RETURN !---  FINAL CALL TO GET THE ERROR MATRIX.
         CALL LINEARBYSQUARE(M,BETA,ALPHA,DA)!---  EVAL PARAM INCREMENTS
         CHISQ1 = CHISQR
         DO 200 J = 1, M
            A(J) = A(J) + DA(J)              !---  INCR TO NEXT SOLUTION
 200     CONTINUE
         CHISQR = CALCCHISQ()
         IF ( CHISQR .LE. CHISQ1 + XICUT ) RETURN
         DO 300 J = 1, M
            A(J) = A(J)-DA(J)     !---  RETURN TO PREV SOLUTION
 300     CONTINUE
         CHISQR = CALCCHISQ()
         LAMBDA = 10*LAMBDA  !---  AND REPEAT THE CALC, WITH LARGER LAMBDA
      END DO
      END

C PROGRAM 8.5:  \CHAPT-8\FITFUNC8.FOR
C USES FITVARS
C -THE FOLLOWING ROUTINES ARE GENERAL FOR FITTING ANY FUNCTION-
      SUBROUTINE CALCULATEY
      REAL YFUNCTION
      INCLUDE '\CHAPT-6/FITVARS.FOR'
      DO 100 I = 1, NPTS
         YCALC(I) = YFUNCTION(X(I))
 100  CONTINUE
      RETURN
      END

      REAL FUNCTION CALCCHISQ()
      REAL  CHI2, YFUNCTION
      INCLUDE '\CHAPT-6/FITVARS.FOR'
      CHI2=0.
      DO 100 I = 1, NPTS
         CHI2 = CHI2 + ( (Y(I)-YFUNCTION(X(I)))/SIGY(I))**2
 100  CONTINUE
      CALCCHISQ = CHI2
```

```
      RETURN
      END

C -STANDARD DEVIATION CALC'D FROM CHISQ CHANGE OF 1 (PARABOLA FIT)
      REAL FUNCTION  SIGPARAB(J)
      INTEGER J
      REAL CALCCHISQ
      INCLUDE '\CHAPT-6/FITVARS.FOR'
      CHISQ2 = CALCCHISQ()
      A(J)   = A(J) + DELTAA(J)
      CHISQ3 = CALCCHISQ()
      A(J)   = A(J) - 2*DELTAA(J)
      CHISQ1 = CALCCHISQ()
      A(J)   = A(J) + DELTAA(J)
      SIGPARAB = DELTAA(J)*SQRT(2/(CHISQ1-2*CHISQ2+CHISQ3))
      RETURN
      END

C -STANDARD DEVIATION CALC'D FROM DIAGONAL TERMS IN ERROR MATRIX
      REAL FUNCTION SIGMATRX(J)
      INTEGER J
      REAL SIG
      INCLUDE '\CHAPT-6/FITVARS.FOR'
      SIG = SQRT(ABS(ALPHA(J,J)))
      IF (ALPHA(J,J) .LT. 0 ) SIG = - SIG  !---  NOTE- AN ERROR
      SIGMATRX = SIG
      RETURN
      END

C PROGRAM 8.6: \CHAPT-8\MAKEAB8.FOR
C MATRIX SET-UP FOR NON-LINEAR FITS
C USES FITFUNC8, NUMDERIV
C
      SUBROUTINE MAKEBETA     !---MAKE BETA MATRICES FOR NON-LINEAR FITTING
      INTEGER  J
      INCLUDE '\CHAPT-6/FITVARS.FOR'
      DO 100 J = 1, M
         BETA(J) = -0.5*DXISQ_DA(J)
100   CONTINUE
      RETURN
      END

      SUBROUTINE MAKEALPHA  !---  ALPHA MATRICES FOR NON-LINEAR FITTING
      INTEGER  J, K
      INCLUDE '\CHAPT-6\FITVARS.FOR'
      DO 100 J = 1, M
         ALPHA(J,J) =  0.5 * D2XISQ_DA2(J)
         IF (ALPHA(J,J) .EQ. 0 ) THEN
            PRINT *, 'DIAGONAL ELEMENT IS ZERO, J =',J
         STOP
         ENDIF
```

```
            IF (J .GT. 1 ) THEN
                DO 200 K = 1, J-1
                    ALPHA(J,K) = 0.5*D2XISQ_DAJK(J,K)
                    ALPHA(K,J) = ALPHA(J,K)
 200            CONTINUE              !--- DO K
            ENDIF                     !--- IF J
 100        CONTINUE                  !--- DO J
            DO 300 J = 1, M
                IF (ALPHA(J,J) .LT. 0 ) THEN
                ALPHA(J,J) = -ALPHA(J,J)
                IF (J .GT. 1 ) THEN
                    DO 400 K = 1, J-1
                        ALPHA(J,K) = 0
                        ALPHA(K,J) = 0
 400                CONTINUE          !--- DO K
                ENDIF                 !--- IF J
            ENDIF                     !--- IF ALPHA
 300    CONTINUE                      !--- FOR J
        RETURN
        END
```

E.5 Routines from Chapter 9

```
C PROGRAM 9.1: \CHAPT-9\LORENFIT.FOR
C MAIN CALLING ROUTINE FOR FIT TO LORENTZIAN + POLYNOMIAL
C USES FITFUNC9, MARQFIT, MATRIX, NUMDERIV, MAKEAB8, FITUTIL
        PROGRAM LORENFIT
        CHARACTER*40 TITLE
        INTEGER TRIAL, J
        REAL XSHIFT, CHISQR, LAMBDA, YFUNCTION
        REAL STEPSCALE(4)/ 0.49999, 0.99999, 0.001, 0.001/
        INCLUDE 'C:\F\CHAPT-6\FITVARS.FOR'
        CHICUT = 0.01
        LAMBDA = 0.001              ! FOR MARQUARDT METHOD ONLY
        STEPSIZE = STEPSCALE(4)     ! SCALES DELTAA[J]
        CALL FETCHDATA('\F\CHAPT-9\SINGLE.HST',TITLE)
        XSHIFT = (X(2)- X(1))/2
        DO J = 1, NPTS
            X(J) = X(J) + XSHIFT    ! MOVE TO BIN CENTER
        ENDDO
        CALL FETCHPARAMETERS        ! USES NPTS, MUST FOLLOW FETCHDATA
        TRIAL = 0
        CHISQR = CALCCHISQ()
        CHIOLD = CHISQR + CHICUT +1
        DO WHILE (ABS(CHIOLD - CHISQR) .GT. CHICUT)
            CHIOLD = CHISQR
            PRINT *,'TRIAL #',TRIAL,' CHISQ = ',CHISQR
            PRINT *, (A(J), J = 1, M)
            CALL MARQUARDT(CHISQR, CHICUT, LAMBDA)
            TRIAL = 1 + TRIAL
```

```
      ENDDO
      CALL CALCULATEY
      CALL MARQUARDT(CHISQR,CHICUT,0)       ! GET ERROR MATRIX
      DO J = 1, M
         SIGA(J) = SIGMATRX(J)                ! ERROR MATRIX
      ENDDO
      CALL OUTPUT(.TRUE., 'CON', CHISQR,TITLE) ! WITH ERROR MATRIX
      DO J = 1, NPTS
         X(J) = X(J) - XSHIFT                 ! RESTORE TO LEFT EDGE
      ENDDO
      CALL PLOTIT('LORENFIT.SCR',.FALSE.,.TRUE.,! SCRIPT FILE, LOG?, SPLINE?
    1 'H', 0.0,                  ! HIST, 0(NOT USED)
    2 0.0, 0.0, 3.0, 220.0,          ! X1, Y1, X2, Y2 FOR PLOT
    3 6, 6,                      ! NUM GRID MARKS X,Y
    4 'E (GEV)', 'NUMBER OF COUNTS')     ! LABELS
C -PLOT THE BACKGROUND-
      A(4) = 0.0
      A(7) = 0.0
      DO J = 1, NPTS
         YCALC(J) = YFUNCTION(X(J))
      ENDDO
      CALL SPLINEMAKE(NPTS,0,0,X,YCALC)
      CALL SCURVE(1, 40, 5, 0.025, X)    ! SPLINE CURVE
      CALL CLOSEGRAPHICS
      END

C LORENTZIAN PEAK ON A QUADRATIC BACKGROUND
      REAL FUNCTION YFUNCTION(XX) ! LORENTZIAN ON POLYNOMIAL
      REAL XX
      REAL YY, PI/3.1415927/
      INCLUDE '\F\CHAPT-6\FITVARS.FOR'
      YY = A(1) + A(2)*XX + A(3)*XX**2 + A(4)*A(6)/(2*PI)
    1 /((XX-A(5))**2 + A(6)**2/4)
      YFUNCTION = YY
      RETURN
      END
      INCLUDE '\F\CHAPT-6\FITUTIL.FOR'
      INCLUDE '\F\CHAPT-9\FITFUNC9.FOR'
      INCLUDE '\F\CHAPT-8\MARQFIT.FOR'       ! MARQUARDT METHOD
      INCLUDE '\F\CHAPT-8\MAKEAB8.FOR'       ! USED BY MARQFIT
      INCLUDE '\F\CHAPT-8\NUMDERIV.FOR'      ! USED BY MARQFIT
      INCLUDE '\F\APPEND-B\MATRIX.FOR'       ! USED BY MARQFIT
```

E.6 Routines from Chapter 10

```
C PROGRAM 10.1: \CHAPT-10\MAXLIKE.FOR
C DIRECT MAXIMUM LIKELIHOOD EXAMPLE
C  USES FITUTIL, QUIKSCRP
      PROGRAM MAXLIKE
      REAL    SIGTAU, TAUMAX, MAXM  !---  M IS LOG OF LIKELIHOOD FUNCTION
      INCLUDE '\CHAPT-10\MAXLINCL.FOR'
```

```
      CALL GETDATA('\CHAPT-10\TEST.DAT')      !--- WAS DA50
      CALL SEARCH(TAUMAX, MAXM)
      CALL WRITEOUTPUT(SIGTAU, TAUMAX, MAXM)
      CALL PLOTLIKECURVE(TAUMAX, SIGTAU, MAXM)
      CALL CLOSEGRAPHICS
      END

      SUBROUTINE GETDATA(INFILE)
      INTEGER  IEVNUM
      CHARACTER*(*) INFILE
      CHARACTER TITLE(80)
      INCLUDE '\CHAPT-10\MAXLINCL.FOR'
      C = 3.00
      LOSEARCH = 0.50
      HISEARCH = 1.5              !--- SEARCH RANGE
      TAUSTEP  = 0.01
      XLO     = 0.50              !---  PLOT RANGE
      XHI     = 1.2
      YLO     = 0.0
      YHI     = 1.2
      NTRIALS = (HISEARCH - LOSEARCH)/TAUSTEP
      OPEN(5, INFILE)             !---  INPUT DATA FILE
      READ(5, *) TITLE
      PRINT *,' ',TITLE
      READ(5, *) NEVENTS, MASS, D1, D2
      IEVNUM = 1
      NEVENTS = 0
      DO WHILE (IEVNUM .GT.0)
         READ(5, *) IEVNUM, XPRODUCTION, PLAB, XDECAY
         IF (IEVNUM .GT.0) THEN
            IF ((XDECAY .GE. D1) .AND. (XDECAY .LT. D2)) THEN
               NEVENTS = 1 + NEVENTS
               LTOTSCALE =  MASS/(C*PLAB)  !---  = 1/(C*BETA*GAMMA)
               TIMES(NEVENTS)=(XDECAY - XPRODUCTION)*LTOTSCALE !---PROPER T
C  CONVERT D1 AND D2 TO TIME LIMITS, LOTLIM AND HITLIM,
C   I.E., INTEGRATION LIMITS IN PROPER TIME FROM THE PRODUCTION VERTEX.
               LOTLIM(NEVENTS) = (D1 - XPRODUCTION)*LTOTSCALE
               HITLIM(NEVENTS) = (D2 - XPRODUCTION)*LTOTSCALE
            ENDIF
         ENDIF
      ENDDO
      PRINT *, 'END OF FILE - ', IEVNUM, ' EVENTS READ'
      PAUSE
      RETURN
      END

      REAL FUNCTION LOGPROB(K, TAU)
      INTEGER K
      REAL TAU
      REAL A, B
      INCLUDE '\CHAPT-10\MAXLINCL.FOR'
```

```
C     D1 AND D2 ARE BEGINNING AND END OF THE FIDUCIAL REGION.
C     MUST CVT TO LOTLIM AND HITLIM WHICH ARE INTEGRATION LIMITS IN PROPER
      TIME,
C     MEASURED FROM PRODUCTION VERTEX.
C     NOW, CALC PROBABILITY-
      B = EXP(-HITLIM(K)/TAU)
      A = EXP(-LOTLIM(K)/TAU)
      PROB = EXP(-TIMES(K)/TAU)/(TAU*(A - B))
      LOGPROB = ALOG(PROB)
      RETURN
      END

      REAL FUNCTION LOGLIKE(T)
      REAL T, LOGPROB
      INTEGER  I
      REAL     M, PROB
      INCLUDE '\CHAPT-10\MAXLINCL.FOR'
      M = 0.0
      DO 100 I = 1, NEVENTS
         PROB = LOGPROB(I,T)
         M = PROB + M
  100 CONTINUE
      LOGLIKE = M
      RETURN
      END

      SUBROUTINE SEARCH(TAUATMAX, MAXM)
      REAL TAUATMAX, MAXM
      INTEGER  TRIAL
      REAL M1, M2, M3, DEL1, DEL2, DELTA1, TAU, MLIKELI, LOGLIKE
      INCLUDE '\CHAPT-10\MAXLINCL.FOR'
      M2    =  -1000
      MAXM  =  -1.0E20
      TAU   =   LOSEARCH
      DO 100 TRIAL = 0, NTRIALS
         MLIKELI = LOGLIKE(TAU)
         PRINT *,'TRIAL',TRIAL,' TAU=', TAU,' LOG LIKELIHOOD=',MLIKELI
         M3 = MLIKELI
         IF (M3 .GT. M2 )  THEN   !---  REMEMBER, THESE ARE NEGATIVE
            M1 = M2
            M2 = M3
         ELSE                     !---  LEAVING MAXIMUM
C     FIND MAXIMUM OF PARABOLA DEFINED BY LAST THREE POINTS-
            DEL1     = M2 - M1
            DEL2     = M3 - 2*M2 + M1
            DELTA1   = TAUSTEP * (DEL1/DEL2 + 1.5)
            TAU      = TAU  - DELTA1
            TAUATMAX = TAU
            MAXM     = LOGLIKE(TAU)    !---  AT  MAXIMUM OF PARABOLA
            RETURN
         ENDIF
```

```
      TAU = TAU + TAUSTEP
100 CONTINUE
    RETURN
    END

    REAL FUNCTION ERROR(T, DT)   !--- 1/SQRT(-2ND DERIVATIVE OF LOG(L))
    REAL T, DT
    REAL  T1, T2, D2YDT2, ERR, LOGLIKE
    T1 = T - DT
    T2 = T + DT
    D2YDT2 = (LOGLIKE(T2) - 2*LOGLIKE(T) + LOGLIKE(T1))/DT**2
    ERR    = 1/SQRT(-D2YDT2)
    ERROR  = ERR
    RETURN
    END

C PROGRAM 10.2 \CHAPT-10\MAXLINCL.FOR    (WEBSITE)
C INCLUDE FILE FOR MAXLIKE
```

E.7 Routines from Chapter 11

```
C PROGRAM 11.1: \CHAPT-11\CHI2PROB.FOR
C CALCULATE CHI^2 PROB. DENS. & THE CHI^2 PROB. INTEGRAL
C USES CHIPROBDENS AND CHIPROB
    PROGRAM CHI2PROB
    REAL CHI2,  CHIPROB
    INTEGER NFREE
    PRINT *,'CALCULATE CHI2 PROBABILITY DENSITY FUNCTION & INTEGRAL',
  1 ' PROBABILITY'
    PRINT *, 'TYPE  NUM DEG OF FREEDOM AND  CHI2.  (EXIT ON ^C)'
    READ *,  NFREE, CHI2
    PRINT 1000, CHIPROBDENS(CHI2, NFREE), CHIPROB(NFREE, CHI2)
1000 FORMAT(' CHI^2 PROB. DENS. = ',F7.3,' , CHI^2 PROBABILITY=',F7.3)
    PRINT *,' ***** NOTE THAT TABLE C.4 REFERS TO CHI^2/NFREE****'
    END

C THE FOLLOWING THREE ROUTINES ARE INCLUDED
C   IN THE PROGRAM UNIT  C:\CHAPT-6\FITUTIL.FOR (WEBSITE)
    REAL FUNCTION CHIPROB(NFREE, CHI2)      !--- MAX NFREE =  56
    EXTERNAL CHIX
    COMMON/UTIL/ GLSIMPS
    REAL CHIX, SIMPSON, GLSIMPS
    INTEGER NFREE
    REAL PI, CHI2, CLIM, INTFROMLIM
    DATA   CLIM /2/,               !--- EXPANSION LIMIT FOR NFREE = 1
  1 INTFROMLIM /0.157/,            !--- INTEGRAL FROM CLIM TO INFINITY
  2    DXO /0.2/                   !--- DETERMINES ACCURACY OF INTEGRATION
  3    PI/3.14159/
    INTEGER NINT
    IF (CHI2 .GE. 1) THEN
       NINT = (CHI2+0.0001)/DXO
    ELSE
```

```
         NINT = 5
      ENDIF
      IF (CHI2 .GT. 15*SQRT(NFREE) ) THEN !--- QUICK CUTOUT
         CHIPROB = 0
      ELSE
         GLSIMPS = FLOAT(NFREE)/2          !--- GLSIMPS IS GLOBAL FOR CHIX
        IF (NFREE .EQ. 1 )    THEN
            IF (CHI2 .LT. CLIM ) THEN
          CHIPROB = 1-SQRT(CHI2/2/PI)*
     1     (2 - CHI2*(1/3 - CHI2*(1/20 - CHI2*(1/168 - CHI2/1728))))
            ELSE
              CHIPROB = INTFROMLIM - SIMPSON(CHIX,NINT,CLIM,CHI2)
     1                 /GAMMA(NFREE/2.0)/2.0**(NFREE/2.0)
            ENDIF                      !--- IF (CHI2 ...)
         ELSE IF (NFREE .EQ. 2 ) THEN
            CHIPROB = EXP(-CHI2/2)     !--- INTEGRABLE
         ELSE
            CHIPROB =  1 - SIMPSON(CHIX, NINT, 0, CHI2)
     1              /GAMMA(NFREE/2.0)/2.0**(NFREE/2.0)
        ENDIF                          !--- IF (NFREE ...)
      ENDIF
      RETURN                           !--- IF (NFREE ...)
      END

      REAL FUNCTION CHIPROBDENS(X,NFREE)
      REAL NUM, DEN, H, X
      INTEGER NFREE
      H = NFREE/2.0
      NUM = X**(H-1) * EXP(-X/2)
      DEN =  2**H * GAMMA(H)
      CHIPROBDENS = NUM/DEN
      RETURN
      END

C  USED BY CHIPROB (FOR SIMPSON WHICH ALLOWS ONLY 1 ARGUMENT.)
      REAL FUNCTION CHIX(X)
      COMMON/UTIL/ GLSIMPS
      REAL GLSIMPS
      REAL X
      IF (X.EQ.0) THEN
         CHIX = 0.0
      ELSE
         CHIX = X**(GLSIMPS-1)*EXP(-X/2) !--- GLSIMPS = H = NFREE/2
      ENDIF
      RETURN
      END

C THIS FOLLOWING ROUTINE IS INCLUDED
C   IN THE PROGRAM UNIT \CHAPT-6\FITUTIL.FOR  (WEBSITE)
C APPROXIMATE GAMMA FUNCTION WITH H = NFREE/2
      REAL FUNCTION GAMMA(H)
```

```
      REAL H, PI/3.1415927/
      GAMMA = SQRT(2.0*PI) * EXP(-H)*(H**(H-0.5)) * (1.0 + 0.0833/H)
      RETURN
      END

C PROGRAM 11.2: \CHAPT-11\LCORPROB.FOR
C CALCULATE LINEAR CORRELATION PROBABILITY INTEGRAL
C USES LCORLATE
      PROGRAM LCORPROB
      INTEGER NOBSERV
      REAL LINCORPROB, RCORR
      PRINT *, 'TEST INTEGRAL OF LINEAR CORRELATION FUNCTION'
      PRINT *, 'TYPE-# OBSERVATIONS, LINEAR CORRELATION COEFFICIENT: '
      READ *, NOBSERV, RCORR
      PRINT *, 'INTEGRAL CORRELATION FUNCTION= ',
     1 LINCORPROB(NOBSERV-2, RCORR)
      END
      INCLUDE '\CHAPT-11\LCORLATE.FOR'

C LINEAR-CORRELATION PROBABILITY FUNCTION AND INTEGRAL
C USES FITUTIL
      REAL FUNCTION LINCORPROB(NFREE, HILIM)
      EXTERNAL LINCORREL     !--- FOR USE IN FUNCTION SIMPSON
      INTEGER NFREE
      REAL   HILIM
      REAL DX /0.01/, LOLIM/0.0/, LINCORREL, SIMPSON
      INTEGER  NINT
      COMMON/UTIL/GLSIMPS
      GLSIMPS = NFREE !--- GLOBAL FOR FUNCTION LINCORREL (FOR SIMPSON)
      NINT   = INT((HILIM - LOLIM)/DX)
      LINCORPROB = 1-2*SIMPSON(LINCORREL, NINT, LOLIM,  HILIM)
      RETURN
      END

      REAL FUNCTION LINCORREL(R)
      REAL R
      COMMON/UTIL/GLSIMPS       !--- GLSIMS = NFREE MUST BE GLOBAL FOR
      DATA SQRTPI/1.7724539/    !    FUNCT "SIMPSONS" WHICH ALLOWS ONLY 1 ARG
      LINCORREL = GAMMA((GLSIMPS+1)/2)/GAMMA(GLSIMPS/2)
     1    *EXP( (GLSIMPS-2)/2 * ALOG(1 - R**2))/SQRTPI
      RETURN
      END
```

E.8 Routines from Appendix A

```
PROGRAM A.1 SIMPSON
C THE FOLLOWING ROUTINE IS INCLUDED
C  IN THE PROGRAM UNIT \CHAPT-6\FITUTIL   (WEBSITE)
C -SIMPSON'S RULE FOR "FUNCTX(X:REAL):REAL"
C  IF FUNCTX HAS OTHER PARAMETERS, THEY MUST BE GLOBAL, E.G., GLSIMPS
```

```
      REAL FUNCTION SIMPSON(FUNCTX, NINTS, LOLIM, HILIM) !---  2 CALCS/INTERVAL
      EXTERNAL FUNCTX !---  THIS STATEMENT REQ'D IN CALLING PGM ALSO
      REAL  FUNCTX, SUM, X, DX, LOLIM, HILIM
      INTEGER  NINTS, I
       X = LOLIM
       DX = (HILIM - LOLIM)/(2*NINTS)
       SUM=FUNCTX(X)
       SUM= SUM - FUNCTX(HILIM)
      DO 100 I = 1, NINTS
      X=X+2*DX
        SUM=SUM + 4*FUNCTX(X-DX) + 2*FUNCTX(X)
 100    CONTINUE
     SUM = SUM
       SIMPSON = SUM*DX/3.0
       RETURN
      END
```

PROGRAM A.2 SPLINE INTERPOLATION

```
C PROGRAM A.1: \APPEND-A\SPLINTST.FOR
C TEST CUBIC SPLINE INTERPOLATION
      PROGRAM SPLINTST
      CHARACTER TITLE(80)
      REAL D2A, D2B, XS, X(100), Y(100), SPLINEINT
      INTEGER N, I
      OPEN(5,'\APPEND-A\SPLINE.DAT')      !--- TEST DATA FILE
      READ(5,1000) TITLE
      PRINT 1000, ' ',TITLE
 1000 FORMAT(80A1)
      READ(5,*) N, D2A, D2B !--- NO. OF POINTS, 2ND DERIVATIVES AT BOUNDARY
      PRINT *,'DATA TABLE: N=', N
      PRINT *,'     X                 Y'
      DO 100 I = 1, N
         READ(5,*) X(I), Y(I)
         PRINT *, X(I), Y(I)
 100  CONTINUE
      CALL SPLINEMAKE(N, D2A, D2B, X, Y)
      CLOSE(5)
 200  PRINT *,'TYPE A VALUE OF X  (EXIT WITH ^C)'
      READ *, XS
      PRINT *, 'INTERPOLATED Y = ', SPLINEINT(XS)
      GOTO 200
      END

C ROUTINES FOR CUBIC SPLINE INTERPOLATION.
C CONSTANT INTERVALS IN THE INDEPENDENT VARIABLE ARE ASSUMED.
      SUBROUTINE SPLINEMAKE(NN, D2YDX2A, D2YDX2B, XIN, YIN)
      INTEGER NN
      REAL    D2YDX2A, D2YDX2B, XIN(100), YIN(100)
C -COMMON VARIABLES SET IN SPLINEMAKE, USED IN SPLINEINT-
      COMMON/SPLINES/N, H, XX(100), YY(100), D2YDX2(100)
      INTEGER N
```

```
      REAL H, XX, YY, D2YDX2
      INTEGER I
      REAL  A(100), DELT1(100), DELT2(100), B(100)
      N = NN        !--- USED BY SPLININT, THROUGH COMMON/SPLINES/
      H = (XIN(N) - XIN(1))/(N-1)
      DO 100 I = 1, N
         XX(I) = XIN(I)
         YY(I) = YIN(I)
100   CONTINUE
      D2YDX2(1) = D2YDX2A !---  END VALUES OF 2ND DERIVATIVES FROM INPUT
      D2YDX2(N) = D2YDX2B
      A(2) = 4
      DO 200 I =  3, N-1
         A(I) = 4-1/A(I-1)                !---  COEFFICIENTS
200   CONTINUE
      DO 300 I = 2, N
         DELT1(I) = YIN(I) - YIN(I-1)     !---  1ST DIFFERENCES
300   CONTINUE
      DO 400 I = 2, N-1                   !---  2ND DIFFERENCES X 6
         DELT2(I) = 6*(DELT1(I+1) - DELT1(I))/(H*H)
400   CONTINUE
      B(2) = DELT2(2) - D2YDX2(1)         !---  B COEFFICIENTS
      DO 500 I= 3, N-1
         B(I) = DELT2(I) - B(I-1)/A(I-1)
500   CONTINUE
      B(N-1) = B(N-1) - D2YDX2(N)
      D2YDX2(N-1) = B(N-1)/A(N-1)
      DO 600 I = N-2, 2, -1
         D2YDX2(I) = (B(I) - D2YDX2(I+1))/A(I) !---  2ND DERIVATIVES
600   CONTINUE
      RETURN
      END

      REAL FUNCTION DYDX(I) !---  FIRST DERIVATIVE (WEBSITE)
      INTEGER I
      COMMON/SPLINES/N, H, XX(100), YY(100), D2YDX2(100)
      INTEGER N
      REAL H, XX, YY, D2YDX2
      DYDX = (YY(I+1)-YY(I))/H - H*(D2YDX2(I)/3+D2YDX2(I+1)/6)
      RETURN
      END

      REAL FUNCTION D3YDX3(I) !---  THIRD DERIVATIVE (WEBSITE)
      INTEGER I
      COMMON/SPLINES/N, H, XX(100), YY(100), D2YDX2(100)
      INTEGER N
      REAL H, XX, YY, D2YDX2
      D3YDX3 = (D2YDX2(I+1) - D2YDX2(I))/H
      RETURN
      END
```

```
      REAL FUNCTION SPLINEINT(X) !--- INTERPOLATE IN TABLE (FROM SPLINEMAKE)
      REAL X
      COMMON/SPLINES/N, H, XX(100), YY(100), D2YDX2(100)
      INTEGER N
      REAL H, XX, YY, D2YDX2, DYDX, D3YDX3, DX
      INTEGER I
      I = INT((X-XX(1))/H)+1
      IF (I .LT. 1 )  I = 1
      IF (I .GT. N-1 ) I = N-1
      DX = X -XX(I)
C -INTERPOLATE
      IF (I .EQ. N ) THEN
         SPLINEINT = YY(I)
      ELSE
         SPLINEINT = YY(I) + (DYDX(I) + (D2YDX2(I)/2 +D3YDX3(I)/6*DX)*DX)*DX
      ENDIF
      RETURN
      END
```

E.9 Routines from Appendix B

```
C PROGRAM B.1: \APPEND-B\MATRIX.FOR
C INVERT A SQUARE MATRIX
C USES FITVARS
      SUBROUTINE MATINV(M, MARRAY, DET)
      INTEGER M
      REAL MARRAY(10,10), DET
      INTEGER IK(10), JK(10)
      INTEGER I, J, K, L
      REAL AMAX, SAVE
      DET=0
C -FIND LARGEST ELEMENT
      DO 100  K = 1, M
         AMAX=0
 1500    DO 200  I = K, M
            DO 300 J = K , M
               IF ( ABS(MARRAY(I,J)) .GT. ABS(AMAX) )  THEN
                  AMAX = MARRAY(I,J)
                  IK(K) = I
                  JK(K) = J
               ENDIF
 300        CONTINUE        !---  DO J
 200     CONTINUE           !---  DO I
         IF (AMAX .EQ. 0 )  RETURN  !---  WITH 0 DETERMINANT AS SIGNAL
         DET = 1
C -INTERCHANGE ROWS AND COLUMNS TO PUT AMAX IN MARRAY(K,K)
         I = IK(K)
         IF (I .LT. K ) THEN
            GOTO 1500
         ELSEIF (I .GT. K )  THEN
            DO 400 J = 1, M
```

```
               SAVE = MARRAY(K,J)
               MARRAY(K,J) = MARRAY(I,J)
               MARRAY(I,J) = -SAVE
 400        CONTINUE                 !---  DO J
         ENDIF                    !---  IF I
         J = JK(K)
         IF (J .LT. K ) THEN
            GOTO 100
         ELSEIF (J .GT. K ) THEN
            DO 500 I = 1, M
               SAVE = MARRAY(I,K)
               MARRAY(I,K) = MARRAY(I,J)
               MARRAY(I,J) = -SAVE
 500        CONTINUE  !---  DO I
         ENDIF  !---  IF J
C -ACCUMULATE ELEMENTS OF INVERSE MATRIX
         DO 600 I = 1, M
            IF (I .NE. K )
  1           MARRAY(I,K) = -MARRAY(I,K)/AMAX
 600     CONTINUE !---  DO I
         DO 700 I = 1, M
            DO 800 J = 1, M
               IF ((I .NE. K) .AND. (J .NE. K) )
  1            MARRAY(I,J) = MARRAY(I,J) + MARRAY(I,K)*MARRAY(K,J)
 800        CONTINUE                     !---  DO J
 700     CONTINUE                        !---  DO I
         DO 900 J = 1, M
            IF (J .NE. K )
  1         MARRAY(K,J) = MARRAY(K,J)/AMAX
 900     CONTINUE                     !---  DO J
         MARRAY(K,K) = 1/AMAX
         DET = DET * AMAX
 100  CONTINUE                        !---  DO K
C -RESTORE ORDERING OF MATRIX
      DO 1000 L = 1, M
         K = M + 1 - L
         J = IK(K)
         IF (J .GT. K ) THEN
            DO 1100 I = 1, M
               SAVE = MARRAY(I,K)
               MARRAY(I,K) = -MARRAY(I,J)
               MARRAY(I,J) = SAVE
 1100       CONTINUE                  !---  DO I
         ENDIF                        !---  IF J
         I = JK(K)
         IF (I .GT. K ) THEN
            DO 1200 J = 1, M
            SAVE = MARRAY(K,J)
            MARRAY(K,J) = -MARRAY(I,J)
            MARRAY(I,J) = SAVE
 1200    CONTINUE !---  DO J
```

```
      ENDIF                         !--- IF I
1000 CONTINUE                       !--- DO L
     RETURN
     END

     SUBROUTINE LINEARBYSQUARE(M, A, B, C)  !--- MATRIX PRODUCT
     INTEGER M
     REAL A(10), B(10,10), C(10)
     INTEGER I,J
     DO 100 I = 1, M
        C(I)=0
        DO 200 J = 1, M
        C(I)=C(I) +A(J)*B(I,J)
200 CONTINUE
100    CONTINUE
     RETURN
     END
```

E.10 Routines from Appendix C

```
C PROGRAM C.1: \APPEND-C\STUDENTST.FOR
C CALCULATES BOTH THE GAUSSIAN PROBABILITY
C   AND THE STUDENT'S T PROBABILITY FOR EXCEEDING  A GIVEN VALUE
C   OF (MU-X)/SIGMA, WHERE MU IS THE MEAN VALUE OF X AND SIGMA IS
C   THE UNCERTAINTY IN THE MEAN.
C FOR SPEED, AND TO REDUCE POSSIBILITY OF OVERFLOW, WE
C   CALCULATE THE RATIO OF THE GAMMA FUNCTIONS DIRECTLY
C   IN FUNCTION GAMMACONST.
C TO IMPROVE SPEED AND ACCURACY BY USING SIMPSON'S FOR INTEGRATION
C
      PROGRAM STUDENTS_T
      REAL GP,TP, T
      INTEGER NU
      PRINT  *,  'TYPE NDOF AND T = |MU - X|/SIGMA  '
      READ *, NU, T
      CALL GTPROB(GP, TP, NU, T)
      PRINT 1100, 100*TP, 100*(1-TP)
      PRINT 1200, 100*GP, 100*(1-GP)
1100 FORMAT(' PROB (STUDENT''S T)  = ',F5.2,'%, 1-PROB = ',F5.2, '%')
1200 FORMAT(' PROB (GAUSSIAN)      = ',F5.2, '%, 1-PROB = ',F5.2,'%')
      END

      REAL FUNCTION STUDENTST(NU, T, G)     !STUDENT'S T DISTRIBUTION
      INTEGER NU
      REAL T, G, X
C X = (1/SQRT(NU*PI)  * (GAMMA((NU+1)/2)/GAMMA(NU/2))*(1+T^2/NU)^(-(NU+1)/2)
      X  = G*EXP( -(NU+1)/2)* ALOG(1+T*T/NU))
      STUDENTST = X
      END

      REAL FUNCTION GAUSS(X)
```

```
      REAL PI/3.14159/,X
      GAUSS = EXP(-X*X/2)/SQRT(2*PI)
      RETURN
      END

C GAUSSIAN AND STUDENT'S T PROBABILITIES
      SUBROUTINE GTPROB(GPROB, TPROB, N, T) !INTEGRAL FROM -T TO +T
      REAL GPROB, TPROB,T
      INTEGER N
      REAL GAM, T1, SUMT, SUMG, DT
      GAM  = GAMMACONST(N)        !RATIO OF GAMMAS - FOR SPEED
      DT   = 0.0001                !INTEGRATION STEP
      T1   = 0
      SUMT = 0
      SUMG = 0
      DOWHILE ((T1 .LT. T) .AND. (SUMT*DT .LT. 0.5)) !SIMPLE INTEGRATION.
C  REPLACE BY SIMPSON'S RULE FOR BETTER SPEED AND ACCURACY
         SUMT  = SUMT + STUDENTST(N,T1,GAM)
         SUMG  = SUMG + GAUSS(T1)
         T1    = T1 + DT
      ENDDO
      TPROB  = 2*SUMT*DT
      GPROB  = 2*SUMG*DT
      RETURN
      END

      REAL FUNCTION GAMMACONST(N)
C  G = GAMMA((H+1)/2)/GAMMA(H/2)/SQRT(H*PI)
C  PRE-CALCULATE RATIO FOR SPEED AND TO  AVOID OVERFLOW
      INTEGER N
      REAL PI/3.14159/
      REAL H, Y1,Y2, G
      H = N
      Y1 = -0.5*(H+1) + 0.5*(H)  *ALOG(0.5*(H+1))
      Y2 = -0.5*H     + 0.5*(H-1)*ALOG(0.5*H)
      G  = EXP(Y1-Y2)*(1+0.0833/(0.5*(H+1)))/((1+0.0833/(0.5*H))
     1  *SQRT(H*PI))
      GAMMACONST = G
      RETURN
      END
      END
```

E.11 Routines from Appendix D

```
C PROGRAM D.1: \APPEND-D\QUIKSCRP.FOR
C CREATE A SCRIPT FILE TO DISPLAY SIMPLE GRAPHS AND HISTOGRAMS
C THE FILE IS READ AND INTERPRETED BY \APPEND-D\QDISPLAY.EXE

C PROGRAM D.2: \APPEND-D\QUIKHIST.FOR
C ASSIGNS DATA TO HISTOGRAM BINS AND PLOTS HISTOGRAM EITHER
C  AS SCREEN CHARACTERS OR IN SCREEN GRAPHISC THROUGH QUIKSCRP
```

REFERENCES

Anderson, R. L. and E. E. Houseman, *Tables of Orthogonal Polynomial Values Extended to N = 104,* Research Bulletin 297, Agricultural Experimental Station, Iowa State University (April, 1942).

Arndt, R. A. and M. H. MacGregor, Nucleon-Nucleon Phase Shift Analysis by Chi-Squared Minimization, in *Methods in Computational Physics,* vol. 6, pp. 253–296, Academic Press, New York (1966).

Baird, D. C., *Experimentation: An Introduction to Measurement Theory and Experiment Design,* Prentice-Hall, Englewood Cliffs, N.J. (1988).

Bajpai, A. C., I. M. Calus, and J. A. Fairley, *Numerical Methods for Engineers and Scientists,* Wiley, Chichester (1977).

Beers, Y., *Introduction to the Theory of Error,* Addison-Wesley, Reading, Mass. (1957).

Box, G. E. P. and M. E. Müller, A Note on the Generation of Random Normal Deviates, *Ann. Math. Statist.,* vol. 29, pp. 610–611 (1958).

David, F. N., *Tables of the Correlation Coefficients,* Cambridge University Press, London (1938).

Dixon, W. J. and F. J. Massey, Jr., *Introduction to Statistical Analysis,* McGraw-Hill, New York (1969).

Eadie, W. T., D. Drijard, F. E. James, M. Roos, and B. Sadoulet, *Statistical Methods in Experimental Physics,* North-Holland, Amsterdam (1971).

Hamilton, W. C., *Statistics in Physical Science,* Ronald Press, New York (1964).

Hamming, R. W., *Numerical Methods for Scientists and Engineers,* McGraw-Hill, New York (1962).

Handbook of Chemistry and Physics, Chemical Rubber Co., Cleveland, Ohio (1973).

Hildebrand, F. B., *Introduction to Numerical Analysis,* McGraw-Hill, New York (1956).

Hoel, P. G., *Introduction to Mathematical Statistics,* Wiley, New York (1954).

IBM, *System/360 Scientific Subroutine Package, Programmer's Manual* (360A-CM-03X).

Knuth, D. E., Seminumerical Algorithms, in *The Art of Computer Programming,* vol. 2, pp. 29ff., Addison-Wesley, Reading, Mass. (1981).

Marquardt, D. W., An Algorithm for Least-Squares Estimation of Nonlinear Parameters, *J. Soc. Ind. Appl. Math.,* vol. II, no. 2, pp. 431–441 (1963).

Melkanoff, M. A., T. Sawada, and J. Raynal, Nuclear Optical Model Calculations, in *Methods in Computational Physics,* vol. 6, pp. 2–80, Academic Press, New York (1966).

Merrington, M. and C. M. Thompson, Tables of Percentage Points of the Inverted Beta (*F*) Distribution, *Biometrica,* vol. 33, pt. 1, pp. 74–87 (1943).

Orear, J., Notes on Statistics for Physicists, UCRL-8417, University of California Radiation Laboratory, Berkeley, Calif. (1958).

Ostle, B., *Statistics in Research,* Iowa State College Press, Ames, Iowa (1963).

Pearson, K, *Tables for Statisticians and Biometricians,* Cambridge University Press, London (1924).

Press, W. H., B. P. Flannery, S. A. Teukolsky, and W. T. Vetterling, *Numerical Recipes, The Art of Scientific Computing,* Cambridge University Press, New York (1986).

Pugh, E. M. and G. H. Winslow. *The Analysis of Physical Measurements,* Addison-Wesley, Reading, Mass. (1966).
"Review of Particle Physics" The European Physical Journal C, vol. 15, p. 193 (2000).
"Review of Particle Properties" Physics Letters, vol. 170B, p. 53 (1986).
Taylor, J. R., *An Introduction to Error Analysis,* University Science Books, Mill Valley, Calif. (1982).
Thompson, W. J., *Computing in Applied Science,* Wiley, New York (1984).
Wichmann, B. and D. Hill, Building a Random Number Generator, *Byte Magazine,* March, p. 127 (1987). *Applied Statistics,* vol. 31, pp. 188–190 (1982).
Young, H. D., *Statistical Treatment of Experimental Data,* McGraw-Hill, New York (1962).
Zerby, C. D., Monte Carlo Calculation of the Response of Gamma-Ray Scintillation Counters, in *Methods in Computational Physics,* vol. 1, pp. 90–133, Academic Press, New York (1963).

ANSWERS TO SELECTED EXERCISES

Chapter 1

1.1. (*a*) 5 (*b*) 2 (*c*) 2 (*d*) 5 (*e*) 4
(*f*) 1 (*g*) 3 (*h*) 3 (*i*) 3 (*j*) 4

1.3. (*a*) 980. (*b*) 84,000 (*c*) 0.0094 (*d*) 3.0×10^2
(*e*) 4.0 (*f*) NA (*g*) 5300 (*h*) 4.0×10^2
(*i*) 4.0×10^2 (*j*) 3.0×10^4

1.5. Mean = 73.48; median = 73; most probable value = 70

1.7. Standard deviation = 15.52

Chapter 2

2.2. (*a*) 20 (*b*) 6 (*c*) 120 (*d*) 270, 725

2.3. For $p = 1/2$, 0.015625, 0.093750, 0.234375, 0.31250, 0.234375, 0.093750, 0.015625

2.6. 4.1 for one lemon; 37 for two lemons; 1000 for three lemons

2.9. (*a*) $2.3 \simeq 2$ students (*b*) 8%

2.13. (*a*) 0.0011 (*b*) $\sim 3 \times 10^{-20}$

2.15. Mean number hitting counter in the 200-ns time interval:
$\bar{x} = \sum_{x=0}^{\infty} x P_P(x; \mu) = \mu$; mean number recorded $= \sum_{x=1}^{\infty} 1 P_P(x; \mu) =$
$1 - P_P(0, \mu) = 1 - e^{-\mu}$; Efficiency $= (1 - e^{-\mu})/\mu$.

2.17. $\bar{r} = \int_0^{\infty} r P(r) dr = 6CR^4$; $\int_0^{\infty} P(r) dr = 1$, so $C = 1/(2R^3)$ and $\bar{r} = 3R$

Chapter 3

3.3. The relative uncertainty in r should be one-half the relative uncertainty in L.

3.5. 1.503 ± 0.024

3.7. (*a*) 15300 ± 6700 (*b*) 165 ± 11

3.9. $\bar{n} = 3.61$; $s = 1.88$

Chapter 4

4.1. $s = 2.18$; $\sigma_\mu = 0.44$
4.3. Fig. 2.3: $\chi^2 = 1.39$ for 5 bins; $\chi^2_\nu = 0.35$
Fig. 2.4: $\chi^2 = 4.88$ for 7 bins; $\chi^2_\nu = 0.81$
4.7. Mean total counts in 1-min interval = 123.2; $\sigma = 9.4$; $\sigma_\mu = 3.0$
(*a*) Background counts in 1-min interval = 11.6; $\sigma = 1.5$
(*b*) Difference = 111.6 ± 3.3 counts per minute from the source
4.9. 32.81 ± 0.46
4.11. (*a*) $1.96\sigma = 31.0$ or 3.1%
(*b*) $1.96s = 30.1$ or 3.0%.
4.13. (*c*) $\chi^2 = 14.7$ (calculated with σ)
(*d*) $\langle\chi^2\rangle = \nu = N - 1 = 12$

Chapter 5

5.10. For 6 rows: (*b*) 8, 48, 120, 160, 120, 48, 8 (*c*) $\sigma = 1.22$

Chapter 6

6.1. $a = 114.3 \pm 9.6$; $b = 9.58 + 0.89$, $\chi^2 = 10.1$
6.4. $b = 3.60 \pm 0.03$; $\chi^2 = 11.9$

Chapter 7

7.2. $a_1 = 512.0 \pm 45.9$; $a_2 = 348.3 \pm 21.8$; $\chi^2 = 13.2$
$\alpha_{11} = 21.09$; $\alpha_{12} = \alpha_{21} = -147.1$; $\alpha_{22} = 476.1$
7.4. All terms: $\chi^2 = 17.21$ for 12 degrees of freedom
Even terms: $\chi^2 = 17.59$ for 14 degrees of freedom
$a_1 = (849.6 \pm 15.4) - (335.5 \pm 85.7)x^2 + (847.3 \pm 87.8)x^4$ with $x = \cos(\theta)$
7.10. $a_1 = 0.0001 \pm 0.0009$; $a_2 = v_0 = 0.871 \pm 0.018$
$a_1 = g/2 = 4.870 \pm 0.057$ (after iterating)

Chapter 8

8.3. (*a*) $\mu = 1.8741 \pm 0.0005$; $\chi^2 = 13.70$
(*b*) $\mu = 1.8471 \pm 0.0005$; $\Gamma = 0.0555 \pm 0.0008$; $\chi^2 = 13.3$
8.4. $a_1 = 148.6 \pm 31.0$; $a_2 = 31.0 \pm 1.1$; $\chi^2 = 13.0$
$\epsilon_{11} = 65.6$; $\epsilon_{12} = \epsilon_{21} = -6.26$; $\epsilon_{22} = 1.156$

Chapter 9

9.4. (*b*) $\chi^2 = 34.2$ for 24 degrees of freedom
Fitted parameters a_1 through a_6:
−2.2 136. −31.6 79.8 0.098 0.20
Uncertainties s_1 through s_6:
2.6 8 3.1 7.0 0.007 0.02

Chapter 10

10.1. $a_1 = 4.16$; $a_2 = 22.8$ at the maximum of the likelihood function

Chapter 11

11.4. Approximately 10% probability
11.5. Approximately 0.1% probability; not a very good fit
11.9. 0.9985
11.10. 0.9729
11.12. 0.9997
11.14. $F \simeq 10$ for $\nu_1 = 1$; $F \simeq 5$ for $\nu_1 \leq \nu_2$
11.18. $\Delta\chi^2 = 2.7$; $a_4 = 3.4^{+4.5}_{-3.1}$; $a_5 = 205^{+70}_{-30}$

INDEX

N

O

P

Q

R

S

T

U

V

W